Images of Science

SCIENCE AND ITS CONCEPTUAL FOUNDATIONS

David L. Hull, Editor

Edited by
PAUL M. CHURCHLAND
and
CLIFFORD A. HOOKER

Images of Science

Essays on Realism and Empiricism,
with a Reply from
BAS C. VAN FRAASSEN

The University of Chicago Press

Chicago and London

The University of Chicago Press, Chicago 60637
The University of Chicago Press, Ltd., London
© 1985 by The University of Chicago
All rights reserved. Published 1985
Printed in the United States of America

94 93 92 91 90 89 88 87 86 85   5 4 3 2 1

Library of Congress Cataloging in Publication Data

Main entry under title:

Images of science.

   (Science and its conceptual foundations)
   Includes bibliographies.
   1. Science—Philosophy.   2. Realism.
3. Empiricism.   4. Science—History.
5. Van Fraassen, Bastiaan C., 1941–   .
I. Van Fraassen, Bastiaan C., 1941–   .
II. Churchland, Paul M., 1942–   . III. Hooker,
C. A. (Clifford Alan), 1941–   . IV. Series.
Q175.I46 1985   501   85-1128
ISBN 0-226-10653-5
ISBN 0-226-10654-3 (pbk.)

Contents

Preface

The occasion of this volume is the renewal of an old debate. The dominant metaphysical and epistemological position among philosophers of science has recently been subjected to its most searching critique in more than two decades. In his book *The Scientific Image* (Oxford, 1980), Bas C. van Fraassen has addressed and argued against most of the tenets of the widely embraced position called *scientific realism*, and he there proposes a systematic alternative conception of science called *constructive empiricism*. His criticisms of realist orthodoxy are urged both with vigor and with charm, and the positive elements of his constructive empiricism constitute important philosophical theories in their own right, independently of their forming a system adequate to challenge realism.

This volume contains an equally vigorous set of papers written in response to van Fraassen's penetrating and challenging work, papers in defense of realism, written by scientific realists of all varieties. Six of the contributions are entirely new, including van Fraassen's extensive replies. Two of the papers (Musgrave's and Churchland's) are substantially expanded versions of earlier publications. And the papers by Hacking, Glymour, and Gutting are reprinted in their initial form, being included here because they address important elements of van Fraassen's position with especial force and clarity.

As the reader will soon determine, these replies are not at all uniform in their responses to the challenges laid down, which reflects the fact that the orthodoxy at issue is not nearly so uniform as one might have supposed. Scientific realists are arrayed on a multidimensional spectrum of positions. Some realists are conservative in their conception of rational cognition; some are radical. Some have a very high estimate of the actual achievements of science; some are decidedly guarded on this point. Some see the development of science as a convergent process likely to reach a unique truth, at least in the long run; others see a different dynamic.

All of this represents a very healthy state of affairs. At an absolute minimum, the challenge of van Fraassen's constructive empiricism constitutes an important test case in which to discriminate the competing forms and grades of realism. Some forms may be less vulnerable to his criticisms than others. Or they may possess virtues equal or superior to the virtues of his own view. In any case, a thorough reassessment of realism is inescapable, and more enlightened versions are already being articulated. Constructive empiricism will do at least this much.

And it may do more. Constructive empiricism, it appears, is prepared to stand its ground against even the most carefully tuned versions of scientific realism. This volume contains extensive replies from van Fraassen, addressed to the various attempts to proof realism against his critique. It also contains replies to the many criticisms that have been leveled against his own view. But here the reader's judgment must begin to supply its own commentary: the debate is joined in the following pages.

PAUL M. CHURCHLAND
CLIFFORD A. HOOKER

Part I

Essays Addressed to
Bas C. van Fraassen

Lex Orandi est Lex Credendi

Richard N. Boyd

> But whilst we are destitute of senses acute enough to discover the
> minute particles of bodies, and to give us ideas of their mechanical
> affections, we must be content to be ignorant of their properties
> and ways of operation; nor can we be assured about them any fur-
> ther than some few trials we make are able to reach. But whether
> they will succeed again another time, we cannot be certain. This
> hinders our certain knowledge of universal truths concerning natu-
> ral bodies: and our reason carries us herein very little beyond par-
> ticular matters of fact. Locke, *Essay*.

Van Fraassen's *The Scientific Image* (1980) represents a sophisticated and
brilliant defense of empiricist philosophy of science. Among the many interesting features of this work are van Fraassen's rebuttals to a class of arguments for scientific realism which other philosophers and I have offered
over the past decade or so and which may be characterized as efforts to
defend scientific realism as an empirical hypothesis which is justified because it provides the best scientific explanation for various facts about the
ways in which scientific methods are epistemically successful. These arguments capture an important source of the plausibility of scientific realism.
For that reason, it is important to assess the extent to which these arguments are successful against sophisticated empiricism. Van Fraassen's rebuttals represent just the sort of sophisticated empiricist criticisms against
which these arguments should be tested. The aim of the present paper is to
show that these rebuttals are ultimately unsuccessful and that the arguments for realism can be sustained.

In section 1 of this paper, I present in some detail empirical arguments
for realism of the sort in question. In section 2, I discuss criticisms of

A version of this paper was presented at the philosophy colloquium at The Johns Hopkins
University. I thank the audience, especially Peter Achinstein, Gary Hatfield, and David
Zaret, for their helpful comments and criticisms. I have also profited greatly from discussions
with John Bennett, Philip Gasper, Kristin Guyot, Harold Hodes, Christopher Hughes,
Norman Kretzmann, Scott MacDonald, Robert Rynasiewicz, Sydney Shoemaker, Alan
Sidelle, and Allen Wood. Rega Wood provided very helpful bibliographical advice.

those arguments which are presented by van Fraassen or are suggested by points which he makes. In section 3, I draw some lessons about philosophical and scientific methodology.

I. Realism and the Theory-Dependence of Experimental Design

In several papers (Boyd 1983, 1981, and esp. 1973), I have argued that scientific realism provides the only scientifically reasonable explanation for the reliability of certain important features of scientific methodology which are crucial in experimental design and in the assessment of experimental evidence. Roughly speaking, these are the features of scientific methodology relevant to the assessment of the "degree of confirmation" of a proposed theory, given a body of observational evidence (if we choose to employ the standard empiricist terminology). In the present section, I want to expand upon this claim and to offer arguments for it in somewhat greater detail.

To begin with, it is important to understand what sort of reliability of methodology is to be explained. If scientific realism is true, then the methodological practices of science provide a reliable guide to approximate truth about theoretical matters and, no doubt, only scientific realism could provide a satisfactory explanation for this fact. But it would be question-begging to suggest that this provides any good reason to accept scientific realism; after all, only realists believe that the methodology of science is reliable in this sense, anyway. What I propose to do is to take advantage of the fact that antirealists in the philosophy of science are typically selective in their skepticism and to define the reliability of the methods of science in such a way that no questions are begged against the position of the typical antirealist. Call a theory instrumentally reliable if it makes approximately true predictions about observable phenomena. Call a methodology instrumentally reliable if it is a reliable guide to the acceptance of theories which are themselves instrumentally reliable. For the antirealist against whom my arguments are directed, it is uncontroversial that the actual methods of science are instrumentally reliable in this sense, although it may of course be a matter for philosophical dispute just which features of actual scientific practice explain this reliability. The arguments I am discussing here are directed against only the selectively skeptical antirealist; I have nothing to say to "the Skeptic."

Let us suppose that some scientific theory T has been proposed and that a body E of experimental results has been obtained which is consonant with the predictions of T. Imagine that T is an ordinary medium-sized theory of the sort which scientists routinely confirm or disconfirm in the course of what Kuhn calls "normal science." (Scientific realism must have something to say about the acceptance of large-scale paradigm-fixing theories as well, but it is a matter of controversy whether there is a reliable methodology for such cases, and so I want here to examine the more commonplace instances of theory testing. For a realist treatment of the other cases, see Boyd 1979, 1981, 1983.) Questions regarding the extent to which T is confirmed by the evidence E may fruitfully divided into three categories:

1. *The question of "projectability."* One of the things which Goodman (1973) has taught us is that something important about inductive inference can be learned by examining the *unrefuted* inductive generalizations which no one ought to accept. Goodman formulates the issue in terms of the projectability of predicates in simple inductive generalizations, but it is clear that the issue he raises is more general. We can think of any sufficiently general theory as representing the proposal to consider as projectable certain possible patterns in observable data, *viz.*, those patterns which the theory predicts. In general, the methodological acceptability of such a proposal will not depend soley upon the projectability of the individual predicates contained in the theory in question considered in isolation, but also on the structure of the theory itself (see Boyd 1981, sec. 2.3; Boyd 1983, sec. 8, esp. p. 41). T will receive significant evidential support from E only if T represents a projectable pattern in possible observational data.

2. *The question of experimental controls and experimental artifacts.* Suppose that T represents a projectable pattern in possible observational data, and suppose further that the data in E represent apparently confirming evidence for T, whatever this latter constraint might come to. It will still be methodologically inappropriate to accept T as well confirmed unless there is reason to believe that the experiments involved in the production of these data were well designed. There must have been experimental controls for the influence of factors irrelevant to the assessment of T; in particular, the data which appear to confirm T must not be artifacts of the design of the experiments in question rather than genuine tests of the empirical adequacy of T. The analogous constraint applies, of course, to the case in which T is apparently disconfirmed by E.

3. *The question of "sampling."* Suppose that T represents a projectable pattern and that the experiments whose results are reflected in E are individually well designed. If we now ask how well, or to what extent, T is confirmed by E, we face head on the methodological analogue of the pure epistemologist's problem of induction. T will typically have infinitely many different observational consequences, and the problem of assessing the extent to which E confirms T comes down to the question of which

(typically relatively small) finite subsets of those consequences are such that their confirmation bestows significant confirmation on all the rest. (For the realist, of course, the problem is broader; one needs to know which such subsets bestow significant confirmation on the theory taken literally as a description of [partly] unobservable reality. Here, as in the case of the definition of reliability of methodology, I frame the issue in a way which does not beg the question against the antirealist.)

We may, I think, frame this question in a revealing way. The question is whether the consequences of T which have been tested are—in an epistemically appropriate sense—*a representative sample* of all the observational consequences of T. We cannot have checked out all of the (epistemically) possible ways in which T could "go wrong" with respect to observational prediction; there are, after all, infinitely many such ways. What we want to know is whether the experimental studies in question involve a representative sample of those ways, so that, if T hasn't gone wrong where we've tested it, then we can be justified in believing that it isn't going to go (very far) wrong at all. (Actually, this description is somewhat idealized; in the actual history of science, well-established theories have often turned out to be very wrong indeed in some of their empirical predictions. What is important is that we—rightly—expect experimental confirmation of a theory to warrant our belief that it will prove instrumentally reliable in a wide range of applications whose limits we cannot set in advance. The problem of identifying a relevantly representative sample of the observational predictions of a proposed theory is hardly rendered easier by this complication.)

The theory-dependence of the answers to these questions

The ways in which scientists answer these fundamental questions regarding the assessment of experimental evidence are quite profoundly dependent upon their prior theoretical commitments. That this is generally true of scientific methodology is now uncontroversial; Kuhn, Quine, both H. Putnams, Goodman, Glymour, and van Fraassen have all emphasized this point without, of course, all drawing realist conclusions. It will be important for our purposes to examine in some detail the ways in which theoretical considerations are involved in answering the three questions about experimental evidence which we have just identified.

1. *Projectability*. Kuhn (1970) correctly insists that in mature sciences the basic form of solutions to particular research problems is tightly circumscribed by the theoretical and research tradition (the "paradigm"), and van Fraassen (1980) agrees that the acceptance of particular theories involves the scientist "in a certain sort of research programme" (p. 12). The proposed theory T whose degree of confirmation by E is to be estimated will not be a serious candidate for comfirmation at all unless it arises as a

proposed solution to some problem: the extension of an existing theory to some new area of application, perhaps, or the explanation of some particular phenomena or observations. The theoretical tradition very sharply constrains such proposals; a proposed solution is unacceptable unless it is *theoretically* plausible in the light of existing theories, unless it is one of the solutions suggested by the existing "paradigm." Only those patterns in observable data are considered projectable which correspond to theoretically plausible theoretical proposals. Two facts about the theory-dependence of such projectability judgments are important for our concerns.

In the first place, such judgments sharply limit the generalizations *about observables* which we take to be confirmable. Suppose, as is typically the case, that T is put forward to account for some particular (finite) set of observational data. Of course, there will be infinitely many possible theories which would accommodate those data. Even if we take two such theories to be equivalent if they are empirically equivalent (or, better, if their respective integrations into the existing theoretical tradition would be empirically equivalent), there will remain infinitely many equivalence classes, each representing one possible observational generalization from the initial data. The effect of our theory-dependent judgments of projectability is to restrict our attention to a quite small finite number of these possible generalizations. Only the generalizations in this small set are potentially confirmable by observations, given the prevailing standards for the assessment of scientific evidence.

Secondly, the projectability judgments in question are genuinely *theory*-dependent. The judgments of theoretical plausibility which these projectability judgments reflect depend upon the *theoretical* structure both of the proposed solutions and of the received theoretical tradition. Proposed problem solutions are plausible, for instance, when the unobservable mechanisms they postulate are relevantly similar to the mechanisms postulated in the received theoretical tradition, where the relevant respects of similarity are likewise dependent on the theoretical structures postulated in the tradition. If the received body of theories were replaced by some quite different but empirically equivalent body of theories, then judgments of theoretical plausibility would pick out quite different problem solutions as acceptable and thus typically identify quite different patterns as projectable. As Kuhn insists, the ontology of the received "paradigm" is crucial in determining the range of acceptable problem solutions (and thus the range of projectable patterns in data).

2. *Experimental artifacts.* Suppose that T is theoretically plausible and thus represents a projectable pattern in observable data, and suppose that the experimental results in E appear to support (or refute) T. If these results are really to be evidentially relevant, then there must be reason to think

that the results favorable (or unfavorable) to T were not the result of features of the experimental situation which are irrelevant to the assessment of T. Of course, it is impossible to control for all epistemically possible experimental artifacts (of which there is an infinite number). Instead, we rely upon established theory to indicate the conditions under which the presence of experimental artifacts is to be suspected and the sorts of experimental controls which will permit us to avoid or discount for their effects. This is, I think, uncontroversial. It is also uncontroversial—although it is not much stressed in the literature—that our theory-dependent judgments in this area cut down the number of epistemically possible artifactual effects we actually control for from infinitely many to rather few.

What may be more controversial is whether or not these judgments are theory-dependent in the broader sense that they depend on the *theoretical* structure of the relevant background theories rather than just on their observational consequences. It might seem that they do not. Consider, for example, the commonplace that one must, in experiments involving electrical phenomena, control for the 60Hz hum induced by the alternating current in ordinary electrical wiring. Of course, the background theories which draw our attention to this sort of possible artifact have a complex theoretical structure, postulating electrons and electrical and magnetic fields and so forth. But, in order to know that we must shield various pieces of apparatus, all we need to know is that unless we do there will appear a certain sort of signal in our recording equipment superimposed on whatever signal comes from the preparation we are studying. If this sort of situation always obtains in cases of controlling for experimental artifacts, then it would appear that the methodological judgments which govern such controls do not depend on the theoretical structure of the relevant background theories.

Even if this were the case, there would, of course, be a significant way in which the identification of necessary experimental controls depends on the *theoretical* structure of the theories in the relevant theoretical tradition: the judgments of "projectability" which governed the acceptance of the generalizations about observables reflected in the currently accepted theories would have themselves depended on the *theoretical* structure of the earlier stages in the theoretical tradition. More importantly, it is by no means the case that the identification of relevant possible experimental artifacts depends solely on the observational consequences of the relevant background theories. This is so for two related reasons. In the first place, sound methodology often requires that we control for possible experimental artifacts whose effects are not by any means *predicted* by the received body of the theories but whose interference with the intended function of the experimental apparatus is *suggested* by those theories. Whatever may be the ultimate "rational reconstruction" of our practice, it is true that, in the typical case, the way in which the possible artifactual effects are suggested is that there are (typically unobservable) mechanisms postulated by the received theories about which it is *theoretically* plausible that either these mechanisms or mechanisms similar to them will produce the artifactual effects in question. Thus, we identify relevant possible experimental artifacts by something like "inductive" inference from theoretical premises, and the sorts of possible artifacts which we thereby identify depend dramatically on the theoretical structure of the theories which are the premises of these inferences.

We may see the same sort of theoretical-structure-dependent inferences in another methodologically important strategy for the identification of relevant possible experimental artifacts. Good methodology often requires that we control in one experimental situation E for some possible artifact A because we have already encountered similar artifacts A' in similar experimental situations E'. In the typical case, the possible artifact will be described in partly theoretical language, and the relevant respects of similarity (between E and E' and between A and A') will be determined by theoretical considerations—by considerations about the structure and effects of the unobservable mechanisms which the received theories postulate as operating in the relevant natural systems. In this case, too, whatever the ultimate reconstruction might be, the theoretical structure of the accepted theories, and not just their observational consequences, plays a crucial role in the identification of the relevant possible artifacts.

Two important points of similarity thus emerge about the way in which sound scientific methodology controls for the possibility of experimental artifacts and the way in which the problem of projectability is solved. In the first place, while there are infinitely many epistemically possible experimental artifacts which might affect any given experiment, scientific attention is paid to only a small finite number. In this regard, the identification of relevant possible experimental artifacts resembles the assessment of projectability: from an infinity of epistemic possibilities, the scientific method identifies a small finite number as methodologically relevant. The identification of relevant possible experimental artifacts resembles the solution to the problem of projectability in another crucial way: in each case the relevant methodology depends on the theoretical structure of the currently accepted scientific theories; were those theories replaced by others which are empirically equivalent but theoretically divergent, quite different methodological practices would be identified as appropriate. In both cases, scientists behave as though their methodology were determined by inductive inferences from the theoretical principles embodied in the received theoretical tradition.

3. *Sampling*. The pattern discernible in our examination of the ways in which scientific methodology handles the issues of projectability and experimental artifacts is even more striking in the case of the solution to the problem of "sampling." It is a fair statement of the most basic methodological principle governing the assessment of experimental evidence that a proposed theory T should be tested under conditions representative of those in which it is most reasonable to think that the theory will fail, if it's going to fail at all. The identification of these conditions rests upon *theoretical* criticism of that theory. The proposed theory T will, typically, postulate various mechanisms, entities, processes, etc., as factors in the phenomena to which it applies. Theoretical criticism involves the identification of alternative conceptions of the mechanisms, processes, etc., involved which are theoretically plausible—that is, which are suggested by the sorts of mechanisms, entities, etc., which are postulated by the received body of theories. These theoretically plausible alternatives to T will suggest circumstances in which the observational predictions of T might be expected to be wrong. It is under (representative instances of) these circumstances that T must be tested if it is to be well confirmed. This is the central methodological princple of experimental design.

Plainly, the methodological solution to the problem of sampling is theory-dependent. Moreover, whatever the "rational reconstruction" of this methodology might be, scientists do not in practice distinguish sharply between unobservable mechanisms, processes, entities, etc., and observable ones in identifying ways in which a proposed theory might reasonably be expected to fail. Indeed, inferences which look for all the world like inductive inferences from accepted premises about unobservables to conclusions about unobservables play an absolutely crucial role in the sort of theoretical criticism we are discussing. Thus, in the present case, as in the case of the methodological solutions to the problems of projectability and of experimental artifacts, the ways in which scientific methodology is theory-dependent are such that, if the existing body of theories were replaced by an empirically equivalent but theoretically divergent body of theories, our methodological judgments *regarding the "degree of confirmation" of generalizations about observables* would be profoundly different.

Projectability and induction about unobservables

The ways in which the features of scientific methodology just discussed depend on the theoretical structure of the received body of background theories may be seen more clearly if we consider a standard way in which philosophers in the tradition of logical positivism have treated the feature of theory testing which I have presented under the heading "Sampling." It has been widely recognized that at any given time in the history of science, and for any given problem or issue, there are typically only a very few theories "in the field" and contending for acceptance. The practice of testing a proposed theory against its most plausible rivals might, in this context, be seen as simply an application of the same pragmatic principle which dictates that, if there are only a very few brands of band saw available, one should evaluate each before making a purchase. This sort of description gives the appearance of reducing the methodological principle we have been discussing to a *merely* pragmatic level, denying it any special epistemic relevance.

Of course, such an interpretation *would not* deprive the practices we have been discussing either of their theory-dependence or of their epistemic importance. So long as the relevant rival theories are identified in the theory-dependent way described in the section on projectability, then the practice would be as theory-dependent as one could wish. Moreover, if testing proposed theories against instrumental rivals in the way suggested represented *the* methological solution to the problem of "sampling," then it has epistemic importance however much it may also have a purely pragmatic justification. What is most interesting in this context, however, is that the actual methodological practice of scientists departs in a revealing way from that suggested by the pragmatic account. In order for the pragmatic picture to have any plausibility, we must think of a proposed theory as competing against other possible predictive instruments roughly as powerful as itself. The rival "theories" against which it must be tested must be theories in the sense of fairly well developed systems with some significant predictive power. One does, after all, test band saws against other band saws.

While it is true that sound scientific methodology does require that a proposed theory should be tested against similarly well articulated rivals which are approximately equally theoretically plausible (roughly, that's what being a rival *theory* means, beyond having been invented in the first place), what is striking about the methodological practices which constitute the solution to the problem of sampling is that they may also require that a proposed theory be tested against a mere hunch, which has no deductive predictive consequences whatsoever. Suppose that a proposed theory T postulates a particular sort of unobservable mechanism as operating in the systems to which T applies, and imagine that T is sufficiently well worked out that (using well-established auxiliary hypotheses) it is possible to obtain experimentally testable predictions from T. Suppose, also, that theoretical criticism of T identifies alternative possible mechanisms, plausible in the light of received theories. Under these circumstances, it becomes necessary to try to pit T's conception of the matter against the alternative in some sort of experimental situation *even if* the alternative account is not nearly so thoroughly worked out as T and even if, for that reason, it yields (together with relevant auxiliary hypotheses) *no* definite predictions about observables at all. Under these circumstances, what sound methodology dictates is the identification of experimental circumstances under which the sorts of observations which it is *theoretically plausible* to expect given the alternative conception are different from those which one would expect given T's account of the relevant mechanisms.

By way of example, suppose that T provides an account of the reaction mechanisms for some biochemical process and that T is worked out in sufficient detail that it has (together with well-confirmed auxiliary hypotheses) significant deductive observational consequences. Suppose that a rival conception of the relevant reaction mechanisms is suggested by theoretically plausible considerations but that this rival conception is insufficiently well developed to have specific testable deductive consequences. It might nevertheless be possible to test T against the rival conception. Suppose that, in the case of better-studied systems, those systems to which mechanisms like those proposed by T and ascribed by the received theories are much more sensitive to some particular class of chemical agent than those to which mechanisms like those proposed in the alternative are ascribed (note here that the relevant respects of likeness will be determined by the content of theoretical descriptions of the systems in question, and the theoretical content of the relevant background theories, *and* that the class of chemical agents in question may similarly be theoretically defined). Under such circumstances, sound methodology will dictate subjecting the biochemical systems to which T applies to chemical agents in the relevant class; data indicating considerable sensitivity to such agents will be especially important for the confirmation of T precisely because they will constitute a test of T against the theoretically plausible rival conception of the relevant reaction mechanisms. Note that in the present case the rival conception need not have any *deductive* observational consequences regarding the experimental situations in question. Instead, reasoning by analogy *at the theoretical level* makes it *theoretically plausible* to expect low sensitivity if the rival conception is true. T is tested against a theoretically plausible hunch about how it might go wrong.

Indeed, the role of considerations of theoretical plausibility in theory testing can go even deeper; a proposed theory may be pitted against a theoretically plausible rival in a particular experimental setting even though neither the rival *nor* the proposed theory have (when taken together with appropriate well-confirmed auxiliary hypotheses) any deductive observational predictions about the results of the experiments in question! In the example we have been considering, the appropriateness of the experimental test in question does not depend on the theory T's having any observational deductive predictions about the results of the experiment. All that is required is that it be *theoretically* plausible that a test of the sensitivity of the relevant biochemical systems to the specific chemical agent will provide an indication of which of the two accounts of reaction mechanisms (if either) is right. We may test T by pitting a hunch about the outcome of experimentation which is theoretically plausible given T (and the body of received theories) against an experimental hunch which is theoretically plausible on the assumption of the rival conception of reaction mechanisms. Even though we have assumed that T makes a significant number of deductive observational predictions, we need not assume that it makes any *deductive* predictions about the outcome of this crucial experimental test! In sciences which deal with complex systems, instances of theory testing which fit the model just presented are by no means uncommon. Indeed, it may be a good idea to ask whether in describing the instrumental application of theories (rather than their confirmation)—when defining empirical adequacy, for example—the idealization that it is the *deductive* observational consequences of a theory (together with auxiliary hypotheses) rather than its *inductive* consequences that are relevant may not be fundamentally misleading; but that is a topic for another paper.

In any event, what we may learn from these examples is that, in practice, inductive inferences in science extend to inferences with theoretical premises and theoretical conclusions. Just as there are theory-dependent judgments about which possible patterns in observables are projectable, so there are judgments about which patterns in the properties or behavior of "theoretical entities" are projectable. Just as there are theory-dependent judgments of the "degree of confirmation" of instrumental claims by empirical data, so there are theory-dependent judgments of the plausibility of various theoretical claims in the light of other considerations both empirical and theoretical. Indeed, whatever the correct philosophical analysis of this matter, scientific methodology does not dictate any significant distinction between inductive inferences about observables and what certainly look like inductive inferences about unobservables. Finally, and most strikingly, the very methodological principles which govern scientific induction about observables are, in practice, parasitic upon "inductive" inferences about unobservables.

An argument for scientific realism

It will be evident how one may argue for scientific realism on the basis of the theory-dependence of experimental methodology. Consider the question, why are the methodological practices of science instrumentally reliable? Both scientific realists and (almost all) empiricists agree that these practices are instrumentally reliable, but they differ sharply in their capacity to explain this reliability. So theory-dependent are the most basic principles for the assessment of experimental evidence that it must be concluded that these are principles for applying the knowledge which is reflected in currently accepted theories as a guide to the proper methods for the evidential assessment of new theoretical proposals; any other conclusion makes the instrumental success of the scientific method a miracle.

According to the empiricist, the knowledge reflected in the existing body of accepted theories at any time in the history of science is entirely instrumental knowledge: the most we know on the basis of experimental evidence is that the existing body of theories is empirically adequate. Thus, the replacement of existing theories by an empirically equivalent set of theories would leave the knowledge they embody unchanged. Thus, the empiricist can explain the epistemic adequacy of only those theory-dependent features of scientific methodology whose dictates are preserved under the substitution, for the actual body of accepted theories, of any other empirically equivalent one. But, as we have just seen, *none* of the central methodological principles which govern the evaluation of scientific evidence have this property! The consistent empiricist cannot explain the instrumental reliability of the methodology which scientists actually employ.

The scientific realist, on the other hand, has no difficulty in providing the required explanation. According to the realist, existing theories provide approximate knowledge not only of relations between observables, but also of the unobservable structures which underlie observable phenomena. In applying theory-dependent evidential standards, scientists use existing theoretical (and observational) knowledge as a guide to the articulation and experimental assessment of new theories. The judgments of projectability, identification of experimental artifacts, and theoretical criticisms of proposed theories which look ever so much like inductive inferences *are* inductive inferences from acquired theoretical knowledge to new theoretical conclusions. When a theoretical proposal is theoretically plausible in the light of the existing theoretical tradition, what that means is that it is supported by an inductive inference at the theoretical level from previously acquired theoretical knowledge.

Judgments of "projectability" are thus just what they look like "pre-analytically": they represent the identification of theoretical proposals for which there are good inductive reasons to believe that they are (approximately) true and thus for which there is good reason to believe that they will eventually be articulated into empirically adequate theories. The role of experimentation is to choose between the various theoretical proposals which pass this preliminary test for probable (approximate) truth.

Similarly, the judgments of theoretical plausibility by which possible experimental artifacts are identified turn out to be inductive inferences from theoretical knowledge which result in reliable assessments of the evidential likelihood that various unobservable factors will influence the outcome of experiments. Finally, the methodological solution to the problem of sampling really does consist in identifying—by reliable inductive inference from theoretical knowledge—the most plausible rivals to a proposed theory and the experimental conditions under which they can be effectively pitted against it. The reliability of scientific methodology in guiding induction about observables turns out to be largely parasitic upon the reliability of the methodology in applying existing theoretical knowledge to guide the establishment of new theoretical knowledge (see Boyd 1973, 1981, 1983). Only this explanation, the realist maintains, can account both for the reliability of the scientific method and for the fact that seemingly inductive reasoning about theoretical matters is so central to it.

Advantages of this defense of scientific realism

The argument for scientific realism just presented has two distinctive advantages when compared with other arguments for realism which emphasize the theory-dependence of scientific practice. In the first place, the standard empiricist response to indications that the methods of science are theory-dependent is to invoke the distinction between the "context of discovery" or "context of invention" on the one hand and the "context of justification" or "context of confirmation" on the other. Theory-dependent features of scientific practice—at least those which depend on the theoretical structure of received theories—are said to be "merely heuristic" features of science which form part of the context of discovery or invention, but which are largely irrelevant to (rationally reconstructed) conformation or justification. The present argument is designed to block such a response by arguing that the relevant theory-dependent features of method are absolutely central to the sort of justification and confirmation which even the empiricist must accept if she holds that scientists can obtain (approximate) instrumental knowledge.

The second advantage to the defense of realism we are discussing is that it provides an epistemologically coherent rebuttal to the empiricist principle that empirically equivalent theories are equally supported or refuted by any body of observations. The evidence for or against a theory is *not* just a matter of the accuracy of its tested empirical predictions; considerations of the theory's own theoretical plausability in the light of received theories, and of the theoretical plausibility of various possible rivals, are essential to its epistemic assessment. The fact (if it is a fact) that a proposed theory (or one of its rivals) is theoretically plausible constitutes inductive warrant for the belief that the theory (or the rival) is (approximately) true. Moreover, the view that considerations of theoretical plausibility are evidential in this way *does not* constitute an abandonment of the doctrine that scientific knowledge is grounded in experiment. The background theories with respect to which theoretical plausibility is assessed are, after all, not *a priori* truths; they have themselves been previously tested by experiment. The theoretical plausibility of a proposed theory thus represents theory-mediated experimental evidence in its favor. Indeed, as we have seen in the discussion of the assessment of experimental evidence, all evidential support which a theory receives from experimental evidence is strongly theory-mediated, even when the evidence involves the confirmation of deductive observational predictions from the theory itself. The empiricist is right that all scientific knowledge is experimental knowledge, but the empiricist conception of experimental evidence fails to include an account of methodologically crucial inductive inferences at the theoretical level; when these are taken into account, the doctrine of the evidential indistinguishability of empirically equivalent theories is evidently false. (For a more elaborate treatment of these issues and of their relation to naturalistic epistemology, see Boyd 1979, 1981, 1983.)

Theoretical inductions and the "unity of science"

Several philosophers have suggested that the so-called "unity of science principle," according to which independently well confirmed theories can be expected to be conjointly empirically adequate, provides a basis for an argument for realism (see, for example, Putnam 1975, chap. 2; Putnam 1978, pt. 3; Boyd 1979, 1981, 1983). The idea here is that, if T and T' are independently tested theories (in the sense that neither has been employed as an "auxiliary hypothesis" in the testing of the other), then the empiricist ought to hold that all that has been tested is the empirical adequacy of each of these theories (together with the auxiliary hypotheses which were used with each in its confirmation). From just the claim that each of the theories is empirically adequate in this sense, it certainly does not follow (even inductively) that the conjunction of the theories will be (even approximately) empirically adequate. Only in the case in which the theoretical vocabularies of the two theories are disjoint would the inference be sound. But the unity of science principle holds (correctly) that the evidence for the two independently tested theories *does* constitute evidence for their conjoint reliability, provided only that no term (theoretical or observational) occurs nonunivocally when the two theories are conjoined.

For the reliability of this principle, the realist offers the explanation that experimental evidence for T and for T' constitutes evidence for the truth of each of the theories and that the judgments of univocality for theoretical terms which are presupposed in the unity of science principle constitute reliable judgments of sameness of reference for the theoretical terms common to T and T . Thus, under the circumstances envisioned for the application of the unity of science principle, we have evidence that T and T' are true theories with relevantly the same subject matter, and, thus, we have evidence that their conjunction is true. It is this consideration which justifies us in taking the empirical adequacy of their conjunction to be confirmed. (For a more careful statement of this argument, see Boyd 1981, sec. 2.4.) Here again realism explains the instrumental reliability of a methodological principle whose reliability the empiricist cannot explain.

It is fairly easy to say why the unity of science principle poses a problem for the antirealist. The fact that scientists accept something like the deductive closure of the various individual theories they accept, together with the fact that many of these theories share common theoretical terms, means that in scientific practice (and, in particular, in scientific inductions about observables) theories are integrated in ways which do not appear to make sense unless theories are taken to reflect knowledge of the entities to which their theoretical terms refer.

In the light of the preceding discussion of the role of inductive inferences at the theoretical level in the assessment of experimental evidence, we may offer an even more powerful version of the argument for realism from the unity of science. Not only does sound scientific methodology dictate the *deductive* integration of theories described by the positivists' unity of science principle, it also dictates the *inductive* integration of theories— the use of individually well confirmed theories (sharing common theoretical terms) as premises in inductive as well as deductive inferences. It is just such inferences which are methodologically crucial in the assessment of experimental evidence, and—as we have seen—these inferences make epistemic sense only if the evidence for particular theories is taken to be evidence for their approximate truth (and if our judgments of univocality for theoretical terms are reliable)—that is, only if a realist conception of scientific inquiry is adopted.

It is thus clear that reflection on the crucial epistemic role of inductive inferences at the theoretical level indicates that scientific theories are even more tightly integrated methodologically than the deductive version of the unity of science principle suggests, and the instrumental reliability of the inductive integration of theories provides even greater evidence for a realist conception of theoretical knowledge and a realist conception of the referential semantics for theoretical terms. (For a suitable realist account of reference, see Boyd 1979, 1981.) If the truth be known, the real unity of science is inductive unity, and inductive unity is no respecter of the observation-theory dichotomy.

II. Constructive Empiricism and the Theory-Dependence of Experimental Design

In the present part of this paper, I want to consider the ways in which the empiricist philosopher of science might reply to the arguments in the preceeding sections. I will take van Fraassen 1980 as a paradigm presentation of the empiricist position, but I will try to focus on features of van Fraassen's views which would be broadly acceptable to empiricists, and I will explore some options open to the empiricist which van Fraassen does not explicitly take. A methodological point is in order here: I will not exactly be speculating about the sorts of replies which van Fraassen might offer to the realist arguments just presented. The argument for realism based on the theory-dependence of experimental design is one which van Fraassen discusses in some detail, and so is the argument for realism based on the deductive version of the unity of science principle. It is nevertheless true that I have had the advantage of reformulating the realist arguments in question after having read van Fraassen's thoughtful and stimulating criticisms of them. Since the method of philosophy is dialectical, it can hardly be expected that my account of how the empiricist might reply to these arguments for realism will be definitive. No doubt van Fraassen and other empiricists will have extremely interesting things to say which are more ingenious than those things I say on their behalf. In the meantime, I hope not to be unfair in my treatment of the arguments which van Fraassen has already presented.

I think that we may fruitfully classify van Fraassen's responses to the arguments for realism into five categories: (1) van Fraassen's explicit reply to the argument for realism based on an examination of the methodological principles which govern experimental design; (2) his explicit reply to the argument for realism based on consideration of the unity of science principle in its deductive version; (3) van Fraassen's appeal to a "Darwinist" explanation of the reliability of the experimental method; (4) his discussion of the role in scientific methodology of "pragmatic" virtues of theories like simplicity and explanatory power; and, finally, (5) van Fraassen's discussion of the quite specific philosophical appeal to explanatory power which is involved in the contention that we should believe scientific realism because only scientific realism adequately explains the instrumental reliability of the scientific method.

Van Fraassen on the theory-dependence of experimental design

Van Fraassen discusses at some length the theory-dependence of experimental design and offers an explicit reply to the resulting argument for realism, responding to the version of that argument presented in Boyd 1973. So effective is van Fraassen in his description of the depth to which the scientific standards of experimental design are theory-dependent that I can't resist the temptation to quote him:

> The real importance of theory, to the working scientist, is that it is a factor in experimental design.
> This is quite the reverse of the picture drawn by traditional philosophy of

science. In that picture, everything is subordinate to the aim of knowing the structure of the world. The central activity is therefore the construction of theories that describe this structure. Experiments are then designed to test these theories, to see if they should be admitted to the office of truth-bearers, contributing to our world-picture.

Whatever the core of truth in that picture (and surely it has some truth to it) it contrasts sharply with the activity Kuhn has termed "normal science", and even with much of what is revolutionary. Scientists aim to discover facts about the world—about the regularities in the observable part of the world. To discover these, one needs experimentation as opposed to reason and reflection. But those regularities are exceedingly subtle and complex, so experimental design is exceedingly difficult. Hence the need for the construction of theories, and for appeal to previously constructed theories to guide the experimental inquiry. . . .

. . . For theory construction, experimentation has a twofold significance: testing for empirical adequacy of the theory as developed so far, *and* filling in the blanks, that is, guiding the continuation of the construction, or the completion, of the theory. Likewise, theory has a twofold role in experimentation: formulation of the questions to be answered in a systematic and compendious fashion, *and* as a guiding factor in the design of the experiments to answer those questions. In all this we can cogently maintain that the aim is to obtain the empirical information conveyed by the assertion that a theory is or is not empirically adequate. (Van Fraassen 1980, 73–74; all subsequent citations are from van Fraassen 1980 unless otherwise indicated.)

When van Fraassen turns his attention to the argument for realism based on the theory-dependence of experimental design, he deals explicitly with a case (from Boyd 1973) of the theory-dependence of what I have here called the solution to the problem of sampling. In that example, a theory L about the mechanisms by which an antibiotic affects a particular bacterial species is proposed and the experimental tests of it which are crucial are identified by considering the sorts of mechanisms which are said by previously accepted theories to operate in, or to interfere with, other antibody-bacteria systems which are (from a theoretical perspective) relevantly similar. The relevant experiments test predictions of L about the effects over time of various doses of the antibiotic on the density of bacterial populations, and the role of theory-dependent considerations is to indicate which of these predictions are most likely to go wrong. The realist argument is that only on a realist understanding both of the theory L and of the knowledge embodied in the relevant background theories can one explain the contribution of these theory-dependent considerations to the reliable testing of L. Here is van Fraassen's rebuttal:

We must admit that this is one explanation: that the collateral theories are believed to be true. But Boyd needs to establish not only that, as a realist, he can explain what is happening, but also that competing explanations are not feasible.

Let us see then, on Boyd's behalf, how an empiricist can render this methodology intelligible. In the above examples, the collateral theories suggested ways in which the function governing population decrease, in terms of drug dosage and time elapsed, might prove to be observably false. Boyd's point is no doubt that the manner in which those theories suggested these consequences, was by suggesting alternative underlying mechanisms which are not directly observable.

I would put this as follows: the models of L are quite simple, and reflection on the models of the collateral theories suggests ways in which the models of L could be altered in various ways. The empirical adequacy of L requires that the phenomena (bacterial population size and its variation) can be fitted into some of its models. Certain phenomena do fit the suggested altered models and not the models of L as it stands. Thus a test is devised that will *favour L* (or not favour it) *as against one of those contemplated alternatives*. But it is easy to see that what such a test will do is to speak for (or against) the empirical adequacy of L in those respects in which it differs from those alternatives.

The talk of underlying causal mechanisms can be construed therefore as talk about the internal structure of the models. In contrast with the logical, syntactic construal of theories which Boyd used in the discussion of what he called principle 1, we must direct our attention to the family of models of the theory to make sense of the pursuit of empirical adequacy through total immersion (for practical purposes) in the theoretical world-picture. (P. 80)

I take it that van Fraassen is here making a very important point: Once the empiricist abandons the verifiability theory of meaning, it is prefectly possible for her to acknowledge that background theories *suggest* new possibilities to scientists in virtue of the theoretical claims which they make and to describe the reasoning involved as inductive inference from theoretical premises. Acknowledging that scientists make theoretical inductions does not, however, commit the empiricist to holding that the premises of such inductions are possible objects of (noninstrumental) knowledge. ". . . [I]mmersion in the theoretical world-picture does not preclude 'bracketing' its ontological implications" (p. 81). Thus, if the realist were to argue for realism merely on the basis of the fact that scientific practice involves genuine inductive inferences at the theoretical level, she would perhaps refute the verifiability theory of meaning, according to which purely theoretical premises are meaningless, but she would not thereby refute the sophisticated empiricist position advocated by van Fraassen. Quite so. But the problem raised by Boyd 1973 is a different one.

What the realist asks for is an explanation of the contribution of theoretical inductions to the identification of the appropriate experimental tests for proposed theories. The problem is not how background theories can *suggest* to us alternatives to proposed theories or how we can design experiments which test proposed theories against the suggested alternatives. The problem is why the alternatives suggested in this way have a privileged epistemic status, why it is against them and not against other logically possible alternatives that a proposed theory must be tested if it is to receive significant evidential support. Experimental design is "exceedingly difficult" as van Fraassen says, but it would be substantially less difficult if one could legitimately test proposed theories against just any logically possible alternatives. One could then solve the problem of sampling by asking a philosopher to suggest alternatives to the predictions which follow from the theory to be tested (since philosophers are experts on logical possibility). Instead, what good scientific practice requires is that one accept the suggestions which follow by induction from the accepted body of theories. What the empiricist apparently cannot do is to explain why it is *this* solution to the problem of sampling which is instrumentally reliable.

It is unclear whether or not van Fraassen's rebuttal to Boyd 1973 is really intended to address this issue. If so, the best reconstruction of the rebuttal which I can see is this: The relevant background theories are well confirmed, and so there is reason to think that they are empirically adequate. The rivals to L which they suggest are, of course, *similar* to these empirically adequate theories, so there are inductive reasons to think that they, too, are empirically adequate. If any one of them is empirically adequate, then L is not, so it is inductively reasonable to use such theories as a guide to the circumstances in which L is most likely to go wrong (empirically), if it's to go wrong at all. All that's involved is the inductive inference: Such-and-such background theories are empirically adequate; so-and-so alternatives to L are similar to these empirically adequate theories; therefore, it is reasonable to believe that one of them is empirically adequate and (therefore) that L makes a false prediction in one or more of the cases in which its predictions differ from those of the so-and-so theories. This is, so far as I can see, exactly right; but, of course, the respects of similarity between the relevant background theories and the suggested alternatives to L lie in the theoretically relevant similarities between the accounts they offer of unobservable phenomena (that's how the examples are constructed). The problem for the antirealist is then why these theory-determined respects of similarity are (out of the infinitely many possible respects of similarity) the relevant ones. Nothing van Fraassen says provides an antirealist answer to the basic question of Boyd 1973:

> Suppose you always "guess" where theories are most likely to go wrong experimentally by asking where they are most likely to be false as accounts of causal relations, given the assumption that currently accepted laws represent probable causal knowledge. And suppose your guessing procedure works—that theories really are most likely to go wrong—to yield false experimental predictions—just where a realist would expect them to. And suppose that these guesses are so good that they are central to the success of experimental method. What explanation beside scientific realism is possible? (Boyd 1973, 12)

Van Fraassen on unity of science

Van Fraassen explicitly discusses the argument for scientific realism based on the deductive version of the unity of science principle, which he calls "The Conjunction Objection" (pp. 83–87). It is somewhat difficult to diagnose all of the dimensions of his criticisms of the argument, but two claims stand out. In the first place, van Fraassen seems to claim that *even as an idealization* the deductive version of the unity of science principle does not really describe the epistemic judgments dictated by sound methodology:

> [A]s long as we are scientific in spirit, we cannot become dogmatic about even those theories which we whole-heartedly believe to be true. Hence a scientist must always, even if tacitly, reason *at least* as follows in such a case: if I believe T and T' to be true, then I also believe that $(T$ and $T')$ is true, and hence that it is empirically adequate. But in this new area of application T and T' are genuinely being used in conjunction; therefore, I will have a chance to see whether $(T$ and $T')$ really is empirically adequate, as I believe. *That belief is not supported yet to the extent that my previous evidence supports the claims of empirical adequacy for T and T' separately, even though that support has been as good as I could wish it to be.* [Emphasis mine] Thus my beliefs are about to be put to a more stringent test in this joint application than they have ever been before. (P. 85)

It is true that applications of the unity of science principle do put the conjoined theory to an additional test (as do all previously untried applications of any theory). Therefore, *in a sense*, the as yet untested conjunction of T and T' is not as well supported as the two theories are separately, but it would be a mistake to conclude from this fact that the conjunction in question is evidentially dubious prior to the "more stringent" tests which result from its application. It is well to remember that all cases of theory-determined improvement of measurement and instrumentation in science represent applications of the unity of science principle (Boyd 1981, secs. 2.1, 2.4). Such improvements in instrumentation do not, in the typical case, represent especially speculative or methodologically dubious features of the scientific method; instead, they are central to the improvement of scientific knowledge. The realist has no difficulty in explaining the sense in which successful applications of the conjunction of T and T' represent a more stringent test; what the antirealist apparently cannot explain is why the independent confirmation of T and T' should constitute any significant evidence for the empirical adequacy of their conjunction *at all*! Van Fraassen's first response, although perfectly true, does not address the crucial epistemological challenge raised by the realist.

The other response to the argument for realism from the unity of science which is clear in van Fraassen's discussion involves his insistence that, before scientists accept the conjunction of two well-confirmed theories, they often *correct* them first. As Demopoulos (1982) observes, no realist denies that this happens, and the challenge for the nonrealist to explain the epistemic legitimacy of the resulting integration of independently established theories still remains. I believe, however, that van Fraassen has

nevertheless made a very significant point about the nature of the integration of theories. It may well be that, for many cases involving "small"
theories whose adoption does not "make waves," the unity of science principle holds in the strong sense that the conjunction of two well-confirmed
theories can be accepted on the basis of the independent evidence supporting each. But, in cases in which the adoption of a particular theory has a
serious effect on scientific understanding (and this need by no means entail
a "scientific revolution"), some correction of previously accepted theories
is often called for, and the necessity of such corrections is often most evident in the case where previously accepted theories are to be applied conjointly with the newly discovered theory. When theories make waves, the
ripple effect requires the modification of previously adopted theories.

What is important for the debate between realists and nonrealists is that
the modifications in question are themselves theory-determined. From the
infinitely many possible modifications of current theories which might be
occasioned by the adoption of a new theory (and by whatever new data
support it), scientists choose to consider those which are theoretically
plausible—which are suggested by inductive inferences at the theoretical
(as well as the observational) level. It is the *inductive* integration of theories
which is reflected in the sorts of theory modifications which follow a major
theoretical innovation. What van Fraassen's insistence on the role of proper
correction in the formulation of the unity of science principle correctly indicates is that the unity of science is inductive unity; but, as we have already seen, the epistemic appropriateness of the inductive unity of science
is something which only a realist can explain.

Darwinism and the methods of science

Toward the end of chapter 2, after discussing the argument that only
realism makes the success of science nonmiraculous, van Fraassen offers a
charmingly succinct alternative to the realist's explanation for the success
of scientific practice:

> Well, let us accept for now this demand for a scientific explanation of the
> success of science. Let us also resist construing it as merely a restatement of
> Smart's "cosmic coincidence" argument, and view it instead as the question
> why we have successful scientific theories at all. Will this realist explanation
> with the Scholastic look be a scientifically acceptable answer? I would like to
> point out that science is a biological phenomenon, an activity by one kind of
> organism which facilitates its interaction with the environment. And this makes
> me think that a very different kind of scientific explanation is required.
> I can best make the point by contrasting two accounts of the mouse who
> runs from its enemy, the cat. St. Augustine already remarked on this phenome-
> non, and provided an intensional explanation: the mouse *perceives* that the cat is
> its enemy, hence the mouse runs. What is postulated here is the "adequacy" of
> the mouse's thought to the order of nature: the relation of enmity is correctly

reflected in his mind. But the Darwinist says: Do not ask why the *mouse* runs from its enemy. Species which did not cope with their natural enemies no longer exist. That is why there are only ones who do.

In just the same way, I claim that the success of current scientific theories is no miracle. It is not even surprising to the scientific (Darwinist) mind. For any scientific theory is born into a life of fierce competition, a jungle red in tooth and claw. Only the successful theories survive—the ones which *in fact* latched on to actual regularities in nature. (Pp. 39–40)

In order to assess the plausibility of this alternative explanation, we need to understand more about what biological Darwinism accomplished and to explore the relation between biological Darwinism and van Fraassen's methodological Darwinism. What Darwin and later Darwinists did was to show that the observed adaptation of organisms to their respective environments can be satisfactorily explained without postulating purposive or teleological forces in nature. What van Fraassen's methodological Darwinism aims to show is that the observed adaptation of theories to observable phenomena (i.e., the instrumental reliability of scientific method) can be explained without postulating theoretical knowledge. In the case of biological Darwinism, research attention has, from Darwin 1859 to the present, focused on the *details* of the *mechanisms* of evolution: the mechanisms which produce and constrain heritable variations in phenotypic traits and the various mechanisms—individual selection, sexual selection, kin selection, (perhaps) group selection, genetic "drift", etc.—which produce changes in gene frequency and which are involved in speciation. Thus, to cite two examples among very many, quite special selection mechanisms (kin selection) and quite particular and unusual features of genetic mechanisms (haplodiploidy) are crucial in prominent evolutionary accounts of the evolution of social castes in the Hymenoptera (see Hamilton 1964), and the most significant current debate in the foundations of evolutionary theory is about whether the integration in the 1930s of Mendelian genetics and Darwin's specifically gradualist conception of natural selection (the "modern synthesis") is compatible with fossil data regarding the tempo of evolution and with what is now known about the genetic mechanisms of speciation (see Gould and Eldredge 1977).

What is important for our discussion is that, not only do evolutionary theorists study the particular mechanisms which underlie heritable variation, selection, and speciation, but the success of Darwin's antiteleological project depends upon the results of such studies. As Darwin was acutely aware, there are a number of possible ways in which the mechanisms of heritable variations or of selection might conceivably work such that, if they actually obtained, the evidence would favor a teleological conception of evolution rather than a materialist conception. There are three significant ways in which facts about genetic variation and evolutionary selection mechanisms *might* have undermined Darwin's antiteleological position. For each of these there is an analogous way in which facts about variation and selection in the evolution of scientific theories might tell against the antirealist. If we follow the example of evolutionary biologists and examine the details of the mechanisms of the evolution of theories, then, I believe, we will find that van Fraassen's antirealist conception of the evolution of theories is refuted at just the points suggested by analogy with the ways in which Darwin's antiteleological position *might* have been refuted. I turn now to a consideration of those points:

1. *Directed variation*. The heritable variations upon which the mechanisms of natural selection operate are limited: not all conceivable phenotypic variations occur, and not all of those which occur are heritable. In general, limitation on heritable variation reduces the effectiveness of natural selection (or kin selection, etc.) in producing adaptations of organisms to their environments. Darwin emphasizes that evolution is opportunistic, "seizing" the opportunities provided by actually occurring variations rather than achieving the perfection in design which might result if heritable variations in the relevant traits were less limited. It *could be*, however, that the natural constraints on heritable variation contribute to, rather than diminish, the effectiveness of evolution in producing adaptations. Heritable variations might be "directed" in the sense that variations are more likely to occur when they would be useful to the relevant organisms. Were this the case, there would be *prima facie* evidence for the operation of teleological mechanisms in evolution, and Darwin is at pains to deny that heritable variation is typically directed in this sense. He proposes that all actual cases of directed variation can be explained by the inheritance of acquired characteristics and other similar mechanisms of a nonteleological sort. If facts had been other than they are—if biological limitations on heritable variation had turned out to contribute to the effectiveness of selection, and if no mechanism like inheritance of acquired characteristics had been sufficient to explain this phenomenon—then there would have been good evidence for the existence of teleological factors in nature, and Darwin's materialist position on this issue would have been *prima facie* refuted.

What is the analogous issue for the antirealist conception of the evoluion of theories? The analogue in the case of the evolution of theories to the biological limitations on heritable variation are whatever methodological limitations act to constrain the range of possible theories which are subjected to "selection," that is, to experimental testing and to other sorts of methodologically appropriate evaluation. The analogue of the issue of directed variation is the issue of whether these methodological limitations act to increase the effectiveness of selection—that is, to increase the instrumental reliability of scientific methodology—in ways which cannot be explained without postulating theoretical knowledge. As we have seen in section 1, the methodological limitations in question are just those which represent the methodological solution to the generalized problem of "projectability" for patterns in observable data, and indeed these limitations *do* contribute to the instrumental reliability of scientific methods in a way which cannot be explained in a nonrealist fashion.

2. *The efficiency of selection.* Darwin 1859 introduces the idea of natural selection by considering variation and selection of organisms under domestication. The purposive selection involved in careful breeding of plants and animals, Darwin argues, can lead to the establishment of strains which are so phenotypically different that they appear to represent different species or even different genera and which are "adapted" to the needs and interests of breeders. Given the evidence for the great antiquity of fossil organisms and of the earth, Darwin suggests, it is reasonable to hold that the nonpurposive and much less efficient mechanisms of natural selection could have produced even more substantial changes and even more various and refined adaptations (in this case, to the demands of the environment). Even if we assume that Darwin was right that species originated by selection operating on heritable variations, and even if we assume that he was right that heritable variations are not teleologically determined, it remains true that Darwin's antiteleological position would have been refuted if the data had shown that selection had been *too* efficient to be explained by any plausible materialist theory. Suppose, for example, that new data about the age of the earth or about the actual rate of speciation, together with findings about genetic mechanisms and about the effects of various natural mechanisms of selection, had made it impossible to explain the extent of adaptation or of speciation merely in terms of the interaction of the relevant genetic mechanisms with naturally occurring mechanisms of selection. In that case, the evidence would have favored a teleological conception according to which something rather like artificial selection guided biological evolution. Darwin's position would have been (*prima facie*) refuted by an observed efficiency of evolutionary selection mechanisms too great to be explained in materialist terms.

What is the analogous issue for the antirealist Darwinian conception of the evolution of theories? The analogues to biological selection mechanisms are the various mechanisms for theory evaluation which are dictated by scientific method. The analogue to the issue of efficiency of the mechanisms of biological selection is the issue of whether or not the important contribution of the methodological principles which govern theory testing to the instrumental reliability (efficiency) of scientific practice can be explained without postulating theoretical knowledge. But these are just the theory-dependent mechanisms by which the problems of experimental artifacts and of "sampling" are solved, and we have already seen that only a realist can explain their contribution to the instrumental reliability of scientific methods. Once again, if we follow the example of biologists and examine closely the mechanisms of theory selection, we see that the antirealist view of the evolution of theories is not sustained by the data.

3. *Future-directed selection.* It is a consequence of any Darwinian conception of evolution that, insofar as natural selection tends to establish adaptive traits, the relevant measure of adaptiveness is the reproductive fitness of the particlar organisms which have the traits in question *rather than* the long-term survivability of the species. It is commonplace, for example, to cite cases of selection for fecundity which eventually helps to create conditions of overpopulation leading to extinction. The conception that selection cannot be directed toward the long-term needs of species except insofar as those needs are captured by present reproductive needs is, of course, *prima facie* dictated by a materialist and antiteleological conception of evolution. If it had turned out that evolutionary mechanisms tended to establish traits which reduced immediate reproductive fitness but which served instead the long-run interest of species survival, then, barring some materialist account of this remarkable phenomenon, there would have been strong *prima facie* evidence for a teleological conception of evolution and against Darwin's conception.

Is there an analogous issue for the theory of the evolution of scientific theories? I think the issue is this: Do the actual mechanisms of theory evaluation permit us to assess the *future* empirical reliability of the theories we test (*future* in the sense of going beyond the reliability actually demonstrated in the outcomes of the relevant experiments) in a way which cannot be explained without postulating theoretical knowledge? Here again, the answer discovered by a careful examination of the methods of science refutes the nonrealist interpretation of the evolution of scientific theories. Scientific methodology does indeed allow us to assess the future empirical reliability of theories by solving the problems of projectability, experimental artifacts, and sampling, but the ways in which it does so are inexplicable except from a realist perspective. The unity of science principle illustrates what is essentially the same phenomenon. The independent testing of particular scientific theories serves to establish their approximate instrumental reliability with respect to both deductive and inductive applications in future (conjoint) contexts quite unlike those in which their reliability has been "directly" tested. Here, too, the only explanation for this future-regarding feature of the methodological selection of theories is one which postulates theoretical knowledge.

I conclude that van Fraassen's appeal to the analogy between theory testing and natural selection provides a valuable perspective from which to study the epistemology of science but that—when we carry out the scientific investigations suggested by the analogy—it turns out that a realist rather than an antirealist conclusion is dictated by the available evidence.

There is one more lesson to be learned from the analogy with evolutionary theory. Van Fraassen's antirealist appeal to a Darwinian conception of theory evolution is strikingly reminiscent of an approach to biological evolution called (by its followers) "optimization theory" or (by its critics) "adaptationism." Roughly, the idea of this approach is that, relatively independently of considerations of the details of evolutionary and genetic mechanisms, one may generally expect natural selection to produce near optimal "solutions" to the "problems" set for organisms by their environments. If this is so, then within certain limits one can take the problem-solving capacity of natural selection for granted without examining the details of the underlying mechanisms. By analogy, one might think it appropriate in epistemological contexts to take the instrumental reliability of scientific methodology for granted, subsuming it under the broad category "induction" and taking the notion of induction pretty much for granted. Gould and Lewontin (1979) have offered serious criticisms of adaptationism in evolutionary theory. Whatever may be the merits of their criticisms, there seems little doubt that analogous arguments of Goodman (1973) show that one cannot take the foundations of induction for granted in epistemology.

> Consider the question: How is reasonable expectation about future events possible? ("Future" may be replaced by "unobserved" for generality.) The recurrent idea that there is some rational form of simple extrapolation from the past, something like rules of induction, may be especially appealing to empiricists because it holds out hope for a presuppositionless, non-metaphysical answer. But it is an idea that goes into bankruptcy with every new philosophical generation. (Van Fraassen 1982, 26)

Van Fraassen on pragmatic virtues

According to the realist conception of scientific knowledge defended in section 1, the fact that a proposed theory is inductively supported at the theoretical level on the basis of already confirmed theories constitutes (some) evidence in favor of its approximate truth. The sort of inductive support at issue is often described in the literature in terms of explanatory power. Typically, when there are good theoretical reasons to believe a theory, then there are good reasons to believe that it provides the right explanation for whatever phenomena it describes; typically, when a theory is said to provide a good (or the best) explanation of some observable phenomenon, what is being reported is that there are *theoretical* reasons (and often experimental reasons as well—but these are theory-dependent, as we have seen) to believe that it provides an approximately true account of how those phenomena come about. I have here avoided reference to the principle of "inductive inference to the best explanation" (Harman 1965) largely because I think that this formulation carries the implicature that there is some conception of explanation, like the "common-cause principle" (pp. 25–31; 118–33), which can be established prior to theoretical understanding and which can serve to justify inferences from observations to theoretical conclusions. In the view expressed here, what are important are inductive inferences from (partly or wholly) theoretical premises to (partly or wholly) theoretical conclusions. The "rules" governing such inductive inferences (judgments of projectability for properties of various sorts of "theoretical entities," for example) are themselves theory-determined. There are no significant pretheoretical rules of inductive inference at either the theoretical or the observational level in science. Thus, insofar as inductive theoretical reasoning can often be described in terms of reasoning about explanation, the relevant notions of explanation should themselves be theory-dependent; our theoretical knowledge tells us what sorts of explanations are possible and what standards are to be used to judge them.

Despite the divergence between the conception of theoretical reasons defended here and the conception of inductive inference to the best explanation considered by van Fraassen, it seems reasonable to take what van Fraassen says about explanatory power, simplicity, elegance, capacity for unification of disparate phenomena, and other such virtues of theories as the basis for a possible empiricist reply to the arguments for realism rehearsed in section 1. Van Fraassen calls these virtues the "pragmatic virtues," and, when he characterizes the dimension which they add to theory acceptance, he does indeed seem to have in mind the sort of theoretical-tradition-dependent evaluative considerations on which the argument for realism depends:

> *Theory acceptance has a pragmatic dimension.* While the only belief involved in acceptance, as I see it, is the belief that the theory is empirically adequate, *more than belief is involved.* To accept a theory is to make a commitment, a commitment to the further confrontation of new phenomena within the framework of that theory, a commitment to a research programme, and a wager that all relevant phenomena can be accounted for without giving up that theory. That is why someone who has accepted a certain theory, will henceforth answer questions *ex cathedra*, or at least feel called upon to do so. (P. 88)

Van Fraassen's treatment of the pragmatic virtues is just what the term *pragmatic* suggests:

> In so far as they go beyond consistency, empirical adequacy, and empirical strength, they do not concern the relation between the theory and the world, but rather the use and usefulness of the theory; they provide reasons to prefer the theory independently of questions of truth." (P. 88)

Here, I believe, van Fraassen makes an extremely important point. If we think of scientists as somehow already knowing about certain scientific theories that they are empirically adequate, then it is an uphill battle for the realist to have to argue that the additional fact that these theories are also explanatory, elegant, simple, harmonious with previous theories, etc., makes it likely that they are (even approximately) true. As I have argued elsewhere (Boyd 1981, 1983), the realist who takes it for granted that (s)he and the empiricist are each already able to assess the experimental evidence for and against the empirical adequacy of scientific theories will find it impossible to use the standard arguments for realism against the powerful epistemological arguments which appear to support empiricist anti-realism. Van Fraassen seems well aware of this difficulty for realists when he characterizes as "rock-bottom criteria of minimal acceptability" "consistency, internally and with the facts," while denying that explanatory power is a rock-bottom virtue in this sense (p. 94).

The defense of realism presented in section 1 does not, of course, face these difficulties. The proposal to understand the theory-dependent virtues as merely pragmatic is subject to the same objections raised in section 1 to the treatment of those virtues as "merely heuristic." Either treatment requires that assessments of the "rock-bottom" virtue of empirical adequacy be independent of theoretical knowledge. But they are not. It is only because our *theoretical* commitments reflect approximate knowledge of unobservables that we are able to assess the empirical adequacy of scientific theories. It *may* be true that our knowledge of the observable results of particular experiments is "rock-bottom" in a way that theoretical knowledge is not, but our knowledge about scientific theories that they are empirically adequate is typically parasitic on our knowledge of "theoretical entities."

Kuhn and others have shown that the justifications scientists actually give for the inductive generalizations they make about observables (their justifications for particular inductive strategies, experimental designs, patterns of inference, etc.) are profoundly theory-dependent. What the realist investigation of the problems of experimental design shows is that this situation cannot be remedied by "rational reconstruction": the only good justifications there are for the inductive practices of scientists are theoretical justifications of the sort only a realist can accept. In the light of these findings, the view that theoretical considerations are merely pragmatic or merely heuristic dictates the absurd conclusion that the inductive inferences about observables which scientists make are without justification (see Boyd 1983, especially part 8).

Inductive inference to philosophical explanations

I have just argued that various empiricist rebuttals to the argument for realism indicated in section 1 are inadequate; the empiricist cannot escape the conclusion that only from a realist perspective is it possible to explain the instrumental reliability of the actual methods of science or the legitimate role of theoretical considerations in those methods. Suppose (as is unlikely) that the empiricist grants this conclusion. What argumentative resources does the empiricist still have at her disposal? One response is clearly suggested by van Fraassen's treatment both of the arguments for realism in particular and of the issue of inductive inference to the best explanation in general: The empiricist must, anyway, reject the epistemic legitimacy of inferences to the best explanation when the conclusions of those inferences are about unobservables; the empiricist may then in particular reject the realist conclusion, even granting that realism provides the best explanation for the instrumental reliability of the methods of science (see also Fine 1984). If the empiricist adopts this response, she will of course still need to retain the doctrine that the methods of science are instrumentally reliable. This is so for two reasons.

First, the doctrine that the methods of science are instrumentally reliable is a generalization *about observables* which is quite evidently true; any philosophical position which abandoned it would be *prima facie* refuted. Secondly, it is the business of philosophers of science, empiricist or otherwise, to identify and discuss the principles which constitute good scientific methodology. The identification of such principles depends on ascertaining which features of scientific practice contribute to the epistemic reliability of science. For the empiricist, the identification of such features will depend precisely on generalizations of the sort in question regarding the contribution of such features to the instrumental reliability of scientific practice. It is also true that, if the empiricist can inductively establish the generalization that the methods of science are instrumentally reliable, then she will have the resources to reply to the claim of the preceding section that the empiricist cannot explain why scientists are justified in making inductive generalizations about observables. The empiricist can offer, on behalf of the scientist, an inductive justification of the inductive practices of science: It is inductively confirmed that the theory-dependent methods of science are instrumentally reliable. Therefore, in any particular case, scientists are justified in accepting as empirically adequate a theory established by those methods, even though they are not justified in adopting any beliefs about unobservables.

I have elsewhere discussed this empiricist strategy in detail (Boyd 1983, pt. 8); I conclude there that the consistent empiricist cannot justify the inference from the observed history of scientific practice to the generalization that the methods of science are instrumentally reliable. I will summarize here one of the arguments which supports this conclusion. In order to establish inductively the instrumental reliability of the scientific method, one must know that it has been appropriately reliable in the past. In order to know this, one must know about various theories which have been accepted by scientists in the past that *they* are approximately empirically adequate. As Kuhn has shown, scientists cannot offer justifications for such conclusions which do not depend upon theoretical beliefs which the empiricist cannot accept. As the argument for realism discussed in section 1 shows, this problem for the empiricist cannot be solved by "rational reconstruction." The inductive justification for theory-dependent inductions about observables cannot be invoked by the empiricist, because the generalization whose justifiability we are discussing is a *premise* for that inductive justification. Therefore, the *consistent* empiricist cannot even justifiably conclude that the methods of science have been instrumentally reliable in the past, much less that they will be reliable in the future.

I conclude that the consistent empiricist can justify neither the methods and empirical findings of science nor the methods and findings of empiricist philosophy of science.

III. Credo: *Lex Orandi est Lex Credendi* (*Analogice Acceptum*)

In chapter 7, "Gentle Polemics," van Fraassen treats the reader to a clever and engaging comparison of various arguments for scientific realism with Aquinas's arguments for the existence of God. The reader, unconvinced by the Thomistic arguments, is supposed to reject the realist's arguments as well. If, as I argued in the previous section, the consistent empiricist cannot justify either the sorts of empirical generalization represented by scientific theories or the basic methods of the philosophy of science, then it is perhaps fair to suggest that the empiricist's position has come to resemble that attributed to Tertullian (160–220), "*Credo quia absurdum.*" In any event, the initial comparison of the realist's arguments with those of Aquinas cannot be sustained in the case of the arguments for realism presented here.

If one looks for precedents for those arguments, they are best found in the persistent "flirtation" with atomism or metaphysical materialism which has characterized the empiricist tradition from Locke to Carnap or in the dialectical materialism of the Marxist tradition. If a theological precedent must be found, the obvious choice is the dictum attributed to Pope St. Celestine I (422–32; the attribution may be faulty—see Buchberger 1954, vol. 6, col. 1001): "*Lex orandi est lex credenti*," "the rule for praying is the rule for believing," or (in a freer translation) "believe what is necessary to 'rationally reconstruct' liturgical practice." For "liturgical practice" put "scientific practice" and you get the strategy for the defense of realism employed here.

Drawing the analogy between this strategy and the principle of theological methodology attributed to Celestine is helpful in understanding the former. The Celestine methodological principle differs sharply from the inferential principles involved in Aquinas's "proofs" of the existence of God. Aquinas appeals to extremely general principles about explanation and causation which are apparently supposed to be basically *a priori* and to be universally applicable. The Celestine principle, by contrast, is not *a priori* and is (even in the view of its defenders) not universally applicable. It is supposed to apply only with respect to the right sort of liturgy, so that a theist considering employing the principle would not know *a priori* whether or not her application of the principle would be epistemically reliable. The Celestine principle is a methodological principal *within* theology, not prior to it, and it is understood to be reliable only within a particular theological tradition.

The analogous claims are true of the methodological principles underlying the defense of realism discussed here. What the realist proposes is to use the ordinary methods of science to investigate the question of why the methods of science are instrumentally reliable. The philosophical methods here are not conceived of as *prior to* scientific methods in any sense. Moreover, according to the realist's own account, the reliability of the scientific methods in question depends on the approximate truth of the background theories in the theoretical tradition; thus, the reliability of the realist's *philosophical* methods depends on logically and epistemically contingent facts about the actual scientific tradition (Boyd 1981, 1983). The realist's philosophical methods are *in that sense* not universally applicable. The principles of inference by which the realist defends realism will be no more stringent than the principles of inference whose reliability the realist is trying to explain. Indeed, if the realist is right, then the principles of scientific methodology to which the realist appeals are tacitly but ineliminably realist themselves. The realist—in insisting that the methods of the philosophy of science should be the methods of science—cannot not offer to defend realism by appeal to inferential principles which are, in the final analysis, themselves neutral with respect to the realism-empiricism controversy. This may seem a grave defect in the realist's program (see, for example, Fine 1984, fn. 7). The realist is content to reply that it is also impossible to defend the inductive inferences which scientists make about observables if one insists on inferential principles neutral with respect to the realism-empiricism controversy; it is likewise impossible thus to defend the philosophical methods of even empiricist philosophers of science (Boyd 1983). Apparently both in science and in the philosophy of science we must make do with Celestine rather than Thomistic principles. For the philosopher who—in the name of science—objects to the *a priorism* of Thomistic reasoning, this may not be an entirely unwelcome conclusion.

References

Boyd, R. 1973. "Realism, Underdetermination and a Causal Theory of Evidence." *Nous* 7:1–12.

————. 1979. "Metaphor and Theory Change." In *Metaphor and Thought*, edited by A. Ortony, 356–408. Cambridge: Cambridge University Press.

————. 1981. "Scientific Realism and Naturalistic Epistemology." In *PSA 1980*, vol. 2, edited by P. D. Asquith and R. N. Giere, 613–62. East Lansing, Mich.: Philosophy of Science Association.

————. 1983. "On the Current Status of the Issue of Scientific Realism." *Erkenntnis* 17:135–69.

Buchberger, M., ed. 1954. *Lexikon für Theologis und Kirche*. Freiburg.

Darwin, C. 1859. *On the Origin of Species by Means of Natural Selection*. London: John Murray.

Demopoulos, W. 1982. Review of *The Scientific Image*, by Bas C. van Fraassen. *Philosophical Review* 91:603–7.

Fine, A. 1984. "The Natural Ontological Attitude." In *Essays on Scientific Realism*, edited by J. Leplin.

Goodman, N. 1973. *Fact, Fiction and Forecast*. 3d ed. Indianapolis and New York: Bobbs-Merrill.

Gould, S. J., and N. Eldredge. 1977. "Punctuated Equilibria: The Tempo and Mode of Evolution Reconsidered." *Paleobiology*, 1977, 115–51.

Gould, S. J., and R. Lewontin. 1979. "The Spondrels on San Marco and the Panglossian Paradigm: A Critique of the Adaptationist Programme." *Proc. R. Soc. Lond.* 3.205, 581–98.

Hamilton, W. D. 1964. "The Genetic Theory of Social Behavior," I, II. *Journal of Theoretical Biology* 7 (1): 1–52.

Harman, G. 1965. "The Inference to the Best Explanation." *Philosophical Review* 74:88–95.

Kuhn, T. 1970. *The Structure of Scientific Revolutions*. Chicago: University of Chicago Press.

Putnam, H. 1975. *Mind, Language and Reality: Philosophical Papers*, vol. 2. Cambridge: Cambridge University Press.

————. 1978. *Meaning and the Moral Sciences*. London: Routledge and Kegan Paul.

Van Fraassen, B. 1980. *The Scientific Image*. Oxford: Clarendon Press.

————. 1982. "The Charybdis of Realism: Epistemological Implications of Bell's Inequality." *Synthese* 52:25–38.

2 The Ontological Status of Observables: In Praise of the Superempirical Virtues

Paul M. Churchland

At several points in the reading of van Fraassen's book, I feared I would no longer be a realist by the time I completed it. Fortunately, sheer doxastic inertia has allowed my convictions to survive its searching critique, at least temporarily, and, as we address you today, van Fraassen and I still hold different views. I am a scientific realist, of unorthodox persuasion, and van Fraassen is a constructive empiricist, whose persuasions currently define the doctrine. I assert that global excellence of theory is the ultimate measure of truth and ontology at all levels of cognition, even at the observational level. Van Fraassen asserts that descriptive excellence at the observational level is the only genuine measure of any theory's truth and that one's acceptance of a theory should create no ontological commitments whatever beyond the observational level.

Against his first claim I will maintain that observational excellence or 'empirical adequacy' is only one epistemic virtue among others of equal or comparable importance. And against his second claim I will maintain that the ontological commitments of any theory are wholly blind to the idiosyncratic distinction between what is and what is not humanly observable, and so should be our own ontological commitments. Criticism will be directed primarily at van Fraassen's _selective_ skepticism in favor of observable ontologies over unobservable ontologies and against his view that the 'superempirical' theoretical virtues (simplicity, coherence, explanatory power) are merely pragmatic virtues, irrelevant to the estimate of a theory's truth. My aims are not merely critical, however. Scientific realism does need reworking, and there are good reasons for moving it in the direction of van Fraassen's constructive empiricism, as will be discussed in the closing section of this paper. But those reasons do not support the skeptical theses at issue.

Previously published in a shorter form as "The Anti-Realist Epistemology of van Fraassen's _The Scientific Image_," _Pacific Philosophical Quarterly_ 63 (July 1982): 226–36. Reproduced by permission. I thank Hartry Field, Michael Stack, Bas van Fraassen, Clark Glymour, Barney Keaney, Stephen Stich, and Patricia Churchland for helpful discussion of the issues here addressed.

35

I. Observation and Ontological Commitment

Before pursuing our differences, it will prove useful to emphasize certain convictions we share. Van Fraassen is already a scientific realist in the minimal sense that he interprets theories literally and he concedes them a truth value. Further, we agree that the observable/unobservable distinction is entirely distinct from the nontheoretical/theoretical distinction, and we agree as well that all observation sentences are irredeemably laden with theory.

Additionally, I absolutely reject many sanguine assumptions common among realists. I do not believe that on the whole our beliefs must be at least roughly true; I do not believe that the terms of 'mature' sciences must typically refer to real things; and I very much doubt that the reason of *homo sapiens*, even at its best and even if allowed infinite time, would eventually encompass all and/or only true statements.

This skepticism is born partly from a historical induction: so many past theories, rightly judged excellent at the time, have since proved to be false. And their current successors, though even better founded, seem but the next step in a probably endless and not obviously convergent journey. (For a most thorough and insightful critique of typical realist theses, see the recent paper by Laudan [1981].)

Evolutionary considerations also counsel a healthy skepticism. Human reason is a hierarchy of heuristics for seeking, recognizing, storing, and exploiting information. But those heuristics were invented at random, and they were selected for within a very narrow evolutionary environment, cosmologically speaking. It would be *miraculous* if human reason were completely free of false strategies and fundamental cognitive limitations, and doubly miraculous if the theories we accept failed to reflect those defects.

Thus some very realistic reasons for skepticism with respect to any theory. Why, then, am I still a scientific realist? Because these reasons fail to discriminate between the integrity of observables and the integrity of unobservables. If anything is compromised by these considerations, it is the integrity of theories generally. That is, of *cognition* generally. Since our observational concepts are just as theory-laden as any others, and since the integrity of those concepts is just as contingent on the integrity of the theories that embed them, our observational ontology is rendered *exactly as dubious* as our nonobservational ontology.

This parity should not seem surprising. Our history contains real examples of mistaken ontological commitments in both domains. For example, we have had occasion to banish phlogiston, caloric, and the luminiferous ether from our ontology—but we have also had occasion to banish witches and the starry sphere that turns about us daily. These latter items were as 'observable' as you please and were widely 'observed' on a daily basis. We are too often misled, I think, by our casual use of *observes* as a success verb: we tend to forget that, at any stage of our history, the ontology presupposed by our observational judgments remains essentially speculative and wholly revisable, however entrenched and familiar it may have become.

Accordingly, since the skeptical considerations adduced above are indifferent to the distinction between what is and what is not observable, they provide no reason for resisting a commitment to unobservable ontologies *while allowing* a commitment to what we take to be observable ontologies. The latter appear as no better off than the former. For me, then, the 'empirical success' of a theory remains a reason for thinking the theory to be true and for accepting its overall ontology. The inference from success to truth should no doubt be severely tempered by the skeptical considerations adduced, but the inference to *un*observable ontologies is not rendered *selectively* dubious. Thus, I remain a scientific realist. My realism is highly circumspect, but the circumspection is uniform for unobservables and observables alike.

Perhaps I am wrong in this. Perhaps we should be selectively skeptical in the fashion van Fraassen recommends. Does he have other arguments for refusing factual belief and ontological commitment beyond the observational domain? Indeed he does. In fact, he does not appeal to historical induction or evolutionary humility at all. These are *my* reasons for skepticism (and they will remain, even if we manage to undermine van Fraassen's). They have been introduced here to show that, while there are some powerful reasons for skepticism, those reasons do not place unobservables at a selective disadvantage.

Very well, what are van Fraassen's reasons for skepticism? They are very interesting. To summarize quickly, he does a compelling job of deflating certain standard realist arguments (from Smart, Sellars, Salmon, Boyd, and others) to the effect that, given the aims of science, we have no alternative but to bring unobservables (not just into our calculations, but) into our literal ontology. He also argues rather compellingly that the superempirical virtues, such as simplicity and comprehensive explanatory power, are at bottom merely pragmatic virtues, having nothing essential to do with any theory's truth. This leaves only empirical adequacy as a genuine measure of any theory's truth. Roughly, a theory is empirically adequate if and only if everything it says about *observable* things is true. Empirical adequacy is thus a necessary condition on a theory's truth.

However, claims van Fraassen, the truth of any theory whose ontology includes unobservables is always radically underdetermined by its empirical adequacy, since a great many logically incompatible theories can all be empirically equivalent. Accordingly, the inference from empirical adequacy to truth now appears presumptuous in the extreme, especially since it has just been disconnected from additional selective criteria such as simplicity and explanatory power, criteria which might have reduced the arbitrariness of the particular inference drawn. Fortunately, says van Fraassen, we do not need to make such wanton inferences, since we can perfectly well understand science as an enterprise that never really draws them. Here we arrive at his positive conception of science as an enterprise whose sole intellectual aims are empirical adequacy and the satisfaction of certain human intellectual needs.

The central element in this argument is the claim that, in the case of a theory whose ontology includes unobservables, its empirical adequacy underdetermines its truth. (We should notice that, in the case of a theory whose ontology is completely free of unobservables, its empirical adequacy does not underdetermine its truth: in that case, truth and empirical adequacy are obviously identical. Thus van Fraassen's *selective* skepticism with respect to unobservables.) That is, for any theory T inflated with unobservables, there will always be many other such theories incompatible with T but empirically equivalent to it.

In my view, the notions of "empirical adequacy" and its cognate relative term "empirically equivalent" are extremely thorny notions of doubtful integrity. If we attempt to explicate a theory's 'empirical content' in terms of the observation sentences it entails (or entails if conjoined with available background information or with possible future background information or with possible future theories), we generate a variety of notions which are variously empty, context-relative, ill defined, or flatly incompatible with the claim of underdetermination. Van Fraassen expresses awareness of these difficulties and proposes to avoid them by giving the notions at issue a model-theoretic rather than a syntactic explication. I am unconvinced that this improves matters decisively (on this issue, see Wilson 1980; also Musgrave [chap. 9], Hooker [chap. 8], Glymour [chap. 5], and Wilson [chap. 10], this volume). In particular, I think van Fraassen has not dealt adequately with the problem of how the so-called 'empirical equivalence' of two incompatible theories remains relative to *which* background theories are added to the evaluative context, especially background theories that in some way revise our conception of what humans can observe. I intend to sidestep this issue for now, however, since the matter is complex and there is a much simpler objection to be voiced.

Let me approach my objection by first pointing out that the empirical adequacy of any theory is itself something that is radically underdetermined by any evidence conceivably available to us. Recall that, for a theory to be empirically adequate, what it says about observable things must be true—*all* observable things, in the past, in the indefinite future, and in the most distant corners of the cosmos. But, since any actual data possessed by us must be finite in its scope, it is plain that we here suffer an underdetermination problem no less serious than that claimed above. This is Hume's problem, and the lesson is that even observation-level theories must suffer radical underdetermination by the evidence. Accordingly, theories about observables and theories about unobservables appear on a par again, so far as skepticism is concerned.

Van Fraassen thinks there is an important difference between the two cases, and one's first impulse is to agree with him. We are all willing to concede the existence of Hume's problem—the problem of justifying the inference to unobserv*ed* entities. But the inference to entities that are downright unobserv*able* appears as a different and *additional* problem.

The appearance is an illusion, as the following considerations will show. Consider some of the different reasons why entities or processes may go unobserved by us. First, they may go unobserved because, relative to our natural sensory apparatus, they fail to enjoy an appropriate spatial or temporal *position*. They may exist in the Upper Jurassic period, for example, or they may reside in the Andromeda galaxy. Second, they may go unobserved because, relative to our natural sensory apparatus, they fail to enjoy the appropriate spatial or temporal *dimensions*. They may be too small or too brief or too large or too protracted. Third, they may fail to enjoy the appropriate *energy*, being too feeble or too powerful to permit useful discrimination. Fourth and fifth, they may fail to have an appropriate *wavelength* or an appropriate *mass*. Sixth, they may fail to 'feel' the relevant fundamental forces our sensory apparatus exploits, as with our inability to observe the background neutrino flux, despite the fact that its energy density exceeds that of light itself.

This list could be lengthened, but it is long enough to suggest that being spatially or temporally distant from our sensory apparatus is only one among many ways in which an entity or process can fall outside the compass of human observation, a way distinguished by no relevant epistemological or ontological features.

There is clearly some *practical* point in our calling a thing "observ*able*" if it fails *only* the first test (spatiotemporal proximity) and "*un*observable" if it fails any of the others. But that is only because of the contingent practical fact that humans generally have somewhat more *control* over the spatiotemporal perspective of their sensory systems than they have over their size or reaction time or mass or wavelength sensitivity or chemical constitution. Had we been less mobile than we are—rooted to the earth like Douglas firs, say—yet been more voluntarily plastic in our sensory constitution, the distinction between the 'merely unobserved' and the 'downright unobservable' would have been very differently drawn. It may help to imagine here a suitably rooted arboreal philosopher named (what else?) Douglas van Fiirrsen, who, in his sedentary wisdom, urges an antirealist skepticism concerning the spatially very *distant* entities postulated by his fellow trees.

Admittedly, for any distant entity, one can in principle always change the relative spatial position of one's sensory appratus so that the entity is observed: one can go to it. But equally, for any microscopic entity, one can in principle always change the relative spatial *size* or *configuration* of one's sensory appratus so that the entity is observed. Physical law imposes certain limitations on such plasticity, but so also does physical law limit how far one can travel in a lifetime.

To emphasize the importance of these considerations, let me underscore the structure of my objection here. Consider the distinction between

(1) things observed by some human (with unaided senses),
(2) things thus observable by humans, but not in fact observed, and
(3) things not observable by humans at all.

Van Fraassen's position would exclude (3) from our rational ontology. This has at least some initial plausibility. But his position would not be at all plausible if it were committed to excluding both (3) *and* (2) from our rational ontology. No party to the present discussion is willing to restrict rational ontology to (1) alone. Van Fraassen's position thus requires a *principled* distinction between (2) and (3), a distinction *adequate* to the radical difference in epistemic attitude he would have us adopt toward them. The burden of my argument is that the distinction between (2) and (3), once it is unearthed, is only very feebly principled and is wholly inadequate to bear the great weight that van Fraassen puts on it.

The point of all this is that there is no special or novel problem about inferences to the existence of entities commonly called "unobserv*ables*." Such entities are merely those that go unobserved by us for reasons *other* than their spatial or temporal distance from us. But whether the 'gap' to be bridged is spatiotemporal or one of the many other gaps, the logical/epistemological problem is the same in all cases: ampliative inference and underdetermined hypotheses. I therefore fail to see how van Fraassen can justify tolerating an ampliative inference when it bridges a gap of spatial distance, while refusing to tolerate an ampliative inference when it bridges a gap of, for example, spatial size. Hume's problem and van Fraassen's problem collapse into one.

Van Fraassen attempts to meet such worries about the inescapable ubiquity of speculative activity by observing that "it is not an epistemological principle that one may as well hang for a sheep as for a lamb" (1980, 72). Agreed. But it is a principle of *logic* that one may as well hang for a sheep as for a sheep, and van Fraassen's lamb (empirical adequacy) is just another sheep.

Simply to hold *fewer* beliefs from a given set is of course to be less adventurous, but it is not necessarily to be applauded. One might decide to relinquish all one's beliefs save those about objects weighing less than five hundred kilograms, and perhaps, one would then be logically safer. But, in the absence of some relevant epistemic difference between one's beliefs about such objects and one's beliefs about other objects, that is perversity, not parsimony.

Let me summarize. As van Fraassen sets it up, and as the intrumentalists set it up before him, the realist looks more gullible than the nonrealist, since the realist is willing to extend belief beyond the observable, while the nonrealist insists on confining belief within that domain. I suggest, however, that it is really the nonrealists who are being the more gullible in this matter, since they suppose that the epistemic situation of our beliefs about observables is in some way superior to that of our beliefs about unobservables. But in fact their epistemic situation is not superior. They are exactly as dubious as their nonobservational cousins. Their *causal history* is different (they are occasioned by activity in the sensory pathways), but the ontology they presuppose enjoys no privilege or special credibility.

II. Beliefworthiness and the Superempirical Virtues

Let me now try to address the question of whether the theoretical virtues such as simplicity, coherence, and explanatory power are *epistemic* virtues genuinely relevant to the estimate of a theory's truth, as tradition says, or merely *pragmatic* virtues, as van Fraassen urges. His view promotes empirical adequacy, or evidence of empirical adequacy, as the only genuine measure of a theory's truth, the other virtues (insofar as they are distinct from these) being cast as purely pragmatic virtues, to be valued only for the human needs they satisfy. Despite certain compelling features of the account of explanation that van Fraassen provides, I remain inclined toward the traditional view.

My reason is simplicity itself. Since there is no way of conceiving or representing 'the empirical facts' that is completely independent of speculative assumptions, and since we will occasionally confront theoretical alternatives on a scale so comprehensive that we must also choose between competing modes of conceiving what the empirical facts before us *are*, then the epistemic choice between these global alternatives cannot be made by comparing the extent to which they are adequate to some common touchstone, 'the empirical facts'. In such a case, the choice must be made on the comparative global virtues of the two global alternatives, T_1-plus-the-observational-evidence-therein-construed, versus T_2-plus-the-observational-evidence-therein-(differently)-construed. That is, it must be made on *superempirical* grounds such as relative coherence, simplicity, and explanatory unity.

Van Fraassen has said that to 'save the appearances' is to exhibit them as a fragment of a larger unity. With this I wholly agree. But I am here pointing out that it is a decision between competing 'larger unities' that determines what we count as "the true appearances" in the first place. There is no independent way to settle that question. And, if such global decisions can only be made on what van Fraassen calls 'pragmatic' grounds, then it would seem to follow that any decision concerning what the *observable* world contains must be essentially 'pragmatic' also! Inflationary metaphysics and 'pragmatic' decisions begin, it seems, as soon as we open our eyes.

Global issues such as these are reminiscent of Carnap's 'external' questions, and I think it likely that van Fraassen, like Carnap, does not regard them as decidable in any but a second-rate sense, since they can be decided only by second-rate (i.e., by 'pragmatic') considerations. If so, however, it is difficult to see how van Fraassen can justify a selectively realist attitude toward 'observables', since, as we have seen, pragmatic considerations must attend their selection, also. (These issues receive extended treatment in Churchland 1979, sec. 2, 3, 7, and 10.) What all of this illustrates, I think, is the poverty of van Fraassen's crucial distinction between factors that are 'empirical, and therefore truth-relevant', and factors that are 'superempirical, and therefore *not* truth-relevant'.

As I see it, then, values such as ontological simplicity, coherence, and explanatory power are some of the brain's most basic criteria for recognizing information, for distinguishing information from noise. And I think they are even more fundamental values than is 'empirical adequacy', since collectively they can overthrow an entire conceptual framework for representing the empirical facts. Indeed, they even dictate how such a framework is constructed by the questing infant in the first place. One's observational taxonomy is not 'read off' the world directly; rather, one comes to it piecemeal and by stages, and one settles on that taxonomy which finds the greatest coherence and simplicity in the world and the most and the simplest lawful connections.

I can bring together my protective concerns for unobservables and for the superempirical virtues by way of the following thought experiment. Consider a man for whom absolutely *nothing* is observable. All of his sensory modalities have been surgically destroyed, and he has no visual, tactile, or other sensory experience of any kind. Fortunately, he has mounted on top of his skull a microcomputer fitted out with a variety of environmentally sensitive transducers. The computer is connected to his association cortex (or perhaps the frontal lobe or Wernicke's area) in such a way as to cause in him a continuous string of singular beliefs about his local environment. These 'intellectual intuitions' are not infallible, but let us suppose that they provide him with much the same information that our perceptual judgments provide us.

For such a person, or for a society of such persons, the *observable* world is an empty set. There is no question, therefore, of their evaluating any theory by reference to its 'empirical adequacy', as characterized by van Fraassen (i.e., isomorphism between some observable features of the world and some 'empirical substructure' of one of the theory's models). But such a society is still capable of science, I assert. They can invent theories, construct explanations of the facts-as-represented-in-past-spontaneous-beliefs, hazard predictions of the facts-as-represented-in-future-spontaneous-beliefs, and so forth. In principle, there is no reason they could not learn as much as we have (cf. Feyerabend 1969).

But it is plain in this case that the global virtues of simplicity, coherence, and explanatory unification are what *must* guide the continuing evolution of their collected beliefs. And it is plain as well that their ontology, whatever it is, must consist entirely of *un*observable entities. To invite a van Fraassenean disbelief in unobservable entities is in this case to invite the suspension of all beliefs beyond tautologies! Surely reason does not require them to be so abstemious.

It is time to consider the objection that those aspects of the world which are successfully monitored by the transducing microcomputer should count as 'observables' for the folk described, despite the lack of any appropriate field of internal sensory qualia to mediate the external circumstance and the internal judgment it causes. Their tables-and-chairs ontology, as expressed in their spontaneous judgments, could then be conceded legitimacy.

I will be the first to accept such an objection. But, if we do accept it, then I do not see how we can justify van Fraassen's selective skepticism with respect to the wealth of 'unobservable' entities and properties reliably monitored by *our* transducing measuring instruments (electron microscopes, cloud chambers, chromatographs, etc.). The spontaneous singular judgments of the working scientist, at home in his theoretical vocabulary and deeply familiar with the measuring instruments to which his conceptual system is responding, are not worse off, causally or epistemologically, than the spontaneous singular judgments of our transducer-laden friends. If skepticism is to be put aside above, it must be put aside here, as well.

My concluding thought experiment is a complement to the one just outlined. Consider some folk who observe, not less of the world than we do, but more of it. Suppose them able to observe a domain normally closed to us: the microworld of virus particles, DNA strands, and large protein molecules. Specifically, suppose a race of humanoid creatures each of whom is born with an electron microscope permanently in place over his left 'eye'.

The scope is biologically constituted, let us suppose, and it projects its image onto a human-style retina, with the rest of their neurophysiology paralleling our own.

Science tells us, and I take it that van Fraassen would agree, that virus particles, DNA strands, and most other objects of comparable dimensions count as observable entities for the humanoids described. The humanoids, at least, would be justified in so regarding them and in including them in their ontology.

But we humans may not include such entities in our ontology, according to van Fraassen's position, since they are not observable with our unaided perceptual apparatus. We may not include such entities in our ontology *even though we can construct and even if we do construct electron microscopes of identical function, place them over our left eyes, and enjoy exactly the same microexperience as the humanoids.*

The difficulty for van Fraassen's position, if I understand it correctly, is that his position requires that a humanoid and a scope-equipped human must embrace *different* epistemic attitudes toward the microworld, even though their causal connections to the world and their continuing experience of it be identical: the humanoid is required to be a realist with respect to the microworld, and the human is required to be an antirealist (i.e., an agnostic) with respect to the microworld. But this distinction between what we and they may properly embrace as real seems to me to be highly arbitrary and radically undermotivated. For the only difference between the humanoid and a scope-equipped human lies in the *causal origins* of the transducing instruments feeding information into their respective brains. The humanoid's scope owes its existence to information coded in his genetic material. The human's scope owes its existence to information coded in his cortical material or in technical libraries. I do not see why this should make any difference in their respective ontological commitments, whatever they are, and I must decline to embrace any philosophy of science which says that it must.

III. Toward a More Rational Realism

I now turn from critic of van Fraassen's position to advocate. One of the most central elements in his view seems to me to be well motivated and urgently deserving of further development. As he explains in his introductory chapter, his aim is to reconceive the relation of theory to world, and the units of scientific cognition, and the virtue of those units when successful. He says, "I use the adjective 'constructive' to indicate my view that scientific activity is one of construction rather than discovery: con-

struction of models that must be adequate to the phenomena, and not discovery of truth concerning the unobservable" (1980, 5).

The traditional view of human knowledge is that the unit of cognition is
the sentence or proposition and the cognitive virtue of such units is truth.
Van Fraassen rejects this overtly linguistic guise for his empiricism. He
invites us to reconceive a theory as a set of models (rather than as a set of
sentences), and he sees empirical adequacy (rather than truth) as the principal virtue of such units.

Though I reject his particular reconception and the selective skepticism
he draws from it, I think the move away from the traditional conception is
entirely correct. The criticism to which I am inclined is that van Fraassen
has not moved quite far enough. Specifically, if we are to reconsider truth
as the aim or product of cognitive activity, I think we must reconsider its
applicability right across the board and not just in some arbitrarily or idio-
syncratically segregated domain of 'unobservables'. That is, if we are to
move away from the more naïve formulations of scientific realism, we
should move in the direction of *pragmatism* rather than in the direction of a
positivistic instrumentalism. Let me elaborate.

When we consider the great variety of cognitively active creatures on
this planet—sea slugs and octopi, bats, dolphins, and humans; and when
we consider the ceaseless reconfiguration in which their brains or central
ganglia engage—adjustments in the response potentials of single neurons
made in the microsecond range, changes in the response characteristics of
large systems of neurons made in the seconds-to-hours range, dendritic
growth and new synaptic connections and the selective atrophy of old connections effected in the day-upwards range—then van Fraassen's term
"construction" begins to seem highly appropriate. There is endless construction and reconstruction, both functional and structural. Further, it is
far from obvious that truth is either the primary aim or the principal product of this activity. Rather, its function would appear to be the ever more
finely tuned administration of the organism's *behavior*. Natural selection
does not care whether a brain has or tends toward true beliefs, so long
as the organism reliably exhibits reproductively advantageous behavior.
Plainly, there is going to be *some* connection between the faithfulness of
the brain's 'world-model' and the propriety of the organism's behavior. But
just as plainly the connection is not going to be direct.

While we are considering cognitive activity in biological terms and in all
branches of the phylogenetic tree, we should note that it is far from obvious that sentences or propositions or anything remotely like them constitute the basic elements of cognition in creatures generally. Indeed, as I
have argued at length elsewhere (1979, chap. 5; 1981), it is highly unlikely
that the sentential kinematics embraced by folk psychology and orthodox

epistemology represents or captures the basic parameters of cognition and learning even in humans. That framework is part of a commonsense theory that threatens to be either superficial or false. If we are ever to understand the *dynamics* of cognitive activity, therefore, we may have to reconceive our basic unit of cognition as something other than the sentence or proposition, and reconceive its virtue as something other than truth.

Success of this sort on the descriptive/explanatory front would likely have normative consequences. Truth, as currently conceived, might cease to be an aim of science. Not because we had lowered our sights and reduced our epistemic standards, as van Fraassen's constructive empiricism would suggest, but because we had *raised* our sights, in pursuit of some epistemic goal even *more* worthy than truth. I cannot now elucidate such goals, but we should be sensible of their possible existence. The notion of 'truth', after all, is but the central element in a clutch of descriptive and normative *theories* (folk psychology, folk epistemology, folk semantics, classical logic), and we can expect conceptual progress here as appropriately as anywhere else.

The notion of truth is suspect on purely metaphysical grounds, anyway. It suggests straightaway the notion of The Complete and Final True Theory: at a minimum, the infinite set of all true sentences. Such a theory would be, by epistemic criteria, the best theory possible. But nothing whatever guarantees the existence of such a unique theory. Just as there is no largest positive integer, it may be that there is no best theory. It may be that, for any theory whatsoever, there is always an even better theory, and so *ad infinitum*. If we were thus unable to speak of *the* set of all true sentences, what sense could we make of truth sentence by sentence?

These considerations do invite a 'constructive' conception of cognitive activity, one in which the notion of truth plays at best a highly derivative role. The formulation of such a conception, adequate to all of our epistemic criteria, is the outstanding task of epistemology. I do not think we will find that conception in van Fraassen's model-theoretic version of 'positivistic instrumentalism', nor do I think we will find it quickly. But the empirical brain begs unraveling, and we have plenty of time.

Finally, there is a question put to me by Stephen Stich. If ultimately my view is even more skeptical than van Fraassen's concerning the relevance or applicability of the notion of truth, why call it scientific *realism* at all? For at least two reasons. The term *realism* still marks the principal contrast with its traditional adversary, positivistic instrumentalism. Whatever the integrity of the notion of truth, theories about unobservables have just *as much* a claim to truth, epistemologically and metaphysically, as theories about observables. Second, I remain committed to the idea that there exists a world, independent of our cognition, with which we interact and of which we construct representations: for varying purposes, with varying penetration, and with varying success. Lastly, our best and most penetrating grasp of the real is still held to reside in the representations provided by our best theories. Global excellence of theory remains the fundamental measure of rational ontology. And that has always been the central claim of scientific realism.

References

Churchland, Paul M. 1979. *Scientific Realism and the Plasticity of Mind*. Cambridge: Cambridge University Press.

————. 1981. "Eliminative Materialism and the Propositional Attitudes." *Journal of Philosophy* 78, no. 2.

————. 1982. "The Anti-Realist Epistemology of van Fraassen's *The Scientific Image*." *Pacific Philosophical Quarterly* 63, no. 2.

Feyerabend, Paul K. 1969. "Science without Experience." *Journal of Philosophy* 66, no. 22.

Laudan, Larry. 1981. "A Confutation of Convergent Realism." *Philosophy of Science* 48, no. 1.

Van Fraassen, Bas C. 1980. *The Scientific Image*. Oxford: Clarendon Press.

————. 1981. Critical notice of *Scientific Realism and the Plasticity of Mind*, by Paul Churchland. *Canadian Journal of Philosophy* 11, no. 3.

Wilson, Mark. 1980. "The Observational Uniqueness of Some Theories." *Journal of Philosophy* 77, no. 4.

3   What Science aims to Do

Brian Ellis

In *The Scientific Image*, Bas van Fraassen has mounted an impressive challenge to scientific realism and argued strongly for a new empiricism which
he calls "constructive empiricism." I agree with van Fraassen that the
form of scientific realism which he challenges is untenable. However, I
cannot accept his constructive empiricism as a satisfactory alternative. It
is, undoubtedly, much more acceptable than earlier, cruder forms of empiricism, since it concedes much that has been found objectionable in
other empiricist theories. But, in the end, it suffers from some of the same
basic defects as they do and must, therefore, be rejected along with them.

Van Fraassen sees empiricism and scientific realism as being opposed to
each other. He locates the basic difference between them as a difference of
view about the *aims* of science. Empiricists, he says, see the aim as being
the anticipation of nature, of producing theories which yield successful
predictions—predictions which are borne out by experience. Scientific realists, on the other hand, see it as being to describe things as they really are
or have come to be—as Duhem says (of explanation), it is "to strip reality
of the appearances covering it like a veil, in order to see the bare reality
itself" (1954, 7).

Traditionally, empiricism has been opposed, not to scientific realism,
but to rationalism, and to revelationism, as a theory about the *sources* of
our knowledge. And most scientific realists would agree with the empiricists on this issue. For example, Galileo was an empiricist about the
sources of our knowledge, and Cardinal Bellarmino, a revelationist. But,
by van Fraassen's criteria, it is Bellarmino who was the empiricist and
Galileo who was not. So van Fraassen's concept of empiricism is not the
traditional one. Nevertheless, the disagreement *within* the empiricist tradition about the aim of science is an important one, and I see no point in
arguing about labels.

The different views about the aim of science lead naturally to different
views concerning its theoretical achievements. For scientific realists, well-
established and accepted theories must be considered to be literally true—
not just useful models for predicting what will occur or, to use the old-
fashioned term, 'saving the phenomena'. For empiricists, such theories are

48

just models of reality which are, and which we believe will continue to be, empirically adequate. Whether they are true or not, van Fraassen maintains, is of no importance—although he does not deny that they are either true or false.

According to van Fraassen, the scientific realist's position is this:

> (1) Science aims to give us, in its theories, a literally true story of what the world is like; and acceptance of a scientific theory involves the belief that it is true. (Van Fraassen 1980, 8)

Van Fraassen's concept of literal truth is a correspondence concept: a statement is literally true if, literally interpreted, it accurately describes or corresponds to reality. The rules for literal interpretation are not clearly specified, but he has in mind at least this: any apparent reference to a theoretical entity is to be construed as a genuine attempt to refer, unless there are good specific reasons for not so construing it.

Against the scientific realists' conception of the aim of science, van Fraassen argues that it is not required either to motivate science or to explain its practice. It is enough for these purposes to see the aim as being the provision of empirically adequate theories. Moreover, on the basis of a general thesis of empirical underdetermination of theories, he argues that we cannot rationally demand more of any theory than that it be empirically adequate. For, once we get down to those theories which are acceptable by his criterion of empirical adequacy, the choice between them cannot be made on grounds which have any bearing on whether or not they are true. The choice has now to be a pragmatic one. So, if we required belief in the *truth* of a theory as a condition for its acceptance, we could not rationally accept any theories at all, even though we had good empirical *and* pragmatic reasons for doing so. Conversely, if we did accept any theories as being literally true, we should have to allow that the pragmatic reasons (simplicity, elegance, and the like) which we have for preferring them to other empirically equivalent theories are actually grounds for believing them to be true.

Van Fraassen thus arrives at the following position:

> (2) Science aims to give us theories which are empirically adequate; and acceptance of a theory involves as belief only that it is empirically adequate. (Van Fraassen 1980, 12)

I assume that van Fraassen intends this statement to be understood normatively and that the same applies to the statement of position he attributes to scientific realists, for there is no sociological investigation of what scientists themselves see as the aims of science or of what beliefs they have concerning the theories they accept. So, presumably, it is a question of what we ought to consider the aims of science to be and of what beliefs ought to be involved in accepting a scientific theory. Nevertheless, since the principal theses are not presented as normative judgments, and there is a lot of discussion of modern physical theories, I cannot help seeing van Fraassen's position as being, at least in part, a reflection of what he thinks are the attitudes of scientists to the dominant physical theories of today.

At the time when the dominant theories in science were mechanistic, it was easy to see the aim of science as being to discover and describe the underlying mechanisms of nature. Think of nineteenth-century chemistry. Of course, the atomic-molecular theories of the time were usually seen as describing basic chemical processes; and to accept them was to believe that they truly described these processes. But the image of science has changed a great deal since then, and the dominant theories are no longer mechanistic. Think now of quantum mechanics and geometrodynamics. Is scientific realism any longer the philosophy of science which we feel naturally compelled to accept? I should think that many space-time and quantum physicists would be quite puzzled by the suggestion that the theories they accept, and work with, might literally be true, since they have no clear conception at all of the reality with which these theories might correspond. And I can well see many of them agreeing with van Fraassen that the aim of science is only to give us theories which are empirically adequate. I think that van Fraassen, rightly or wrongly, draws some comfort from this and sees scientific realism as a philosophy of science more appropriate to another age. On this point I would agree with him, given his characterization of the position.

In considering van Fraassen's position and his arguments for it, I find myself also in agreement with much of what he says in detail. I am convinced, at least, that any scientific realist who accepts a correspondence theory of truth and van Fraassen's distinction between empirical and pragmatic considerations is in serious trouble defending his position. For, given that the only considerations relevant to the truth or falsity of a claim are empirical or logical and that such considerations alone can never determine what theories we should accept as being true (because of underdetermination), it follows that we can never have any good reason to believe that a theory is true. So anyone who believes in the literal truth of the theories they accept must do so irrationally. To counter this argument, it would be necessary to challenge either van Fraassen's empiricial underdetermination thesis, or the correspondence theory of truth upon which the whole argument is based.

However, most scientific realists are in a serious bind, for they see acceptance of a correspondence theory of truth as being essential to their position. Coherence and pragmatic theories, they think, go together not with realism but with idealism. So most scientific realists are likely to focus on either the empirical underdetermination thesis or the empirical/pragmatic distinction as the weak point of the argument. However, the empirical underdetermination thesis is very widely accepted and seldom challenged; and most scientific realists would be reluctant to give up the empirical/pragmatic distinction which gives the argument for underdetermination such force, since doing so would seem to undermine the correspondence theory of truth which they believe to be essential to their position. Nevertheless, I think that they *must* do one of these things if they wish to defend scientific realism. The correspondence theory of truth is *not* essential to their position; nor is van Fraassen's empirical/pragmatic distinction. For scientific realism can be combined with a pragmatic theory of truth; and, given such a theory of truth, all of the criteria we use for the evaluation of theories, including the so-called pragmatic ones, can be seen as being relevant to their truth or falsity.

In opposition to van Fraassen's constructive empiricism, I would propose the pragmatist thesis:

> (3) Science aims to provide the best possible explanatory account of natural phenomena; and acceptance of a scientific theory involves the belief that it belongs to such an account.

Now, I think that (3) is nearly right, and empiricists and scientific realists can both accept it. Empiricists can accept it by allowing that the best possible account is just any that is empirically adequate. Scientific realists can accept it by agreeing that the best possible account, if it exists, is *necessarily* the true one. I suppose not many scientific realists would accept this pragmatist defense of their position, since most of them would accept a correspondence theory of truth—not a pragmatic theory. Nevertheless, I do not think they have much choice. They have either to reject the correspondence theory of truth, and accept the position known as internal realism, or to follow van Fraassen along the road to his constructive empiricism.

My aim in this paper is to defend scientific realism from the perspective of an internal realist. From this perspective I shall argue that van Fraassen's case against scientific realism fails. Many of the arguments I shall present should be acceptable to scientific realists whatever their metaphysical persuasions. And these arguments should be of independent interest to scientific realists. In particular, I shall argue that the arguments for conventionalism and strong underdetermination are not as good as they are usually thought to be. However, in the end, I think that van Fraassen's arguments must defeat the metaphysical realist version of scientific realism, for there remains a skeptical challenge to metaphysical realism, to which there is no adequate reply. The final choice, therefore, is between constructive empiricism and internal scientific realism.

I. Scientific Realism

We can isolate a number of strands in the thought of scientific realists, apart from those already mentioned; and perhaps it will be useful to do so, for most scientific realists see them as going together as a package deal.

First, there is a commitment to a physicalist ontology. The precise nature of this ontology is rarely spelled out, and there are disagreements between scientific realists about what it includes. It would be almost universally agreed that it includes the fundamental particles, but it would be disputed whether properties, relationships, forces, sets, or qualia should also be included. Some would include universals (properties, relationships) in their ontology but deny the existence of sets or other abstract particulars.[1] Others are happy about sets but unhappy about universals.[2] Some would offer a Humean account of causation, and hence of forces (as I did once [1965]), while others would accept some doctrine of natural necessitation (see Harré and Madden 1975). And there are some who would admit the existence of epiphenomenal qualia while yet claiming to be scientific realists (as Keith Campbell does [1976]).

A second strand of scientific realism concerns the status of scientific laws and theories. All would say that these are to be understood realistically rather than instrumentally. The laws and theories of science are genuine claims about reality, not mere instruments of prediction which are more or less useful for this purpose. This is the *central thesis of scientific realism*. Every scientific realist subscribes to some version of it. Yet this thesis, like the doctrine of physicalism, is also subject to various interpretations. Basically, the idea is that there are things in the world to which our laws and theories refer and of which they are true or false. They are, that is, to be understood as referring to real existents and ascribing genuine properties to them. It is not clear, though, how even this claim is to be interpreted. On a naïve interpretation, we must suppose that there are Hilbert spaces, perfect gases, inertial systems, and ideal incompressible fluids in steady flow in uniform gravitational fields, for it is to things like these that many of our laws seemingly refer, if they are not just vacuously true. But my impression is that most scientific realists do not really want to take the ontological commitments of laws or theories of science in such a

1. This, as I understand it, is the position taken in Armstrong 1978.
2. The position is Quine's, although Smart and others have adopted it.

naïvely realistic way, but only to those which can be so understood conformably with their ontology. The rest would have to be suitably reduced and the apparent references parsed away to avoid any unwanted commitments.[3]

The third strand of scientific realism is the *objectivity thesis* that the laws and theories of science are objectively true or false. Every scientific realist is committed to some version of this. The objectivity thesis is rarely distinguished from the central thesis of scientific realism and is often confused with it. But, as I understand them, they are quite distinct. One could hold, for example, contra the conventionalists, that it is objectively true or false whether space is Euclidean without believing that there is such a thing as space or property of Euclideanness. One who thinks that truth is what occurs in the ultimately best theory, for example, might think this. Moreover, one could believe that this was so even though one did not believe that statements about the geometry of space were reducible to statements about, say, spatial relationships between physical entities.

The two theses are often confused, because most scientific realists accept some form of the correspondence theory of truth. They hold that a statement is objectively true iff it corresponds to reality. Of course, they do not hold that all objectively true statements correspond to reality in the same way. Some do so more directly than others. Some statements, e.g., some of those ascribing physical properties to physical objects, have a kind of *direct* correspondence to reality; and for these *basic statements* a Tarski *T*-sentence provides a sufficient explication of the correspondence relationship. Other statements, however, are related to reality in more complex ways, and to understand how they may correspond, or fail to, we must have analyses of their truth conditions which will enable us to understand them in terms of direct correspondences. Now, assuming that a statement is objectively true or false iff there exists such an analysis, and that the existence of such an analysis is both a necessary and sufficient condition for the possibility of interpreting a statement realistically, the distinction between the central thesis of scientific realism and the objectivity thesis collapses. A statement is capable of a realistic interpretation iff it is objectively true or false. However, anyone who would reject the correspondence theory of truth can continue to make the distinction.

The main difficulty with the objectivity thesis derives from the arguments of the conventionalists which appear to establish a general thesis of empirical underdetermination. The thesis in question is not the weak one that our theories are inductively underdetermined, but the strong one that, even if *all* the empirical consequences of a theory could be checked and it could be *known* that they have all been checked, it would still be

---

3. For example, some of those who admit universals in their ontologies have tried to construe laws as relationships between universals (e.g., Armstrong 1978).

possible to construct an alternative theory which has a different set of ontological commitments, or is otherwise incompatible with the given theory,
but which has precisely the same observational consequences. If this strong
thesis is true, then it threatens the realists' ontology, or at least the claim
that they would all certainly make that the ontology is empirically well
supported.

A fourth tenet of scientific realism is the correspondence theory of
truth. Most scientific realists believe that we must have such a theory if we
are to be realists about scientific entities—or, for that matter, about anything else nonmental. In championing the correspondence theory of truth,
realists thus see themselves as being opposed to idealism—to the view that
reality is a construct out of experience rather than something existing independently of it. In believing that there is this connection, I think they are
profoundly mistaken. We need not have a correspondence theory of truth
to accept a physicalist ontology or to believe in an independently existing
reality. On the contrary, I think that anyone who accepts a physicalist ontology should not also hold a correspondence theory of truth, although
this is not a claim that I wish to defend here.[4]

Clearly, I cannot discuss all of these issues properly in one paper. So I
shall focus on those questions which seem to be most at issue between scientific realists and constructive empiricists. The question of what ontology
one should have if one is a scientific realist can safely be set aside. The
main question concerns how scientific theories are to be understood—
realistically or otherwise.

II. Theories and Explanations

One thing you learn about the United States when you visit that country
is that it is dangerous to generalize about it. And one thing philosophers of
science should have learned about scientific theories and explanations is
that it is also dangerous to generalize about them. The danger is that what
seems to be true of theories and explanations in one field, or of some of
them at any rate, may not be true in others, so that different theories of
theories and explanations may seem plausible or not depending on what
examples are taken. I think that van Fraassen is well aware of this danger—probably more so than most of his realist opponents. For scientific
realists do have a rather lamentable tendency to take nice homely examples
of causal explanations as typifying scientific explanations generally, and
simple mechanistic theories taken from nineteenth-century physics or
chemistry, which were obviously intended by their authors to be under-

4. This is a major theme of a book on truth and realism which I am currently writing.

stood realistically, as paradigms of scientific theories. Their philosophical position is thus given a strong flavor of initial plausibility. Van Fraassen's position, on the other hand, probably derives more from his earlier work on space-time theories (e.g., 1969, 1970), where it is not at all obvious that the theories were intended to be anything more than models of some kind which could be used with greater or less facility to 'save the phenomena'.

In fact, there are different sorts of theories and explanations which arise as answers to different sorts of questions. I follow van Fraassen in thinking that any request for explanation is a request for information. A *causal explanation* is information about the causal history of something or about the causal processes which result in something.[5] A *functional explanation* is information about the role of something in some ongoing system—about the contribution it makes to sustaining it. A *model theoretic explanation* is information about how (if at all) the actual behavior of some system differs from that which it should have ideally if it were not for some perturbing influences and, where necessary, includes some information about what perturbing influences may be causing the difference. A *systemic explanation* is information about how the fact to be explained is systematically related to other facts.

Theories, on the other hand, provide us with the general schemata for giving such explanations. *Causal process theories* attempt to describe the basic causal processes of nature. *Functionalist theories* are concerned with ongoing systems of various kinds and with the kinds of mechanisms, described in terms of their functional roles, necessary for their maintenance. *Model theories* define norms of behavior against which actual behavior may be compared and (causally) explained. *Systemic theories* set forth some general organizational principles adequate to determine the basic structure of some system of relationships between things. Euclidean geometry, for example, is such a theory for the system of spatial relationships.

Now, the argument for scientific realism, insofar as it concerns the reality of theoretical entities, derives whatever force it has from taking causal process theories to be typical of scientific theories generally. For to accept that A is the cause of B is to accept that both A and B are real existents (or events). But no such argument applies to the theoretical entities of model theories, for the hypothetical entities of model theories are not the postulated *causes* of anything. Consequently, there is no parallel argument that to accept a model theory involves the belief that the entities to which it apparently refers really exist. We do not have to believe in the reality of Newtonian point masses or of Einsteinian inertial frames or in the existence of the perfectly reversible heat engines of classical thermodynamics

5. David Lewis has elaborated on this theme in a paper entitled "Causal Explanation," which is to appear in his *Philosophical Papers* to be published by Oxford University Press.

to accept these various theories, for none of these entities has any causal role in the explanations provided by the theories in question. Consequently, it is not necessary to think of these theories as being literally descriptive of the underlying causal processes of nature. And, if we do so, we become committed to the absurd view that many, if not all, of the basic principles of Newtonian dynamics, of special relativity, and of thermodynamics are just vacuously true.

Van Fraassen's view, on the other hand, seems to come from taking model and systemic theories as typical, for the value of such theories derives, not from any insights they may provide about the workings of nature, but from their capacity to systematize our knowledge of it. Consequently, it does not matter whether these theories are literally true or false. What matters is whether they are adequate to the task for which they were devised. Plausibly, therefore, to accept such a theory involves no more than the belief that it is so adequate—that it is, in van Fraassen's sense, an empirically adequate theory.

Scientific realists run into trouble when they try to generalize about scientific theories. To cope with laws which hold only for ideal systems of some kind (and most of the laws of nature are like this) and which would strictly be false if they were meant to apply to actual systems, many scientific realists think of them as being only good *approximations* to the truth (e.g., Boyd 1976, 1980). So, instead of van Fraassen's characterization of their position, they would accept something like this:

> (1a) Science aims to give us, in its theories, a literally true story of what the world is like; and acceptance of a scientific theory involves the belief that it is at least approximately true.

Thus, they would consider such laws as conservation of energy and conservation of momentum, which strictly apply only to closed and isolated systems, as being only good approximations to the truth about actual systems. What is strictly true, they would say, is that the energy and momentum of any more or less closed and isolated system are more or less conserved. And, as science progresses, these essentially 'inaccurate' laws should be replaced by more accurate ones involving less idealization and hence greater faithfulness to the actual relationships among things. Think, for example, of the replacement of the ideal gas laws by van de Waal's equation. Many scientific realists thus envisage the eventual replacement of model theories by systemic ones in which all of the laws and principles are just true generalizations about how actual things behave.

However, if science is aiming to achieve such a result, it seems often to be pointing in the wrong direction, for a great deal of theoretical scientific research goes into devising increasingly abstract model theories, and relatively little into reducing the degree of idealization involved in our theories in order to make them more realistic. It is true that in economic forecasting, and in the applied sciences generally, researchers labor to develop increasingly elaborate computer simulations of real systems, taking into account as many as possible of the relevant variables so as to maximize the accuracy of their predictions. And I do not deny the importance of such research. But basic theoretical development in science tends, if anything, to proceed in the opposite direction—to greater abstraction and generality.

Take the development of Newtonian mechanics from 1700 to about 1900 (see Dugas 1955). In this period no major scientist working in the field of mechanics developed theories which they thought were closer to the truth than Newton's. Most of them would have said that Newton's theory was true already. Nor, for that matter, did they think they were developing theories which were empirically more adequate than Newton's. So it seems wrong to suppose that they were aiming to increase either the realism or the empirical adequacy of the basic theory they were working with. Yet the great works of classical mechanics of Euler, the Bernoullis, d'Alembert, Fermat, Lazare Carnot, Lagrange, Laplace, Gauss, Coriolis, Hamilton, and Jacobi surely contributed *something* to fulfilling the aims of science, for they improved greatly our knowledge and understanding of mechanical processes. They solved many previously unsolved problems, they applied Newtonian theory in new ways, they discovered new principles and unsuspected symmetries, and they invented powerful new mathematical techniques for handling complex mechanical problems. I think we may conclude that scientists, qua pure scientists, are not always greatly concerned to make their model theories more realistic or more adequate empirically. They have other much more interesting things to do.

Van Fraassen's constructive empiricism is in trouble for these reasons. But he adds to his troubles by construing all theories on the model of model theories, for he is now committed to saying that the postulated entities of causal process theories have no more claim to be considered real existents than the theoretical constructs of model theories. Atoms, creatures of the Jurassic period, inertial systems, and possible worlds are all on a par, according to him, and to accept theories, which apparently make reference to such entities, involves only the belief that the theories are empirically adequate. Now, if the theories we are talking about are special relativity and 'possible worlds' semantics, van Fraassen's position is at least plausible. But it loses all plausibility if the theories in question are historical theories or theories of chemical combination. And the reason, I think, why this is so is that the postulated causes of the phenomena must be supposed to exist if the theory is to be accepted as doing what it purports to do; and normally we should expect to be able to find independent confirmation of their existence from various sources. The situation is quite dif-

ferent with the theoretical entities of abstract model theories. Since they
are not postulated as causes, they are not supposed to have any effects. So
we should not expect them to leave any traces or to manifest themselves in
other ways, or indeed in *any* way at all. And that is why, apparently, we can
play with them as we like or assign any properties to them we wish to
produce a better theory. We know that no astronomers are going to discover inertial frames which don't have the properties we assign to them and
that travelers are not going to stumble across other possible worlds where
they shouldn't be and so spike our theories about conditionals.

The status of the theoretical entities of causal process theories is not like
this, however. When the theory is accepted, we think we know only *some*
of their properties. We know we might be wrong about them in some ways,
and we might expect our picture of them to change somewhat. But, typically, we expect to discover more about them—to add to and refine our
knowledge of them and to explain why they have the properties they do.
That is, we expect them to be like other physical things and to participate
in various ways in causal processes, depending on what their properties are
and what the surrounding circumstances are like. In short, we think of
them, and expect them to behave, as real things do.[6] Moreover, our reasons for believing in them are not basically different from our reasons for
believing in more ordinary things. If the existence of atoms or of the
moons of Jupiter were a legal issue, I think almost any jury would find the
case proven by the ordinary rules of evidence.

Now, I think that van Fraassen is aware of all this. I don't think he embraces constructive empiricism out of naïveté or any failure to distinguish
between causal process theories and model theories. His philosophical
position probably arose out of a less ambitious theory about the status of
the laws of special relativity and other abstract model theories (see van
Fraassen 1969, 1970). Originally he argued that many of the principles involved in these theories were conventional. For example, he argued that it
is ultimately a matter of convention whether we say that the one-way speed
of light is the same in all directions. About this he would have said, following Reichenbach and the other positivists, there is no truth of the matter.
However, he has now come to believe that his arguments for the conventionality of such principles apply *right across the board*. So, consistently
with his earlier position, he ought to conclude that *in general* there is no
truth of the matter concerning the fundamental laws and theories of science. But, as I understand him, van Fraassen does *not* accept this general
conclusion. There *is* a truth of the matter, he now wants to say, *but we*

6. According to James Clark Maxwell and P. W. Bridgman, an entity is physically real if it
manifests itself in more than one way. G. Schlesinger (1963) calls this the Maxwell-Bridgman
criterion for physical reality. It is surprising that this important criterion is not more widely
known or discussed.

cannot really know what it is. For the best that we can ever hope to do is construct a system of empirically adequate theories. Anything beyond this is necessarily beyond our grasp.

III. Conventionalism

Conventionalists argue that there are many questions in science which call, not for further empirical investigation, but for decision. Is the one-way velocity of light equal to its round-trip velocity? Is the geometry of space Euclidean? Does a body which is not acted upon by forces continue in its state of rest or uniform motion in a straight line? These questions, and many others, have been said by conventionalists not to be questions about what is true of reality but ones which can be resolved only by stipulation or definition. Concerning these questions, there is said to be no truth of the matter.

There are three main arguments for considering the statement of a law or theory to be conventional. First, there is *the circularity argument*, which was frequently used by Reichenbach. For example, in arguing for the conventionality of the principle that the one-way velocity of light is the same in all directions, he tries to show that we should have to presuppose the principle in order to prove it. He argues that to measure the one-way velocity of light we should need clocks in synchrony at different places, i.e., reading the same at the same time. Consequently, we must know what it is for two events (readings) occurring at different places to be simultaneous. But ultimately, he claims, we cannot determine whether two such events are simultaneous unless we make some prior assumptions about the one-way velocity of light. And he then goes on to say that "the occurrence of this circularity proves that simultaneity is not a matter of knowledge, but of a coordinative definition, since the logical circle shows that a knowledge of simultaneity is impossible in principle" (Reichenbach 1958, 127). He concludes that the law in question is conventional.

It has now, I think, been conclusively established that Reichenbach was wrong in thinking that one has to make some assumption about the one-way velocity of light to determine the simultaneity of distant events. There are, in fact, several procedures which could be used to establish clocks in a relationship of distant synchrony which, logically, do not depend on this assumption (as argued in Ellis and Bowman 1967). Consequently, there are several logically independent criteria for distant simultaneity, and the standard signal synchrony criterion is just one of them.

However, conventionalists have a second argument to fall back on—*the argument from the need for definition*. It may be conceded that distant simultaneity is a multicriterial concept, so that any of a number of different criteria for distant simultaneity might be chosen to define this relationship. Still, a choice has to be made,[7] and the law or principle which underlies the choice will have to be considered to be true by definition. This principle might be the one-way-light principle (that the one-way velocity of light is constant) or the principle of slow clock transport (that locally synchronized identical clocks transported sufficiently slowly remain in synchrony), or it might be some other principle. But surely at least one of these laws or principles must be regarded as conventional if the concept of distant simultaneity is to be well defined. If the one-way-light principle is the chosen one, then it is conventional and all the others are empirical.

It is, however, arbitrary to pick *one* law out of a law cluster like this, and say that it is conventional while the rest are empirical, if no good reason can be given for choosing it. And, if the choice is arbitrary, then we might as well have chosen another one. Consequently, any given law in such a law cluster might arbitrarily be regarded as empirical or conventional, depending on how we choose to axiomatize our system. Assuming that the 'simultaneity' law cluster is like this and that the one-way velocity of light has been set equal to the round-trip velocity by definition, then the principle of slow clock transport is empirical. It is, however, a matter of convention that the one-way-light principle is conventional, and it is also conventional that the principle of slow clock transport is empirical. For we could equally well have reversed the roles of these two principles. Assuming that the special theory of relativity is correct, we know that the 'simultaneity' law cluster exists. In all of the vast literature on distant simultaneity, there is no argument for preferring the standard light signal to the slow-clock-transport definition of distant simultaneity, or conversely; and we may reasonably assume that there is none. Therefore, it is arbitrary which principle we choose to call conventional and which empirical.

This being the case, why should we choose at all? What difference would it make, either to our practice or to our beliefs, if we thought of all the laws in the cluster as empirical laws? Suppose that, contrary to the predictions of the special theory of relativity, we found that clocks synchro-

---

7. The use of this argument is hard to document. But the assumption that there is a need for definition to fix meaning *before* any questions of truth or falsity can arise is a common background assumption of logical positivists. This assumption was challenged, most notably by Quine and Putnam, in the course of their attacks on the analytic/synthetic distinction in the 1950s and early 1960s. Nevertheless, the implications of their work for the fact/convention distinction in the philosophy of science were not understood. In 1967, Bowman and I (pp. 116–36) took the challenge into the center of the philosophy of science camp by arguing specifically that their most cherished convention, distant simultaneity, was not in any interesting sense conventional. Grünbaum, Salmon, and van Fraassen (*Philosophy of Science* 36 [1969]) replied in terms which left no doubt that they still considered the fact/convention distinction to be absolutely indispensable for the analysis of science and stoutly defended the conventionality of distant simultaneity. The arguments which follow are derived from my reply to them (1971, 177–203).

nized by slow transport were not in standard signal synchrony? Would it make any difference to how we should proceed which of the two principles we had chosen to call true by definition or convention? It is my belief that it would make no difference at all. A radically new space-time theory would be needed, and I do not believe that, in constructing it, we should feel in any way constrained to accept what we had earlier said was true by definition or to reject what we had said was empirical.

In general, the empirical/conventional distinction is of no practical importance in science. Scientific practice is not affected by what we might choose to say is conventional or what we might think of as empirical. So we can just as well regard all of the laws and theories of science as empirical— at least in the sense that they are open to revision in the light of experience. The argument from the need for definition is, therefore, no good argument for conventionalism, for, in general, there *is* no need for definition. It may be useful or even necessary to offer a definition of a term when introducing it to the profession or to students for the first time, but the statement of that definition enjoys no privileged immunity to revision or even rejection in the light of further experience.

The third and perhaps most important argument for conventionalism is that from *empirical underdetermination*.[8] We can discover empirically, perhaps, that the 'simultaneity' law cluster exists, but this does not bind us to accepting any or even the conjunction of all of the laws in the cluster as defining the relationship of distant simultaneity. For we are free to adopt some nonstandard signal or transport definition of distant simultaneity, say one which makes the one-way velocity of light a function of direction, or relative clock-rates a function of their relative positions, and use this to determine what all of the other laws in the cluster are. Given such a definition, there will still be a 'simultaneity' law cluster, but it will be a different one. No doubt, such a coordination would seriously complicate our physics, but, so the conventionalist maintains, it cannot be ruled out on any *a priori* or empirical grounds. Consequently, there can be no truth of the matter concerning *any* of the laws in the cluster. They are *all* conventional.

This move is the most serious one for the scientific realist who wishes to retain the objectivity thesis along with the correspondence theory of truth, for it threatens his whole program of interpreting the laws and theories of science realistically. It is the main reason why those positivists, who accepted the correspondence theory, rejected the objectivity thesis for some form of conventionalism; and it is one of the main reasons why some, who wish to retain the objectivity thesis, have rejected the correspondence

8. This is the argument most commonly used by empiricists from Mach and Poincaré onward.

theory of truth in favor of some form of coherence theory. The point is that the coherence theorist can argue that what is true is what occurs in the best (ultimately best) theory and is not embarrassed by this conventionalist argument, for no one would pretend that a nonstandard definition of distant simultaneity would produce a theory which was anything like *as good* as the special theory of relativity. It might be empirically equivalent, but we should have to allow that there are some strange spatial asymmetries in many of our laws of nature for which we have no adequate explanation.

Now, van Fraassen, as I understand him, accepts the main argument for conventionalism—that from empirical underdetermination—and continues to accept the correspondence theory of truth; but he rejects the conventionalist conclusion that there is no truth of the matter concerning those laws and theories which are empirically underdetermined. He agrees with the scientific realists that they are either true or false; but, unlike them, he considers that whether they are true or false is ultimately not an empirical question but a metaphysical one. Truth, for van Fraassen, is a metaphysical concept.

IV. Empirical Underdetermination

The implications of the conventionalist argument from empirical underdetermination are serious for the scientific realist, for empirical underdetermination would appear to be a feature of *all* of our theories. Consequently, it threatens to render even the realists' belief in a physicalist ontology either false or metaphysical. If it is always possible to construct another theory which is incompatible with a given theory, but is empirically equivalent to it, then all theories are empirically underdetermined. Therefore, if one accepts both the correspondence theory of truth and the objectivity thesis, one may be forced to consider all theories to be either false or metaphysical—a conclusion which would be unavoidable if the general principle of empirical underdetermination were accepted.

This thesis of empirical underdetermination is not, of course, the obvious point that our theories are *inductively* underdetermined by the evidence we have for them. Things might turn out to be grue or bleen rather than green or blue. The point is that there are incompatible theories which are empirically equivalent in the sense that *no* evidence could distinguish between them; and the claim is that our theoretical understanding of the world could be underdetermined even by the supposed totality of empirical evidence. This thesis, if it is true, is the real threat to scientific realism; for what reason could we have to believe in the existence of any theoretical entity if we could be assured that there is another theory, empirically equivalent to the one in which it is postulated, which may assume the existence of other, quite different kinds of entities?

Although the thesis of empirical underdetermination is commonly asserted, the arguments for it are not as compelling as they are usually thought to be. First, there is the failure of Carnap's program of reductive analysis (summarized in 1956). All attempts to define the theoretical terms of science in an observational language have met with failure. Even the simplest theoretical predicates have stoutly resisted such an analysis. Consequently, the conviction has grown that theories say more than anything that can be said in an observational language. It is not just that they go beyond the evidence. All inductive generalizations do that. They go beyond anything for which there could be inductive support. For they cannot even be expressed in a language the terms of which are purely observational. Consequently, it is held that all theories must be strongly underdetermined by evidence. Even if all possible observations had been made, and were known to be have been made, and our theories were compatible with this supposed totality of evidence, our theories would still say more than this and so be underdetermined by it. This argument is plausible, I think, but not entirely convincing, for it does seem to rely on a rather naïve inductivism—the crucial premise being that if A cannot be expressed in an observation language L, then A cannot be supported by any evidence which is expressible in L. And I know of no good arguments for this premise.

Second, and I think more importantly for van Fraassen, there are many persuasive examples of empirical underdetermination—ones which make it reasonable to believe that all theories are similarly underdetermined. At any rate, van Fraassen does argue for empirical underdetermination primarily by cases and is not content to rely on sweeping general arguments such as the one I have briefly presented.

Without going into detail, I would admit that there are persuasive arguments for the existence of empirically equivalent theories, but, even if they are sound, they do not establish the general thesis. For one cannot establish that all theories are ultimately underdetermined by showing that some are. One might, for example, try to isolate our dynamical theories and our theories of space and time, for which the most detailed arguments for conventionality exist, and admit that at least some of the entities which occur in these theories, such as forces, may not be real, but preserve a realist interpretation for the entities of atomic and subatomic physics by arguing that the theories which postulate the existence of these entities are not conventional in the same way. This is the strategy I had in mind when I spoke of 'scientific entity' realism (1979, p. 45, n. 15).

Now, I do not particularly wish to defend 'scientific entity' realism in this form, because I would now prefer the stronger internal realist position. However, it is worth pointing out that the main arguments for the conventionality of our space-time and other abstract model theories do not carry over as arguments for the conventionality of our causal process theories. For, if the entities postulated in these theories exist, we should expect them to manifest themselves in various ways and to participate in causal processes *other* than those described in the theories we have so far developed. Consequently, we should expect confirmation to come from unexpected sources as our causal process theories are developed. Hence, theories which cannot be distinguished between empirically in one theoretical context may be distinguished in another. New theoretical developments may show some empirical consideration to be differentially relevant to the truth of the two theories, which previously would have been considered to be empirically equivalent.

It is possible, for example, to construct two theories of chemical combination to explain why one volume of hydrogen combines with one of chlorine to form two volumes of hydrogen chloride (described fully in Ellis 1957). One is the classical theory of Avogadro; the other postulates the existence of certain 'gas numbers' characteristic of elemental and compound gases and certain laws relating these gas numbers to combining volumes. In terms of empirically testable consequences, the theories are equivalent. Nevertheless, they are very different theories. One postulates the existence of atoms and molecules and paramechanical processes of chemical combination; the other does not. We may add the further hypothesis to the second theory that there are no such things or processes—just to make sure that the two theories are incompatible. Now, in spite of the fact that these two theories have the same empirically testable consequences, they clearly are not, and I want to say were never, empirically equivalent. For they offered quite different prospects for development, and, as new theories were developed, new facts became relevant to the acceptance of one of them (Avogadro's theory). The facts of electrolysis, for example, are not empirical consequences of Avogadro's theory, but they certainly supported it, because they could be readily explained on the atomic-molecular model which it used.

Many philosophers think of theories as sets of sentences implying observational conditionals, i.e., sentences of the form

$$C_1 \wedge C_2 \wedge \ldots C_n \supset O$$

where $C_1, C_2, \ldots C_n$ are empirically determinable boundary conditions and O_i is an observation statement. (E.g., an event of kind K is occurring in the space-time region S.) This model of a theory is so widely accepted and used that it may reasonably be called "the standard model". On this model, it is natural enough to say that two theories are empirically equivalent iff they imply the same set of observational conditionals. For only if this is so will they have the same empirical consequences. Yet, according to this criterion, my 'gas number' theory is empirically equivalent to Avogadro's—which in my view is absurd. It is absurd because the evidence in favor of Avogadro's theory and against the gas-number theory, is now over-whelming. This is so because the original theory of Avogadro has become embedded in a very general and powerful theory of chemical combination and has well-established links with (in the sense that it has hypotheses in common with) a wide range of other physical and chemical theories. Avogadro's theory has gained support from these other theories because it has become an integral part of them. The evidence in favor of Avogadro's theory cannot be identified with the confirmation that its observational consequences have received—at least not if "observational consequences" is understood as narrowly as the standard model would suggest.

The point is a Duhemian one. Theories do not normally occur in isolation, and evidence for or against a theory can come from unexpected quarters. This evidence may be unexpected not because we have failed to carry out the relevant deductions from the axioms but *because it is not a consequence of these axioms at all*, at any rate not of these axioms *alone*. It may be evidence which can be seen to be relevant to the theory in question *only* because some new linked theoretical development has occurred. Therefore, unless we can know in advance what theoretical developments might occur, we cannot say in advance what evidence, if any, might distinguish between theories which on present indications would appear to be empirically equivalent. Therefore, assuming that we cannot know what theoretical developments will occur, we cannot ever be fully justified in claiming that two incompatible theories are empirically equivalent in the sense that no evidence could ever distinguish between them. That they cannot be distinguished, given our present theoretical understanding of the world, does not imply that they cannot be distinguished. A scientific realist may therefore be able to defend his position against the empiricist/conventionalist arguments for the empirical underdetermination of theories, for he can argue that the underdetermination thesis *cannot be demonstrated*. It cannot be shown, except by fiat, that there is *any* genuine case of empirical underdetermination.

The reason for this perhaps needs emphasizing. It derives from the *openness of the field of evidence*. By the field of evidence of a theory I mean the set of possible empirical discoveries relevant to its truth or falsity. This set is to be clearly distinguished from the set of *empirical consequences* of the theory. The empirical consequences of a theory are the observational conditionals entailed by it. Hence, the set of such consequences is a subset of the set of logical consequences of a theory. But not so the field of evidence. There can be evidence for a theory which is in no way entailed by it and evidence against it which does not contradict it. Faraday's laws of electrolysis, for example, certainly supported the atomic-molecular theory of Avogadro, since the atomic-molecular model proposed by Avogadro to explain certain facts about chemical combinations could readily be adapted to explain the facts of electrolysis. But Avogadro could not, without making additional assumptions, have deduced Faraday's laws from his original theory. The failure of the Michelson-Morley experiment, on the other hand, surely proved to be evidence against Newtonian mechanics, even though this null result was compatible both with Newton's laws of motion and with his law of gravity. Newton's laws of mechanics simply had *nothing to say* about the behavior of electromagnetic radiation.

This point is of considerable importance in the present context, for it implies that claims of empirical equivalence should be treated with caution or be relativized to a given stage of theoretical development. From the fact that two logically distinct theories have the same empirical consequences (in the sense that they imply the same set of observational conditionals), it does not follow that there is no evidence which could distinguish between them. Nor does this follow from our inability to think of any way in which they might be so distinguished. For, so long as the theories we are concerned with are logically distinct, it is always logically possible that unforeseen theoretical developments will occur which will enable us to devise tests which would decide between them. To justify accepting any empirical equivalence claim, therefore, we should at least need a good theoretical argument that no such developments could occur.

Defenders of the thesis of the conventionality of distant simultaneity claim to have just such an argument. They claim that the facts relating to signal speeds and slow clock transport make it impossible for us ever to distinguish empirically between standard and nonstandard relativity theories, e.g., theories which make the one-way speed of light in an inertial system some kind of function of direction.[9] For reasons I have given at length elsewhere, I am not convinced by their arguments (Ellis 1971). However, even if it were established that there is some significant conventionality in our theories of space and time, this would not establish the *general* thesis of empirical underdetermination which van Fraassen needs for his argument to apply to all theories. For the theoretical entities of causal process theories are different from those of the model theories we have been considering. They are the sorts of thing which, if they exist, we can expect to discover a great deal more about as we improve our understanding of nature. And, because of the openness of the field of evidence concerning them, we can never be sure that logically distinct but empirically equivalent causal process theories cannot ultimately be distinguished empirically. So the general empirical underdetermination thesis van Fraassen needs to establish his position is, as yet, not proven.

As far as I know, the only way of demonstrating the empirical under-

9. See the panel discussion on distant simultaneity with papers by A. Grünbaum, W. C. Salmon, B. C. van Fraassen, and A. Janis (*Philosophy of Science* 36 [1969]) for the detailed argument.

determination thesis for causal process theories is to build empirical equivalence into the specification of the supposed, logically nonequivalent, alternative theories. For example, instead of the theory T, we might have the theory T', where

> T' = Although the world is not as theory T says it is, the world behaves, so far as we can ever tell, as if T were true.

Of course, there are all sorts of variants of this, like

> T'' = Actually, we are all brains in a vat, but . . .

or

> T''' = Actually, the world began five minutes ago (local time), but . . .

But the variations just illustrate the general strategy. So the fundamental issue which needs to be considered is whether the metaphysical realism presupposed by such specifications of alternative theories is acceptable to a scientific realist. If it is, then I think van Fraassen wins, for the following theses cannot be consistently maintained:

1. All theories are empirically underdetermined in the sense that logically nonequivalent but empirically equivalent alternative theories always exist.
2. The choice between empirically equivalent theories cannot be made on any grounds except pragmatic ones (simplicity, symmetry, elegance, explanatory power, etc.).
3. Pragmatic considerations have no relevance to truth or falsity.
4. We have good reason to believe that the world is more or less how our best theories say it is.

In discussing the issue of metaphysical realism, let me contrast it with my own position of internal realism. For I want to show how scientific realism can be defended by rejecting thesis 3 and accepting a pragmatic theory of truth. Also, in retrospect, I think that this has always been the basic issue on which van Fraassen and I have disagreed.

V. Internal and Metaphysical Realism

To know that something is the case, we must somehow be justified in believing it and, moreover, it must be true. I say "somehow" because apparently not every case of justified true belief is a case of knowledge. But the refinements here need not concern us, since the point I want to make is independent of how the concept of justification may be spelled out. So let us just say that knowledge is true and justified belief and ignore the refinements. Then, apparently, there are two kinds of conditions for knowledge: one concerns the way the world is, and the other the justification of our beliefs about it. Let us call them the *ontological* and the *epistemic* conditions, respectively. Now, I think that nearly everyone would agree that satisfaction of the epistemic conditions does not entail satisfaction of the ontological ones, or conversely. We may be justified in believing what is false and not justified in believing what is true. This is common ground between internal and metaphysical realists. Where they differ is in the extrapolation of this point to the limit of what is ultimately justifiable. For the internal realist, what is true, if anything, *is* just what is ultimately justifiable. For the metaphysical realist, however, truth remains independent of rational evaluation, even in the projected limit of this process.

It is important to be clear that the internal realist does *not* equate truth with *warranted assertability* or *reasonable belief*. For these notions are tied to existing or available evidence and perhaps also to certain background assumptions and theories which, in the given context, are not in question. No, for the internal realist, truth is a kind of limit notion of reasonable belief. We all believe that we are limited in what we know and that probably some of the things we believe are false, but we do not, for these reasons, think that it is unreasonable for us to believe what we do. Circumstances can change what we may reasonably believe. If new evidence is discovered, if a better theory is proposed, if results previously accepted are brought into question, then what we may reasonably believe is affected by such changes. Moreover, we should normally consider these changes to be ones for the better—to be changes which *improve* our knowledge and understanding of the world. Truth, for the internal realist, is the supposed perfection of this process. It is what we should believe, if our knowledge were perfected, if it were based on total evidence, was internally coherent and was theoretically integrated in the best possible way. There are, of course, very great problems in trying to explicate this concept of perfection. But for the internal realist this is what truth, if it exists, must be.

The metaphysical realist, on the other hand, has a different concept of truth. For him truth is a relationship between what is true and what it is true of, which holds independently of our epistemic values. Our epistemic values are those which come obviously into play whenever we are involved in evaluating (for truth or falsity) what other people say or believe or in reevaluating our own beliefs. We judge *how well* something has been attested or supported or explained or *how well* it coheres with other things that we think we know. And to do this we must have some system of preferences. The values which give rise to these preferences are our epistemic values. The metaphysical realist normally does, and certainly should, admit that we have such values. But truth, he would say, is not dependent on what these values may be. We may, as a matter of fact, have evolved criteria for epistemic evaluation which are well adapted to the goal of discovering what is true. But, then again, we may not. Even the perfect theory (supposing it existed) might not be true, though the world and our systems of epistemic values might be such that we could never turn up (or have turned up) any evidence which would give us good reasons to doubt it.

One's concept of truth need not greatly affect one's beliefs about reality. Consequently, internal realists and metaphysical realists may have quite similar ontologies. There is, for example, no reason why an internal realist should not be a physicalist, for there is no reason, which is not a reason for anyone else, why an internal realist should not consider the physicalist ontology to be the theory that it is most rational to believe. On the contrary, there would appear to be very good reason why he/she should do so. Moreover, an internal realist can and indeed should be a realist in the quite full-blooded sense of believing that there is a reality which exists independently of anyone's knowledge of it or beliefs about it. For surely, on the best available theories that we have concerning the nature of the physical world, reality would be much the same even if the human race had never existed. Any sensible person must believe that, and internal realists can be eminently sensible. Moreover, the independent reality which the internal realist may believe in is not a featureless noumenon but a world of physical objects with physical properties interacting with each other in the usual ways. If it is right to believe that these things exist, and if truth is what it is right to believe, then, of course, it is true that they do.

Internal and metaphysical realists differ from each other essentially only in that they have different concepts of truth. But this difference gives rise to others. For example, the metaphysical realist is open to a kind of skeptical challenge: if truth is independent of our epistemic values, then what reason do we have to believe that those theories which we should judge to be the best are nearer to the truth? How do we know that our epistemic values are well adapted to discovering what is true? We can investigate nature and develop a theoretical understanding of the world, but we cannot compare what we think we know with the truth to see how well we are doing. We cannot even be assured that science has made progress toward its goal of discovering the true nature of reality. The internal realist, on the other hand, is not plagued by these skeptical doubts. Our epistemic values *must* be adapted to the end of discovering what is true, because *truth* is just the culmination of the process of investigating and reasoning about nature in accordance with these values.

Given these characterizations of the two positions, I think most scientific realists would consider themselves to be metaphysical realists, for this is the position which must be associated with adherence to the correspondence theory of truth. The idea that one could be a scientific realist and yet have a concept of truth so clearly reminiscent of Bradley's would strike most scientific realists as absurd. Nevertheless, I think they are wrong about this. I think they go wrong because they systematically misunderstand the position of the internal realist. They take internal realists to be denying that there is a reality which exists independently of anyone's knowledge or understanding of it. But, of course, they do not, or at least they *need* not, do this. I certainly believe in the existence of such a reality, because that is what I think any sane person *ought* rationally to believe. The position I should defend, therefore, is not an idealist one in the usual sense, even though my concept of truth is similar in some ways to the coherence concept defended by some idealists.

The point that an internal realist does not have to deny the existence of an independent physical reality needs to be stressed. But I fear that the name that Putnam (1979) has given to the kind of position I want to defend will insure that it is constantly misunderstood. There is, however, nothing 'internal' about my ontology, and it is not my view that reality is a construction out of ideas or sense impressions or whatever. Reality exists and has virtually all of the properties that it has independently of anyone's knowledge or belief about it. That is, the following counterfactual conditional is one that I think is true, i.e., right to believe:

If human beings had never existed, the world would not be so very different from the way it is.

Of course, there would be no cities, plowed fields, books, or theories about the world, and maybe some species of animals would not now be extinct, if we had never existed. But, overall, the difference would not be very great. The difference would be even less great if we just had different beliefs about the world. It would still be different, because our beliefs are parts of reality and beliefs have effects, but the basic ontology to which we should subscribe would be unaffected by our having or coming to have different beliefs about it.

An internal realist can thus believe in the existence of an independent reality. That is, he can believe that the ontology to which he subscribes and the basic laws of nature would be the same, whatever he might believe. Indeed, it is part of his understanding of the world that this should be so. The independence here is a *causal* independence. One cannot change the basic structure of the world by changing one's beliefs about it.

But what about changes in our epistemic values? Wouldn't a change in our epistemic values change the way the world is by making it right for us to believe things about the world which it would not now be right for us to believe? Isn't an internal realist not, after all, a kind of idealist?

First, let me distinguish between two sorts of changes. If I change the position of my little finger, then in one sense I change Sirius because I change its relationship to my little finger. But in another sense I do not change Sirius, because there is no causal influence—at least nothing which will be felt on Sirius for a long time. Changes of the former kind are called "Cambridge" changes. What is characteristic of such changes is that they are brought about *without causal influence* on the object that is said to be changed. The causal explanation of the change in the position of Sirius relative to my little finger will refer only to events occurring in me. Now, in this sense a change in our epistemic values effects only a Cambridge change on the world. The explanation of what has occurred will refer only to events occurring in us. So anyone with our present epistemic values would not see the world as having changed. He would see only a set of changes in the way people think about it.

But suppose we all changed so that we all became different sorts of beings equipped with a different system of epistemic values. Then, presumably, there would be a different way the world is for the beings we then become. Yes, I think that is right. The way the world is is relative to the sorts of beings we are. That is one of the consequences of internal realism. It does not make it wrong for us to believe what we do about the world. Nor would it necessarily be wrong for us to believe something different if we were different sorts of beings. Nor is there any third standpoint from which the belief systems of two different sorts of beings could be compared and evaluated, or, if there is, then it enjoys no privileged position. So, according to the internal realist, there is no way that the world is absolutely, only ways in which it is relative to various kinds of beings, none of which can claim absolute priority.

Does this then mean that an internal realist is after all a kind of idealist? Yes, insofar as he believes that the way the world is for us is dependent, although not causally dependent, on the sorts of being we are, an internal realist might fairly be called an idealist of sorts. But he is not, or certainly need not be, an idealist in the sense that he denies the existence of an independent reality. For he may well think, as I do, that it is right to believe in the existence of a reality which is not affected by our beliefs about it and is not dependent on our experience of it. There is, therefore, no strong incompatibility between internal realism and scientific realism. An internal realist can accept most of the principal theses of scientific realism.[10] And this paper has been a defense of scientific realism from an internalist perspective.

But what of the alternative? What is wrong with metaphysical realism? Is there any reason why a scientific realist should not be a metaphysical realist? Well, yes, I think there is. First, metaphysical realism is essentially a skeptical position. If even the perfection of human knowledge by human

10. But not, of course, the correspondence theory of truth.

standards does not necessarily lead to truth, then the truth is essentially unknowable, as van Fraassen has shown. We can have no reason to think that improving our knowledge and understanding of the world brings us any nearer to the truth. Second, there cannot be a good argument for metaphysical realism. If the assumption that there is a way the world is independent of our *epistemic* values had any explanatory power, then it would be right for us to believe in it. But in that case it would be part of our world and not an absolutely independent reality. We do, I think, have good arguments that there are things which exist independently of our *knowledge and understanding* of the world. The internal realist accordingly embraces them. Likewise, if we had good arguments for the existence of a transcendental noumenon, then it would be right for us to believe in it, too. But let us not pretend that we can frame an argument which would be persuasive whatever our epistemic values might be and, so, binding on all thinking creatures. Third, the assumptions of metaphysical realism are unnecessary. Human truth is all we can ever aspire to. If there is another kind of truth for beings of another kind, then that's their problem. There is not and cannot be any absolute truth, and therefore there cannot be any way that the world is independently of how we, or some other kind of creature, would evaluate its beliefs about it. So I conclude that, if you want to be a scientific realist, then you had better be an internal realist.

But isn't there, finally, a crucial objection to this position? I say that there is a way the world is for us, and that perhaps for other sorts of beings there is a way, a different way, it is for them, and that neither can claim priority. But surely there has to *be* a world, a *common* world, to which the members of both species are reacting. Well, yes, I think that is right. I am not denying the existence of a world or asserting the existence of a multiplicity of worlds. What I am denying is that there is any *way* that the world is independently of how it is for various kinds of beings. Different beings may have different perspectives on the world, but there is no reason to think that there is any perspective which can claim priority. But perhaps if we met some aliens who had a different perspective we could explain why they see the world as they do, differently from us. So perhaps in this way we could achieve a better understanding of how the world really is. Maybe so, but we still could not achieve any truly objective stance, for the theory we constructed to explain the belief systems of creatures different from us would still be a theory *of ours* which we should have to evaluate as we would evaluate any other theory. And, if the aliens constructed a theory about how *we* thought about the world and evaluated it by *their* standards, then there is no guarantee that it would be the same. So no objective stance which would define the way the world is independently of our epistemic values seems to be possible. Yet the metaphysical realist believes that there is a way the world is independently of our epistemic values. But I confess that I cannot make much sense of this view. Nor can I see any reason at all for holding it. Why do things behave as if the theory T were true? The perfectly adequate answer is that it is because the world is, for us, a T world.

The concept of absolute or metaphysical realism is like the concept of absolute space. It does no useful work. We can explain anything that can be explained about how bodies move without supposing that there is any absolute motion. Likewise, we can explain anything that can be explained about our knowledge of reality, e.g., the phenomena of epistemic convergence, without supposing that there is any way that the world is absolutely. It is enough if there is a way (for us) that the world is *for us*.

References

Armstrong, D. M. 1978. *Universals and Scientific Realism.* 2 vols. Cambridge: Cambridge University Press.

Boyd, R. 1976. "Approximate Truth and Natural Necessity." *Journal of Philosophy* 73:633–35.

————. 1980. "Scientific Realism and Naturalistic Epistemology." *PSA* 2:613–62.

Bradley, F. H. 1914. *Essays on Truth and Reality*, chap. 7. Oxford: Clarendon Press.

Campbell, K. K. 1976. *Metaphysics: An Introduction.* Encino, Calif.: Dickenson Publishing Co.

Carnap, R. 1956. "The Methodological Character of Theoretical Concepts." In *Minnesota Studies in the Philosophy of Science*, vol. 1, edited by H. Feigl and M. Scriven, 38–76. Minneapolis: University of Minnesota Press.

Dugas, R. 1955. *A History of Mechanics.* Translated by J. R. Maddox. Neuchatel, Switzerland: Editions du Griffon.

Duhem, P. 1954. *The Aim and Structure of Physical Theory.* Translated by P. P. Wiener. Princeton, N.J.: Princeton University Press.

Ellis, B. D. 1957. "A Comparison of Process and Non-Process Theories in the Physical Sciences." *British Journal for the Philosophy of Science* 8:45–56.

————. 1965. "The Origin and Nature of Newton's Laws of Motion." In *Beyond the Edge of Certainty*, edited by R. G. Colodny, 29–68. Englewood Cliffs, N.J.: Prentice Hall, Inc.

————. 1971. "On Conventionality and Simultaneity—A Reply." *Australasian Journal of Philosophy* 49:177–203.

————. 1979. *Rational Belief Systems.* Oxford: Basil Blackwell.

Ellis, B. D., and P. Bowman. 1967. "Conventionality in Distant Simultaneity." *Philosophy of Science* 34:116–36.

Harré, R., and E. H. Madden. 1975. *Causal Powers.* Oxford: Basil Blackwell.

Putnam, H. 1979. "How to Be an Internal Realist and a Transcendental Idealist (at the same time)." *Language, Logic and Philosophy: Proceedings of the Fourth International Wittgenstein Symposium*, 100–108.

Reichenbach, H. 1958. *The Philosophy of Space and Time*. Translated by M. Reichenbach and J. Freund. New York: Dover.

Schlesinger, G. 1963. *Method in the Physical Sciences*. London: Routledge and Kegan Paul.

Smart, J. J. C. 1969. "Quine's Philosophy of Science." In *Words and Objections*, edited by D. Davidson and J. Hintikka, 3–14. Dordrecht, Holland: Reidel. Also see Quine's reply in the same volume, 292–94.

Van Fraassen, B. C. 1969. "Conventionality in the Axiomatic Foundations of the Special Theory of Relativity." *Philosophy of Science* 36:64–73.

———. 1970. *An Introduction to the Philosophy of Time and Space*. New York: Random House.

———. 1980. *The Scientific Image*. Oxford: Clarendon Press.

4   Constructive Realism

Ronald N. Giere

Empiricism, writes van Fraassen, "could not live in the linguistic form the positivists gave it." His own empiricist image of science, therefore, utilizes an alternative linguistic form for scientific theories. In liberating empiricism from its positivist shackles, however, van Fraassen has unintentionally also set free the realism he abhors. That, at least, is the thesis of this paper.[1]

The Scientific Image treats a number of topics that would be central in any theory of science. I shall focus on just two: (1) The *nature* of scientific theories and their relations with the world and (2) the *justification*, or acceptance, of theories. Van Fraassen devotes much more attention to the nature of theories than to their justification. It is his account of what theories are that liberates both empiricism and realism. His arguments for empiricism over realism, however, turn primarily on questions of justification. By adopting the core of his view about theories, I am in the enviable position of being able to employ his best-developed weapons to attack his weakest defenses.

I would not enter this battle if I did not feel strongly that realism is right and empiricism wrong—and, moreover, that the difference matters. It matters because I take our task not merely to engage in scholastic debate, but to construct a general *theory of science* that could provide a theoretical background for diverse studies in the history, philosophy, psychology, and sociology of science. Here, as in science generally, it is important to "get it right."

I. Models

According to van Fraassen, the logical empiricists' preoccupation with the linguistic structure of scientific theories obscured the importance of

---

1. The quote is from van Fraassen 1980, p. 3. It will be obvious to readers of this paper that my debt to van Fraassen is enormous. I have benefited not only from his writings but from more than a decade of correspondence and discussion that was unfailingly gracious and patient. In genesis, at least, without constructive empiricism there could be no constructive realism.

the *models* satisfying those linguistic structures. In science, he claims, it is the models, not the linguistic forms, that occupy center stage. I agree completely.

Van Fraassen draws much of his inspiration from quantum theory. His image of theories, however, is intended to be much more general, and he does employ classical examples. We should not go too far astray, therefore, if we stick to classical mechanics. Consider, then, a one-dimensional linear harmonic oscillator—one of the original exemplars of Newtonian science.

As described in texts for nearly three hundred years, a linear harmonic oscillator is a mass subject to a restoring force proportional to the distance from its rest position. For reasons that will be obvious shortly, I agree with van Fraassen in preferring the Hamiltonian formulation:

A one-dimensional linear harmonic oscillator is a system consisting of a single mass constrained to move in one dimension only. Taking its rest position as origin, the total energy of the system is,

$$H = T + V = p^2/2m + \tfrac{1}{2}kx^2, \text{ where } p = mdx/dt.$$

The development of the system in time is given by solutions to the following equations of motion:

$$dx/dt = \partial H/\partial p, \text{ and } dp/dt = -\partial H/\partial x.$$

This is a capsule version of descriptions found in standard textbooks (see, for example, Marion 1970, chap. 3).

Thinking of k and m as merely constants, and both x and p as simply mathematical functions of t, solutions of the equations in x-p-t space are elliptically shaped spirals moving out along the t axis. Projecting the full x-p-t state-space onto the x-p plane yields an ellipse. Projections onto the x-t and p-t planes yield sinusoidal curves as illustrated in figure 1. Abstracting, then, from the standard meanings of terms like *mass* and *position*, we are left with a family of purely *mathematical* structures determined by the parameters m and k (and by specified values of both x and p for some given value of t).

Van Fraassen follows Patrick Suppes in holding that the proper language for the philosophical study of science is mathematics, not metamathematics. This doctrine frees the philosophy of science from the concerns and methods of the foundations of mathematics, one of the twin original sources of inspiration for logical empiricism—the other being the classical empiricism of Hume, Mill, Russell, and Mach. Suppes, however, retained the idea of a canonical language, namely, set theory. Van Fraassen frees us from even this constraint, letting the appropriate language be dictated by the specific scientific subject under investigation. On his view, interesting foundational problems in the various sciences are generally not

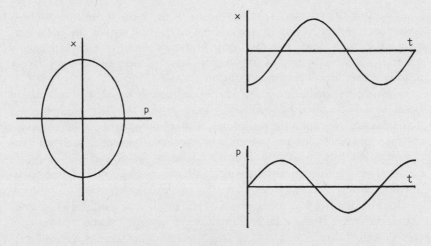

Figure 1

such that they can be removed merely by reformulation in the proper linguistic framework. They reside in the structure of the models employed.[2]

Van Fraassen has another reason for rejecting the set-theoretic framework. It is strongly biased toward purely *extensional* formulations of scientific theories, and van Fraassen wants to represent the *modalities* as well. I will return to this point later.

The desire to free philosophy of science from general questions about language is laudable. Philosophy of science should no more be just a subbranch of the philosophy of language than of epistemology or the foundations of mathematics. One cannot, however, eliminate all questions about language or *interpretation*. Any consistent formal structure has purely mathematical models, say in number theory. Some additional semantic categories, such as meaning or reference, are needed to distinguish masses from numbers—and thus mechanics from pure mathematics.

In earlier publications, van Fraassen introduced "elementary statements" which yield what he then called a "semi-interpreted language" (1970). Interpretation of purely mathematical symbols is thus provided, or at least transmitted, by the elementary statements. I will simply ignore such issues here. The theory of science need not wait on the development of adequate general theories of meaning and reference to proceed. We need not know in detail *how* general terms such as *mass* come to be associated with terms in an abstract mathematical structure. We know *that* it can be done because it *is* done.

2. Suppes's most recent writings (1979) take a more liberal view of the proper language for foundational studies. This liberal view now separates Suppes, Suppe (1973), and van Fraassen from the "structuralist" school of Sneed (1971) and Stegmueller (1976).

Assuming the standard interpretations of the basic terms, we obtain a generalized model (or family of specific models) that we call *the* linear harmonic oscillator. I will use the term *theoretical model* to refer either to a general model or to one of its specific versions obtained by specifying unique values for all parameters and initial conditions.

A theoretical model, then, is a *defined* entity. It has all and only those characteristics explicitly specified. In cases where the definition employs mathematical concepts, the model has all the precision of the corresponding mathematical notions. The mass in *the* linear harmonic oscillator, for example, exhibits a perfectly sinusoidal motion. There can, therefore, be no question whether linear harmonic oscillators perfectly satisfy the equations of motion. That they do is a matter of explicit definition.

On the other hand, if the behavior of the theoretical model is a matter of definition, what is the relationship between theoretical models, so conceived, and real oscillating systems such as bouncing springs, pendulums, and vibrating molecules?

II. Theoretical Hypotheses

The logical empiricist notion of a *correspondence rule* conflated what on a definitional view are two distinct, though related, functions. One is providing a general *interpretation* of theoretical terms such as *mass*, *momentum*, etc. The other is providing the means for *identifying* particular instantiations of these terms. I think van Fraassen would agree that the details of this process are not a matter for philosophical analysis or for armchair psycholinguistics. Rather, the topic calls for deep empirical investigations into how we humans use abstract symbols in describing particular objects in the real world. Here again we do not need a detailed account at hand to construct a useful theory of science. A fairly shallow analysis will do.

Figure 2 pictures a weight suspended on a spring—the kind of thing one finds in elementary physics laboratories. Also pictured is the two-dimensional *x-p* state-space with initial conditions specified. The arrows indicate the evolution of the state of the system in the state-space. Here we make the "standard" identifications. The object on the spring is our mass. The place the mass comes naturally to rest is taken as the origin of our coordinate system. Moving up vertically is the positive *x* direction. We shall assume that the system is set in motion by pulling the weight down a distance *x'* and releasing it with no initial velocity.

In the circumstance indicated, what is the relationship between our *theoretical model*, the ideal linear harmonic oscillator, and this real oscillating system? Here, finally, we confront the difference between constructive empiricism and constructive realism. Let us begin with realism.

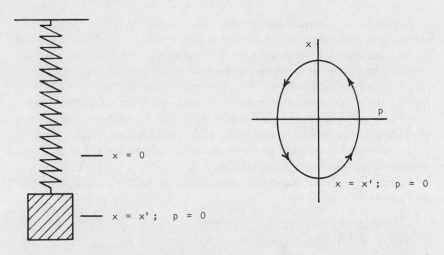

Figure 2

At one point, van Fraassen characterizes the realist as claiming "to have a model which is a faithful replica, in all detail, of our world" (1980, 68–69). In fairness, this is not his most considered statement of a realistic hypothesis, but it provides a useful starting point.

Now, our theoretical model is obviously not a "faithful replica, in all detail," of our bouncing weight. Real springs, for example, have internal friction which eventually brings the system to rest. And there are other apparent forces, centrifugal and coriolis, due to our reference system being on a rotating sphere, i.e., the *earth*. These circumstances insure that the motion of the actual system will differ measurably from that indicated by the model.

The traditional response to such worries is to construct a more detailed model. So the claim would be that there is *some* Newtonian model that *exactly* captures the behavior of the real system. Since Einstein, however, we know that this claim is false. There is *no* Newtonian model that exactly represents any dynamic system. Moreover, even in the classical context, say as of 1875, such a claim could not have been *justified*. Since all evidence is finite and of less than perfect precision, the most that could ever have been justified is that there is some Newtonian model that matches the real system to within then-existing experimental limits of detection. Here I am invoking the principle that one cannot justify a claim by appeal to experiments that have no chance of detecting the falsity of the claim even if it is false. Later I shall provide a basis for this principle.

If we are to have scientific hypotheses which are both realistic and have some reasonable chance of being true, we must avoid claims that any real system is *exactly* captured by some model. Realists typically have adopted this view. Most realists would qualify their claims by saying that the real system is at best *approximately* captured by the model.[3] The difficulty with this formulation is to explain what "approximately" means. Van Fraassen suggests that approximation be explained in terms of a *class* of models, one of which *exactly* fits (1980, 9). This explication, however, must be rejected. Since there is no Newtonian model that exactly captures any real system, van Fraassen's analysis would have us deny that Newtonian models now provide a good approximation for any real world systems. Such a position is just too much at variance with the present convictions of scientists—and of most students of scientific practice.

I propose we take theoretical hypotheses to have the following general form:

The designated real system is *similar* to the proposed model in specified *respects* and to specified *degrees*.

We might claim, for example, that all quantities in our spring and weight system remained within ten percent of the ideal values for the first minute of operation. The restriction to specified respects and degrees insures that our claims of similarity are not vacuous.

The precision associated with any hypothesis cannot justifiably be *greater* than that of the measurement techniques actually employed, though one might in many contexts claim *less* precision than could be achieved. I shall say little more about precision here except to note that this is one example of how the logical empiricists' preoccupation with simple languages obscured what almost everyone knows to be an essential feature of scientific hypotheses. The state-space representation makes possible an intuitive analysis of degrees of approximation in terms of a volume in the state-space which includes the idealized history of the theoretical model.

The above general characterization of a theoretical hypothesis permits a number of variations corresponding to various grades of empiricism and realism. The most extreme realist hypothesis would be that the real system resembles the model in *all* respects. There are many examples showing that this version of realism is too extreme. Here one might mention the phase factor in the wave functions of quantum theory, the imaginary part of complex numbers used in electromagnetic theory, or even negative square roots.

I would therefore recommend a more modest, constructive realism that claims only a similarity for *many* (or perhaps most) aspects of the model. This formulation leaves the residual problem of saying when an aspect of

3. The writings of Putnam (1975) and Boyd (1973, 1981), for example, are full of talk about approximation. But their sentential framework makes it difficult for them to deploy the notion effectively. This makes van Fraassen's criticisms seem more powerful than they are.

the model is to be denied a counterpart in reality. The positron shows us that it sometimes pays to take seriously a negative square root—and thus that there is unlikely to be a simple answer to the residual problem.

How, then, does the above modest realism differ from van Fraassen's constructive empiricism. At first sight, the difference seems obvious. Empiricism is much more restrictive. It enjoins us to limit our claims of approximate similarity to *observable* aspects of our models. The difference between realism and empiricism, therefore, turns on how one specifies the observable aspects.

Van Fraassen is clear in denying the logical empiricist doctrine that the specification of observable aspects can be done in terms of vocabulary or by segregating individual statements. Rather, it is done by designating substructures of one's models as the *empirical substructures*. Empiricists claim only that a model is *empirically adequate*—meaning that there is some empirical substructure which approximately matches the observable facts. How, then, do we pick out the empirical substructures? Here van Fraassen is uncharacteristically unclear in that his examples and theoretical pronouncements do not always cohere.

He considers, for example, Newton's claims about absolute space. Imagine, then, that we complicate our model for the simple harmonic oscillator by adding a constant velocity representing the motion of the solar system relative to absolute space. No matter what velocity we add, the resulting behavior of the system, as measured in our earth-based coordinate system, is unchanged. It is not just that we can't observe any difference. Our model itself tells us that there is no difference we could detect, no matter how sophisticated or how sensitive our instruments. Here the constructive realist could agree completely with van Fraassen that to assert any similarity between the real system and the model with respect to some added absolute velocity term would be gratuitous and totally unjustified.

Van Fraassen, however, appears to claim more. Indeed, he seems to endorse the idea that the appropriate empirical substructures for classical mechanics consist only of relative positions and velocities. Even the standard Hamiltonian state-space, which includes momentum, and thus mass, is too theoretical. Why? I am not sure. The kinds of reasons that have been given by other philosophical students of classical mechanics make strong appeal to the logical structure of mechanics itself.[4] Van Fraassen, however, explicitly denies that his distinction is theory-dependent in this way.[5]

The correct approach, he suggests, is not through philosophical analysis or armchair psychology but through the empirical study of human per-

---

4. Van Fraassen mentions the studies of McKinsey, Sugar, and Suppes (1953); and Simon (1954).
5. He writes, "I regard what is observable as a theory-independent question" (1980, 57).

ceptual capabilities. So determining what is observable depends on scientific theory, but on psychology and physiology, not on physics. Whether this generates a "logical catastrophe" for psychology and physiology is unclear.[6] But it is very difficult to see how the study of human perceptual capabilities could tell us that velocity is observable while mass is not.

The operative scientific notion, I suggest, is not human observability but scientific *detectability*. What is now detectable depends primarily on two things: (1) the structure of our current models, together with our interpretations and identifications, and (2) our present ability to design and build experimental apparatus, which in turn depends on other models. That the ultimate output of our measuring instruments must be something that *humans* can *observe*, e.g., a dial or a computer printout, is a simple consequence of the fact that all the scientists we know are humans. Satisfying this requirement is primarily a matter of engineering.[7]

If the requirement of empiricism is only that our scientific claims be restricted to aspects of our models that are, in a broad sense, *detectable*, then one major difference between constructive empiricism and constructive realism is removed. There remain serious differences regarding *modalities*, that is, what our models tell us about the *possible* behavior of real systems. This is the subject of the following section.

To sum up the present section, van Fraassen is to be thanked for liberating us from one of the two sources of logical empiricism, metamathematics and formal linguistic analysis. But he remains in the clutches of the other source—classical empiricism. His empiricism, therefore, is a vestige of the classical empiricist philosophy that sought to ground all knowledge in *experience*—and thus to make man the *measure*, rather than merely the *measurer*, of all things. Such an empiricism may indeed "deliver us from metaphysics" (van Fraassen 1980, 69), but it delivers us into the hands of an anthroprocentrism that is antithetical to the whole modern scientific tradition.

III. Physical Modality

Most recent versions of the debate between empiricists and realists have focused on the distinction between observational and theoretical aspects of science. Van Fraassen's treatment is no exception. He does, however, give special consideration to the status of the physical *modalities*. This seems to

6. The threatened vicious circle is that we must use psychology and physiology to tell us what are the observable substructures of our models of psychological and physiological systems.

7. The importance of detectability rather than observability has been emphasized recently by Churchland (1979) and Shapere (1982).

me the crucial dividing line between empiricism and realism. Van Fraassen is willing to grant that real systems may in fact exhibit the *theoretical* structure of our models. He merely insists that we cannot justifiably assert this correspondence. Regarding physical *modality*, however, he is not merely agnostic but atheistic. Possibilities and necessities are only figments of our models—useful, perhaps, but not even candidates for reality.[8] Thus, even if we substitute detectability for observability, there remains a vast difference between empiricism and a realism that extends to physical modality. Here I will merely describe the position of modal realism. The following section will treat the justifiability of modal claims.

Van Fraassen's own preferred form for presenting theories, the state-space approach of Herman Weyl and Evert Beth, provides an excellent framework for expressing the claims of modal realism. To keep the discussion as concrete as possible, let us return to the harmonic oscillator and bouncing weight represented in figure 2. I will begin by distinguishing a number of possible claims, of increasing strength, about the relationship between the idealized model and the real system. In each case, an appropriate degree of precision over a suitably restricted interval of time is assumed.

1. (Extreme empiricism) The model agrees with the positions and velocities of the real mass which have been observed up to the present time.
2. (Extended empiricism) The model agrees with all the positions and velocities that ever have been or will be observed.
3. (Actual empiricism) The model agrees with all the actual positions and velocities of the real system, whether they are observed or not.
4. (Modal empiricism) The model agrees with all possible positions and velocities of the real system.
5. (Actual realism) The model agrees with the actual history of all (or most) system variables.
6 (Modal realism) The model agrees with all possible histories of all (or most) system variables.

Constructive empiricism is a version of actual empiricism. One could identify constructive realism with either actual realism or modal realism. I prefer the modal version. In either case, constructive realism is a variety of what is sometimes called *structural realism*.[9]

As Quine delights in pointing out, it is often difficult to *individuate* pos-

---

8. "The locus of possibility is the model, not a reality behind the phenomena" (van Fraassen 1980, 202).

9. Constructive realism is thus a model-theoretic analogue of the view advocated by Grover Maxwell (1962).

sibilities. Often, yes, but not always. Many models in which the system laws are expressed as differential equations provide an unambiguous criterion for individuating the possible histories of the model. They are the trajectories in state-space corresponding to all possible *initial conditions*. Threatened ambiguities in the set of possible initial conditions can be eliminated by explicitly restricting the set in the definition of the theoretical model. If, for example, one wonders whether $x' = -100$ meters is a possible initial condition, we could explicitly restrict x to ± 10 cm. Nor is this criterion limited to models defined in terms of differential equations. A clear distinction between system laws and parameters or initial conditions is generally sufficient. Of course, even classical physics presents cases in which the specification of boundary conditions is not unambiguous. But the ambiguity is not nearly so great as Quine or van Fraassen suggest.

As noted earlier, part of van Fraassen's reason for preferring the state-space representation of theoretical models, rather than a purely set-theoretic representation, is that the state-space presentation emphasizes the *modal* structure. Thus, even for van Fraassen, the modal structure is important for our understanding of reality. And, indeed, for a two- or three-dimensional state-space, one can easily visualize the possible histories. But van Fraassen explicitly denies that the modal structure is, or perhaps even could be, a part of the physical world.

One's attitude toward modalities has a profound effect on one's whole theory of science. Actualists, including actual realists, must hold that the aim of science is primarily to describe the actual history of the world. For modalists, including modal empiricists, the aim is to describe the structure of physical possibility (or propensity) and necessity. The actual history is just that one possibility that happened to be realized. This difference in aims is connected with profound differences in how one understands diverse scientific activities such as causal attribution, explanation, and experimental design.

Our bouncing spring is a *causal* system if anything is. Now, our theoretical model tells us that the frequency of oscillation, f, is *functionally* related to the ratio, k/m, and functionally independent of the amplitude, x'. If the functional relationship obtains merely between the *actual* values of these quantities, it is difficult to see what more there is to *causality* than merely this functional relationship. Empiricists have traditionally sought to ground the causal claim in *universal generalizations*; e.g., all bouncing springs exhibit these relationships. On examination, such generalizations turn out to be either false or vacuous. For the modal realist, the *causal* structure of the model, and thus, to some degree of approximation, of the real system, is identical with the *modal* structure. For any real system, the functional relationships among the actual values of f, m, k, x', etc., are causal not because they hold among the *actual* values in *all* such real systems but because they hold among all the *possible* values in *this* particular system.

Giving *explanations* for particular happenings or citing causes thereof is, as van Fraassen insists, a complex human activity involving theoretical hypotheses. It is thus not to be expected that the correct responses to requests for explanation or for causal antecedents are uniquely determined by the scientific hypotheses themselves. It does not follow, however, that such questions do not have unique answers when applied to *restricted* systems such as our weight and spring, for which we have a complete model. In such cases there is no need for externally imposed relevance conditions. The model itself tells us everything that could be relevant to a system exemplifying its structure. And it gives unique answers for any fully specified set of parameters and initial conditions. What would the system do if we were to change the initial conditions in a specified way? Our hypothesis gives us a unique answer. The uniqueness is lost only when the physical system is embedded in a larger social context for which we have no similarly complete models.

From the standpoint of modal realism, to *understand* a system is to know how it works. And this means knowing how it would behave under conditions other than those which in fact obtain. It is knowing the causal structure. In a curious way, this is also van Fraassen's view. The curiosity is that he would replace "knowing" with "having an empirically adequate model with the given modal structure." But this reduces scientific understanding derived from a theoretical model to understanding conferred by a good historical novel, one which remains faithful to the known historical facts. They both provide a good story, which, however, we have no reason to believe is true. I am not sure what motivates van Fraassen to advocate such a degree of epistemic caution, but I am fairly confident that it is not justified.

IV. Acceptance

Van Fraassen distinguishes between the empiricist *acceptance* of a model as empirically adequate and the realist *belief* that the model provides a true picture of reality. I shall speak of the acceptance of hypotheses in both cases. The difference is in the nature of the hypotheses at issue. Empiricist hypotheses claim only that a real system resembles the model in its observable aspects, while realist hypotheses assert the resemblance for theoretical (and perhaps modal) aspects as well.

In *The Scientific Image*, the main argument for empiricism is that realism is unnecessary. Everything we want to say about science, e.g., its use in giving explanations, can be said on an empiricist understanding of scientific hypotheses. I shall take up this argument, though in an indirect way, in the final section of this paper.

Here I wish to confront a more direct, secondary argument. This argument is based on the logical principle that if H implies H', then, no matter what the evidence, H cannot be better justified than H'. In other words, evidential *safety* is inversely related to logical *strength*. Thus, even if one has other reasons for wanting to be a realist, one's realistic hypotheses could never be better justified than the corresponding empiricist hypotheses.[10] The obvious reply is that one might wish to trade off some evidential security for the logical strength of realistic hypotheses. And van Fraassen cannot totally deny this appeal, since he himself stops short of embracing extreme, or even extended, empiricism. But the issue deserves a deeper investigation.

I shall argue not only that realistic hypotheses can be justifiably accepted but also that, in some circumstances, a realistic hypothesis might be *better* justified than the corresponding empiricist hypothesis. The argument requires appeal to an explicit theory of acceptance. Such an appeal carries no danger of directly begging any questions, since van Fraassen does not himself present an account of acceptance.

Given his generally pragmatic approach to epistemic issues, van Fraassen would agree that the activities of scientists are to be understood in essentially the same way as the activities of other people engaged in other pursuits. Now, our everyday understanding of human behavior is in terms of beliefs and desires. That is, we generally explain why people do what they do in terms of their desires, or goals, and their beliefs both about these goals and about various possible means to their goals. Moreover, there exist fairly well developed theoretical models for describing such behavior. These are supplied by decision theory. Like all theoretical models, decision-theoretic models are only partial and approximate. And there are other kinds of models one might apply.[11] My aim here, however, is only to show that my approach to the justification of realistic hypotheses is sufficient, not that it is the best possible.

Although he does not present a theory of acceptance, van Fraassen seems to hold an implicit view of what such a theory must do. Recast in decision-theoretic terms, van Fraassen's view is that we must choose one hypothesis out of a literally countless set of possible hypotheses. This looks to be a very difficult task given finite evidence and the logical underdetermination of our models by even their complete empirical substructures. I would urge a quite different picture.

10. Van Fraassen himself gives a sharp formulation of the empiricist/realist debate in terms safety and strength in his later paper (1981).
11. The methodologies of Lakatos (1970) and Laudan (1977) may be understood as providing an empirical account of scientific activities such as the pursuit or acceptance of a theoretical program.

	H is true	H is false
Accept H	V = 1	V = 0
Reject H	V = 0	V = 1

Figure 3

Looking at the scientific process, we see scientists struggling to come up with even one theoretical model that might account for phenomena which previous research has brought to light. In some cases we may find two or three different types of models developed by rival groups. These few models may be imagined to be a selection from a vast number of logical possibilities, but the actual choices faced by scientists involve only a very few candidates. Given this already limited context, we sacrifice little by assuming that the choice is simply between a hypothesis and its negation.

The simplest decision-theoretic representation of the acceptance or rejection of a given hypothesis is a two-by-two matrix, as pictured in figure 3. The two possible actions are "Accept H" and "Reject H." The two relevant possible states of the world are that H is true and that H is false. These descriptions of the actions and states require some immediate qualifications.

To accept H means to regard it as being true. But acceptance is not absolute. It is provisional in the sense that it may be revoked in the light of new evidence. And even without further evidence, one acknowledges the *possibility*, indeed, some *probability*, that H is really false. We are all fallibilists.

H has the form earlier proposed for theoretical hypotheses, namely, that the real system is similar to the model in specified respects and to specified degrees. We can therefore use the same matrix for both empiricist and realist hypotheses simply by changing our understanding of what sorts of respects are included. The difference between the empiricist and the realist, in this framework, shows up as a difference in *goals*. The empiricist strives to accept true empiricist hypotheses; the realist strives to accept true realistic hypotheses.

Any decision-theoretic model requires a value input. A full treatment of the decision problem could thus be quite complex, since there is in reality a variety of relevant values associated with the individual, the particular field, and even the institution of science as a whole. For our simplified model of the situation, we can get by assuming only that truth is valued over falsity. We need not even attempt to distinguish the relative value of the two kinds of possible mistakes. Giving correct acceptance or rejection the value 1 and incorrect acceptance or rejection the value 0, we obtain the completed matrix shown in figure 3.

A decision-theoretic framework also requires the adoption of some explicit *decision rules*. I agree with van Fraassen in shying away from Bayesian models of scientific agents. Such models assume scientists to have a belief function over hypotheses, a function having the formal properties of probability. There is little evidence that scientists actually have such nicely ordered beliefs. For this and other reasons, I prefer a version of classical decision theory. In all honesty, I think van Fraassen could better defend his position from a Bayesian standpoint. It is a tribute to his sensitivity to the practice of science that he does not take this route. Virtue, however, is not always rewarded.

From the standpoint of classical decision theory, the decision problem represented in figure 3 has no solution. The matrix is completely symmetrical. The symmetry, of course, is partly the result of our having assigned equal value the two types of successful and mistaken decisions. But this is not essential. No one would argue that H should be accepted or rejected solely on the basis of such value differences. The reason the problem is indeterminate is that it is *incomplete*. We have not yet factored in any *evidence*. Both empiricists and realists agree that the decision is to be made in light of the evidence. The question, therefore, is how the role of the evidence is to be represented in our model of the decision problem.

Evidence enters the decision problem by way of the *decision rule*. Although there are in principle any number of possible rules, the following simple rule is consistent with van Fraassen's presentation: (1) Accept the hypothesis if the known observations are as implied by the hypothesis; reject the hypothesis if the known observations are inconsistent with it. To evaluate this rule, we must compare it to other possible rules. There are only three other rules with the same general form. These are: (2) Accept the hypothesis if the observations are inconsistent with it; reject the hypothesis if they are as it implies. (3) Accept the hypothesis whether or not the observations are consistent with it. (4) Reject the hypothesis regardless. There is no doubt that the first rule is the best of the four. The question is why it is best.

To answer this question we need to consider the *metadecision* problem in which our options are the above four possible decision rules. The relevant states of the world are, as before, the truth or the falsity of the hypothesis. The matrix for this metaproblem is shown in figure 4, where (A, R) represents the first rule. Following standard procedures in classical decision the-

	H is true	H is false
(A, R)	V = 1	?
(R, A)	V = 0	?
(A, A)	V = 1	V = 0
(R, R)	V = 0	V = 1

Figure 4

ory, the value attached to an outcome is the *expected value* of that outcome given the corresponding state of nature and decision rule. Rule 3, for example, yields an expected value of 1 if H is true and 0 if it is false.

The obvious difficulty we now face is that we do not know what is the expected value of adopting rules 1 or 2 in the case where H is *false*. The decision-theoretic approach to the acceptance of hypotheses thus forces us to consider a question that might otherwise go unnoticed. The question will surely go unnoticed if one considers only those aspects of the problem that are determined by logic alone. The negation of H does not logically imply whether or not the observations will agree with H.

To give a decision-theoretic account of why rule 1 is correct requires some additional considerations sufficient to determine the missing expected values in the metadecision matrix of figure 4. The route suggested by classical decision theory, though not the only logically possible route, is to look for some way of determining the *probability* of observations implied by H if H is nonetheless false. Let us postpone questions about the nature of these probabilities and how they might be determined in order first to see why some probabilities of this form do provide a solution to the metadecision problem—and thus to the original problem of deciding whether to accept or reject H.

It is fairly easy to see that what we need is a *high* probability of the observations being *inconsistent* with H if H is false. In that case, using rule 1 results in there being a high probability that we make the correct decision of rejecting H. The missing expected value for the rule, (A,R), is thus close to 1, while the corresponding expected value for the opposite rule, (R,A), is near zero. In this case, choice of the intuitively correct rule, (A,R), could be justified by the minimax principle, since this rule gives a non-zero expected value whether H is true or not.

But minimax justifies too much. In a real scientific context, an expected value minimally greater than zero would not be enough. One wants an expected value that is reasonably high. So the operative decision strategy is something like *satisficing*.[12] Then the (A,R) rule is the correct choice because it gurantees a sufficiently high expected value whether H is in fact true or not. Fixing the value scale as we have, this is equivalent to saying that (A,R) is best because it yields a high probability of a correct decision regarding H whether H is true or not.[13]

There remains the problem of justifying the strategy of satisficing. My view is that satisficing is justified simply on the grounds of being an efficient strategy to follow in a world in which one cannot guarantee correct decisions. This is an empirical claim. I will later reply to those who would at this point raise the specter of a Humean regress.

Nothing in the above decision-theoretic model of acceptance distinguishes empiricist from realist hypotheses. It works equally well for both. If there is a difference, it will have to be found either in the values assigned or in the probabilities which we have thus far left unanalyzed.

At this point I think it best to proceed in the context of a new example. The example has been selected because it nicely illustrates the virtues of a realistic image.

V. The Double Helix

In the fall of 1951, James Watson arrived at the Cavendish Laboratory in Cambridge intent on discovering the chemical structure of DNA. He was hardly starting from scratch. Behind him was Chargaff's determination of the ratios of the four basic nucleotides in DNA and Pauling's discovery of the helical structure of the polypeptide, α-kerotin. There was also much knowledge of the structure of chemical bonds for various atoms and considerable experience with X-ray diffraction. And Watson had inside information both about Pauling's progress on DNA and about X-ray studies going on at King's College in London.[14]

---

12. For an authoritative introduction to uses of satisficing, see Simon 1957.
13. For references and further discussion of a classical decision-theoretic approach to the acceptance of theories, see Giere 1983.
14. My understanding of the history of this case is derived from Watson 1968 and Olby 1974.

Following Pauling's example, Watson and Francis Crick began building models of DNA—not theoretical models, but actual *scale* models using bits of wire, metal, and cardboard. One of the beauties of the model-theoretic approach to scientific theories is that this sort of activity can be seen as continuous with the development of theoretical models using language and mathematics. The scale model provides a satisfactory means of defining a theoretical model. In this case the theoretical model is an idealized nucleic acid with idealized subunits composed of idealized individual atoms in very specific spatial arrangements. Watson's hypothesis was that actual DNA molecules are very similar in composition and structure to the theoretical model. In practice, of course, Watson and Crick referred directly back to the scale model without explicitly invoking the theoretical intermediary.

At the time, the clearest sort of evidence for such structural hypotheses was provided by the techniques of X-ray diffraction. The diffraction patterns and relative intensities produced on photographic plates tend to be quite sensitive to the structure of the material producing the pattern. On the other hand, no one at the time knew enough about the X-ray patterns produced by nucleic acids to be able simply to read off the structure of the material from an X-ray photograph. If that had been possible, the structure would have been known long before Watson arrived on the scene.

Once the double helical structure had been proposed, workers at King's College quickly reexamined photographs already in their possession. The results agreed substantially with those predicted by the Watson-Crick model. Papers discussing this evidence appeared in the same issue of *Nature* as the original Watson and Crick paper.

Because of the complexities of interpreting X-ray photographs and uncertainty about the role played by photographs already in Watson's possession, the confirmation in this case was not as dramatic as in some other recent cases, e.g., the confirmation of continental drift or of big-bang models of the universe. But the DNA case is clear enough and more relevant to an examination of the relative virtues of empiricism and realism. I will proceed first to argue the minimal thesis that a realistic hypothesis can after all be justified.

The realistic hypothesis, of course, is that the molecular structure of DNA is quite similar in the relevant respects to the proposed model. The evidence is the particular pattern of light and dark spots observed on X-ray photographs. The decision rule is to accept the hypothesis if the observed pattern matches that predicted by the model and reject it if the pattern fails to match. Our choice of this decision rule is justified by the principle of satisficing if we can show that the *probability* of getting the pattern is *high* if H is true and *low* if H is false.

Enough was known about the patterns produced by various molecular structures and about X-ray techniques to justify the *conditional* judgment that, if the hypothesis were correct, the specified pattern would very likely result. But the same body of background knowledge also justified the converse judgment. If the real structure were not very similar to the model, this particular pattern would be unlikely to result. One does not have to know just which patterns go with which structures to know that importantly different structures rarely yield similar patterns. Moreover, the theory of X-ray diffraction substantiates this judgment—that is, it tells us why the observed pattern should be quite sensitive to differences in structure. Thus, even if we set a quite high satisfaction level, the standard decision rule is justified—and so, therefore, is the realistic hypothesis.[15]

It might be objected that we were able to justify the realistic hypothesis only because we assumed that some *other* realistic, indeed, *modal*, hypotheses were already justified. In particular, we appealed to our background knowledge to determine what *would* be the likely result if *H* *were* true and if it *were* false. This is correct. But does this circumstance constitute an objection to realism? We have not literally begged the question by assuming that the hypothesis under investigation is justified. We have only assumed that some *other* realistic hypotheses were previously justified.

To argue that the above procedure begs the question is, I think, to misconceive the problem. Our task was to show how realistically interpreted hypotheses are justified. That we have done. What we have not done is show how a realistic hypothesis could be justified from a background consisting only of observed facts—perhaps together with empirical generalizations. This is a very different, and peculiarly philosophical, problem. It might be confused with the old problem of induction, in which case it is simply impossible. Suitably interpreted, however, it can be solved, as I shall shortly attempt to show.

Right now let us see how the empiricist version of the Watson-Crick hypothesis fares in our decision-theoretic framework. The hypothesis is that the model is empirically adequate to the phenomena of X-ray photographs. From this hypothesis it follows immediately that the observed patterns will match. To justify the standard decision rule, however, we need to know the probability of a match if the proposed model is *not* adequate. Now, for the empiricist, our background knowledge consists of other models previously justified as being empirically adequate. In particular, the empiricist knows that models of molecular structure differing substantially from empirically adequate models are rarely themselves empirically adequate. Employing this knowledge, the empiricist concludes that the predicted match is unlikely if *H* is false. This shows that the decision rule is satisfactory and that the observation of the predicted pattern does justify the hypothesis.

What the empiricist cannot do, however, is appeal to knowledge of the

---

15. This holds even for the *modal* version of the realistic hypothesis.

internal causal structure of nucleic acids and their interactions with X rays in order to bolster the judgment that the predicted pattern is indeed unlikely if H is false. The empiricist's knowledge is exhausted by the claim that these other models are empirically adequate. Unjustified claims about causal structure cannot provide justification for their consequences. Thus, the realist claim that the decision rule is sufficiently justified may itself be better justified than the corresponding empiricist claim. And surely the acceptance of a hypothesis using a better-justified decision rule is itself better justified. So the realist hypothesis might after all be *better* justified than the empiricist hypothesis.[16]

I say only that the realist hypothesis "might" be better justified because we have not yet taken account of the logical principle that safety is inversely related to strength. But at least we have shown the possibility that in science, as in diplomacy, there may be security in strength.[17] To pursue this possibility, we must consider two more cases.

First, let us return to the philosophical problem of whether a realistic hypothesis could be justified assuming only a background of empirically interpreted hypotheses. In fact, it is fairly easy. All we need do is substitute the realistic Watson-Crick hypothesis for the empiricist version in the empiricist's decision problems. The decision rule is justified by appeal only to empiricist hypotheses. In the end we accept the realist hypothesis. We reach realist goals using empiricist means.

The reverse strategy, pursuing empiricist goals using realist means, is also possible. That is, we might justify the empiricist version of the Watson-Crick hypothesis using realistic background knowledge to justify our decision rule. Though philosophically unmotivated, the possibility of pursuing such a strategy prepares us to face the issue of safety versus strength.

The decision-theoretic approach to acceptance has introduced a new factor in addition to the principle that safety is inversely related to strength. For a fixed hypothesis, the amount of justification is greater if one begins with a better-justified decision rule. Or, to put it in more familiar terms, the more *severe* the test, the better the justification—where the severity of a test is proportional to the *improbability* of the implied result if the hypothesis is false.[18] And we have argued that using realistic background knowledge provides the basis for a stronger test. Using these two principles, then, we obtain the following partial ordering of justifications.

The *strongest* justification is obtained if we judge the empiricist hy-

16. My point here is similar, though in a different framework, to that argued by Boyd (1973, 1981).

17. I have adapted the formulation "security in strength" from an unpublished paper by William Harper.

18. The classical decision-theoretic approach thus provides an "inductivist" reconstruction of established Popperian dogma.

pothesis using realistic background information to justify the decision rule. The realist hypothesis, being logically stronger, would be less well justified on the same background knowledge. At the other end, the *weakest* justification is obtained if we judge the realist hypothesis using empiricist background information. The empiricist hypothesis is better justified on the same empiricist background knowledge. On the other hand, keeping the hypothesis fixed and varying the background knowledge yields the result that the empiricist hypothesis is better justified on realistic background knowledge than it is on empiricist background knowledge. Similarly, the realist hypothesis is also better justified with a realistic rather than an empiricist background.

All these relationships, however, fail to determine whether the realist hypothesis on realistic background knowledge is better justified than the empiricist hypothesis with an empiricist background. We don't know the rate of exchange between justification due to severity of test and to logical strength of the hypothesis. Nor does it seem that any such rate could be determined without a logical measure of content and confirmation, something van Fraassen and I both reject. So the opposition between constructive empiricism and constructive realism cannot be decided by appeal to considerations regarding safety and strength.

The decision-theoretic framework suggests another sort of appeal— *value*. The realist values realistic hypotheses over empiricist hypotheses, regardless of the background knowledge. This value difference may compensate for the difference in logical strength. The appeal to value, however, obviously cannot decide the issue, since the empiricist has contrary values. But it does put the issue in better focus. The question for van Fraassen is why he prefers empiricist hypotheses. It cannot simply be a matter of epistemic caution, since he values actual empiricism over extreme empiricism. He is willing to take some risks. Why is he unwilling to take a few more? Nor is it sufficient to point to the specter of metaphysics. The physical modalities he would regard as metaphysical are for others the essence of theoretical science.

Perhaps, after all, he simply can't let go of the old Hume in our philosophical tradition. Since the old Hume dies hard, it may help to ease the grief to consider a Darwinian tale. Like van Fraassen, I am a Darwinian. Indeed, I am more so, since I believe not only in the facts of evolution evidenced by the fossil record but also in the reality of the mechanisms— modeled in part by the genetic theory of natural selection—that produced this record.

My tale begins with human beings and human culture evolving together over many centuries. This evolution could not have been successful if the culture did not both generate and transmit a fair amount of pretty reliable, though perhaps mundane, knowledge about the natural world. This knowledge, I claim, must have included some understanding of the causal mechanisms underlying the phenomena. The development of science, solidified in the west in the seventeenth century, consisted both in acquiring the *goal* of learning more details about the deep structure of the natural world and in devising *means* appropriate to this goal. The stock of knowledge developed over the past three hundred years provided the background for the current highly accelerated growth of science.

Filling in the details of this story would not, of course, provide a philosophical *justification* of induction. The tale does, however, provide the outlines of an *explanation* of induction. To provide more is neither necessary nor possible.

VI. The Theory of Science

One way of understanding *The Scientific Image* is as an attempt to construct a *theory of science*.[19] Moreover, in the terminology of some recent sociologists of science, the theory is fully *reflexive*. That is, it has itself as an instance.[20] From this standpoint, we see van Fraassen constructing a model of science and attempting to show that the model is empirically adequate. His arguments, however, proceed at a high level of abstraction. He is concerned to show, for example, that the goal of empirical adequacy is sufficient to account for philosophical theses about methodology, explanation, causation, and probability. Only occasionally, as in the Millikan example, does he argue the case at the level of scientific practice. But it is at this level that empiricism as a theory of science stands or falls. So we must ask, is constructive empiricism empirically adequate?

My introduction of an example from molecular biology was partly motivated by this question. It is very difficult, I think, to save the phenomena of molecular biology, *as a scientific enterprise*, using the empiricist model. The major figures in the 1950s, Pauling, Watson, Crick, et al., could not be described as having the goal of merely accounting for the phenomena of X-ray diffraction and chemical reactions. Watson clearly had little interest in X-ray pictures except as a means to the end of discovering the structures of DNA. Rather, much time in Crick's laboratory was spent

19. The phrase "theory of science" appears several times in *The Scientific Image*, e.g., on p. 196. I am not sure that van Fraassen intended it to be taken as seriously as I am now taking it. In a later paper (1981), however, he states that the goal of the philosophy of science is to "make sense of science." Moreover, he explicitly employs his own empiricist conception of what "making sense" requires. So my description of his enterprise fits his later characterization even if this does not appear explicitly in *The Scientific Image*.

20. The requirement of reflexivity for the sociology of science is discussed, for example, in Bloor 1976.

measuring the angles between the metallic representations of nucleic acids in their scale model to see if the model fit the known bonding angles of various atoms. These are not the actions of people striving merely to account for spots on X-ray photographs.

Nor is the picture much changed if we move from talking about the goals of individuals to considering the goals of the science as a whole. Reading the history of molecular biology and excerpts from papers and letters written at the time, it is difficult to find any evidence of an overriding concern with saving the phenomena. The whole profession acted as if it were after the real molecular structure of real molecules.

Van Fraassen says that scientists live in a life-world created by their models. They talk *as if* they take their models to be descriptive of reality. To adapt what he says about Millikan, "it may be natural to use the terminology of discovery" to describe the achievements of the early molecular biologists, but "the accurate way to describe it is that [they were] writing theory by means of [their] experimental apparatus." We should, van Fraassen urges, take "a purely functional view" of the relation between theory and experiment. "Experimentation is the continuation of theory construction by other means" (1980, 74–77).

Van Fraassen's willingness radically to reinterpret how scientists talk and act raises serious questions about the phenomena which a theory of science is to save. Of course, an individual's words and even deeds may belie his real goals. But are we to be free to reinterpret the words and deeds of a whole scientific community?[21] If so, do there not cease to be any phenomena that could count against the adequacy of any theory of science? If this is so, then the theory of science ceases to be a scientific theory even on empiricist grounds. My worry, then, is that van Fraassen's constructive empiricism avoids empirical inadequacy only by forfeiting any claim to being a genuine theory of science.

Putting aside these global worries, let me close on a more harmonious note. The model-theoretic account of theories permits a theory of science which avoids one of the major methodological defects of much social science—the demand for *universality*. If the empiricist model does not fit molecular biology, that does not mean it is worthless as part of a theory of science. The question is whether there are *any* major sciences, or long periods in the life of some major sciences, that fit the empiricist model. It seems hard to deny that there are. Greek astronomy, thermodynamics in the late nineteenth century, and quantum theory in the twentieth century are obvious candidates. It may be more than coincidence that quantum

21. By describing what he is doing as giving a more "accurate" description of scientific activities, van Fraassen disguises, perhaps even from himself, the degree to which he is *reinterpreting* these activities.

physics is the science van Fraassen knows best. On the other hand, many contemporary sciences, including chemistry, molecular biology, and geology, seem decidedly realistic.

Of course, many disagreements remain. For example, is empiricism or realism the dominant mode in contemporary science? I would guess that realism is dominant. Empiricism is the mode of sciences lacking a solid theoretical tradition, e.g., many of the social sciences, or of sciences whose models are difficult to understand in spite of much empirical success, e.g., quantum theory. Another issue is whether having realistic goals tends to make scientists, or even scientific fields, more successful in discovering new results. Again I would guess that realism is scientifically more fruitful. Though difficult to resolve, these issues, happily, are more scientific than philosophical.[22]

References

Bloor, David. 1976. *Knowledge and Social Imagery*. London: Routledge & Kegan Paul.

Boyd, Richard. 1973. "Realism, Underdetermination and a Causal Theory of Evidence." *Nous* 7 : 1–12.

———. 1981. "Scientific Realism and Naturalistic Epistemology." In *PSA 1980*, vol. 2, edited by P. D. Asquith and R. N. Giere, 613–62. East Lansing, Mich.: Philosophy of Science Association.

Churchland, Paul M. 1979. *Scientific Realism and the Plasticity of Mind*. Cambridge: Cambridge University Press.

Giere, Ronald N. 1983. "Testing Theoretical Hypotheses." In *Testing Scientific Theories: Minnesota Studies in the Philosophy of Science*, vol. 10, edited by John Earman. Minneapolis: University of Minnesota Press.

Lakatos, I. 1970. "Falsification and the Methodology of Scientific Research Programmes." In *Criticism and the Growth of Knowledge*, edited by I. Lakatos and A. Musgrave. Cambridge: Cambridge University Press.

Laudan, Larry. 1977. *Progress and Its Problems*. Berkeley: University of California Press.

Marion, Jerry B. 1970. *Classical Dynamics*. 2d ed. New York: Academic Press.

Maxwell, Grover. 1962. "The Ontological Status of Theoretical Entities." In *Minnesota Studies in the Philosophy of Science*, vol. 3, edited by H. Feigl and G. Maxwell, 3–27. Minneapolis: University of Minnesota Press.

22. The author's research has been in part supported by a grant from the National Science Foundation. The support, and good fellowship, of the Center for the Philosophy of Science, University of Pittsburgh, is also gratefully acknowledged. Clark Glymour, James Woodward, and John Worrall generously read and commented on an earlier draft.

McKinsey, J. C. C., A. C. Sugar, and P. Suppes. 1953. "Axiomatic Foundations of Classical Particle Mechanics." *Journal of Rational Mechanics and Analysis* 2:253–72.

Olby, Robert. 1974. *The Path to the Double Helix*. Seattle: University of Washington Press.

Putnam, Hilary. 1975. *Mathematics, Matter and Method*. Cambridge: Cambridge University Press.

Shapere, Dudley. 1982. "The Concept of Observation in Science and Philosophy." *Philosophy of Science* 49 (4): 485–525.

Simon, H. A. 1954. "The Axiomatization of Classical Mechanics." *Philosophy of Science* 21:340–43.

———. 1957. *Models of Man*. New York: Wiley.

Sneed, J. D. 1971. *The Logical Structure of Mathematical Physics*. Dordrecht, Holland: Reidel.

Stegmueller, W. 1976. *The Structuralist View of Theories*. New York: Springer.

Suppe, F. 1973. "Theories, Their Formulations, and the Operational Imperative." *Synthese* 25:129–64.

Suppes, Patrick. 1979. "The Role of Formal Methods in the Philosophy of Science." In *Current Research in Philosophy in Science*, edited by Peter D. Asquith and Henry E. Kyburg, Jr. East Lansing, Mich.: Philosophy of Science Association.

Van Fraassen, B. C. 1970. "On the Extension of Beth's Semantics of Physical Theories." *Philosophy of Science* 37:325–39.

———. 1980. *The Scientific Image*. Oxford: Clarendon Press.

———. 1981. "Theory Construction and Experiment: An Empiricist View." In *PSA 1980*, vol. 2, edited by P. D. Asquith and R. N. Giere, 663–78. East Lansing, Mich.: Philosophy of Science Association.

Watson, J. D. 1968. *The Double Helix*. New York: Atheneum.

5   Explanation and Realism

Clark Glymour

One way to argue for a theory is to show that it provides a good explanation of a body of phenomena and, indeed, that it provides a better explanation than does any available alternative theory. The pattern of argument is not bounded by time or by subject matter. One can find such arguments in sociology, in psychometrics, in chemistry and astronomy, in the time of Copernicus, and in the most recent of our scientific journals. The goodness of explanations is a ubiquitous criterion; in every scientific subject it forms one of the principal standards by which we decide what to believe. The ambition of philosophy of science is—or ought to be—to obtain from the literature of the sciences a plausible and precise theory of scientific reasoning and argument, a theory which will abstract general patterns from the concreta of debates over genes and spectra and fields and delinquency. A philosophical understanding of science should, therefore, give us an account of what explanations are and of why they are valued, but, most importantly, it should also provide us with clear and plausible criteria for comparing the goodness of explanations. One-half of the subject of this essay concerns a fragment of such a theory. I will try to describe, without gratuitous formality, some features that generate clear and powerful criteria for judging the goodness of explanations. That is half of my subject; the remainder concerns what such criteria determine about what we ought to believe. Both halves are prompted by Bas van Fraassen's delightful book *The Scientific Image*.

Wherever theorists have postulated features of the world that could not at the time be observed, debates have erupted over the scientific credentials of beliefs in the unobserved or unobservable. The scientific debates have more often than not been best articulated in throwaway lines: "If I were the Master," wrote Dumas in the 1830s, "I would ban the word 'atom' from chemistry, for it goes beyond all experience, and never, in chemistry, should we go beyond experience" (1937). And in our own time we have B. F. Skinner's argument by rhetorical question: When one has a secure equational linkage between observed variables, why introduce gra-

Previously published in *Scientific Realism*, edited by J. Leplin (Berkeley: University of California Press, 1984), pp. 173–92. Reproduced by permission.

tuitous unobserved variables? We have more considered, philosophical discussions that lead to the same opinion, namely, that there is no warrant to be found in scientific observations for conclusions that concern unobserved or unobservable features of the world. Call those who hold such views antirealists, and those who hold the contrary view realists. Kepler and Dalton and Spearman were realists. Osiander and Dumas and Skinner were (and in the last case remains) antirealists. The antirealist case has been made with care and with something approaching logical precision by several writers, but still nowhere better than by Hempel (1965), who argued roughly as follows: Make the idealization which supposes that all of the evidence pertinent to our theories can be formulated using only a part of the terminology of science. Call this part the "observational vocabulary." Then full-fledged theories whose vocabulary goes beyond this observational fragment seem to have no epistemological advantage over the collection of their consequences that can be stated in observational terms alone. For example, Hempel views explanation as a kind of deductive systematization, and the "observational consequences" of a theory are as well systematized—by Hempel's criteria, anyway—as is the theory itself. (It must be noted that Hempel himself did not draw the antirealist conclusion, for he hoped that theories might be shown to afford an "inductive systemization" unobtainable without them.) Further, both Duhem (1974) and Quine (1953) have argued that, once we entertain hypotheses that transcend all possible experience, we enter into the realm of convention and arbitrariness, for the evidence and the canons of scientific method can determine only what we ought to believe about the observable, and the unobservable is indeterminate: a plethora of alternative conjectures will explain the evidence equally well. Claims of the same sort were made against Dalton's theory by his antiatomist opponents—the properties of atoms, they said, are indeterminate and arbitrary.

Duhem and Quine and others are right enough if we allow science only a sufficiently impoverished collection of principles of assessment and inference. But even the fragmentary criteria for comparing scientific explanations which I am able to display are strong enough to defeat their antirealist arguments. For, although the criteria make no use of the notion of observability, they yield the result that sometimes the best explanation does go well beyond what is observed or observable. Or, put in something more like Hempel's fashion, they yield the result that sometimes a theory with "non-observation terms" has explanatory virtues which are unobtainable without such terms. Moreover, the criteria for comparing explanations do not, in leading us beyond the observable, enmesh us in indeterminacy. On the contrary, in some contexts they suffice to determine a unique best theory which explains the phenomena. Further, there is nothing about these criteria for the goodness of explanations which requires or presupposes that they be applied only to theories that contain claims about what is unobservable. The very same criteria are used to determine the best of competing theories about observable features of the world. The same criteria sometimes determine—at least so well as I understand the notion of "observability"—that theories that are confined to observable features of the world are better explainers than competitors which postulate unobservable features. That all seems to me exactly as it should be. The result is simply that the same features of inference which lead to general conclusions about the observable also lead in other contexts to determinate conclusions about the unobservable. In consequence, I believe there are only two ways to maintain the antirealist position: either by impoverishing (perhaps I should say emasculating) the methods of science, and disallowing altogether explanatory criteria such as those I will describe, or by arbitrarily (and vaguely) restricting the scope of application of such principles to the realm of the observable.

II

In discussions of scientific explanation and of the grounds such explanations may afford for belief in unobserved features of the world, it is important to keep cases in mind. Eventually I will try to show that all of the following cases have something important in common:

1. The Copernican explanation of the regularities of the superior planets. In particular of the regularity, noted by Ptolemy, which asserts that if, in a whole number of solar years, a superior planet goes through a whole number of oppositions and also a whole number of revolutions of longitude, then the number of solar years equals the number of oppositions plus the number of revolutions of longitude. Copernican theory explains the regularity by noting that the number of solar years is equal to the number of revolutions the earth has made about the sun, that the number of oppositions is equal to the number of times the faster-moving earth has overtaken the superior planet moving more slowly in an orbit outside the earth's orbit, and that the number of revolutions of longitude equals the number of revolutions the superior planet has made about the sun.

2. The Daltonian explanation of the law of definite proportions. The law asserts that, in any two cases in which quantities of the same pure chemical reagents combine to produce quantities of the same products, the ratios of the combining weights of the reactants are the same. Dalton's explanation is that any sample of a pure chemical substance consists of molecules, each having the same weight as any other, and each molecule of a given kind is composed of the same numbers of atoms of various kinds. All atoms of the same kind have the same weight, and a chemical reaction is

just a process in which the constituent atoms of the molecules in the reagents are recombined to form molecules of the products. Thus, the ratios of the numbers of molecules of the various reactants which combine are invariant and characteristic of the reaction. Two different samples of reactants with different weights will differ in the number of molecules they contain, but, since the weight of any pure sample is just the sum of the weights of its constituent molecules, the ratio of the weights of the reactants will equal the ratio of the number of molecules they severally contain.

3. The general relativistic explanation of the anomalous motion of Mercury's perihelion and of the deflection of starlight passing near the limb of the sun. The explanations go roughly as follows: For weak gravitational fields, such as those of the planets at astronomical distances, the theory of relativity closely approximates Newtonian gravitational theory. Hence, in the case of Mercury and in the case of the starlight, the only significant contribution to the gravitational field is that of the sun, which is known to be close to spherical. The gravitational field is therefore to good approximation the gravitational field of a single point mass. From the field equations of the theory the gravitational field, in empty space, of a single point mass is uniquely determined and describes the geometry of the surrounding space-time. The equations of motion of the theory imply that light rays move on null geodesics of this geometry and massive particles move in timelike geodesics. Light from a star may be described as a ray, and the planet Mercury reacts to gravity (to good approximation) as a point mass. Given the initial conditions, the phenomena follow.

Spearman's explanation of the correlations among intelligence tests. Although the history is complicated, it is probably fair to say that Spearman invented factor analysis when, in 1904, he published two papers, one on the measurement of correlations and another on the measurement of intelligence. Spearman obtained several assessments of populations of schoolchildren, including assessments of "intelligence" and of sensory discrimination. Analyzing his data from 1904 and later, he argued in the *Analysis of Human Abilities* that for any four such measures, say X_1, X_2, X_3, X_4, the correlations among the measures have vanishing "tetrad differences." That is to say, empirically it is found that for any such quadruple of measures for his samples

$$\rho_{12}\,\rho_{34} = \rho_{13}\,\rho_{24}$$
$$\rho_{13}\,\rho_{24} = \rho_{14}\,\rho_{23},$$

where ρ_{ij} is the correlation between x_i and x_j, Spearman claimed that the best explanation of these equations is that there is a common factor which determines the outcomes of scores on all of the measures, and he termed that factor "General Intelligence."

These cases are not essential ones for any theory of explanation, but as cases they have important virtues. They represent a range of subject matter, they are explanations that have a genuine historical role, and they were not used to satisfy idle curiosity—as with questions about soap bubbles—but were instead crucial pieces of the arguments given for the respective theories. They have no uniform connection with distinctions between what is observable and what is not. The Copernican theory postulated features—the motions of the planets in three-dimensional space—that could not be observed in the sixteenth century. So did competing theories. Yet some of these features might arguably be regarded as observable today. Dalton's atomic theory certainly did concern features of the world that have served almost as paradigm cases of the unobservable. I have no idea whether the metric field of space-time ought to count as observable or as unobservable. Spearman's General Intelligence is surely an unobservable feature of persons—more carefully, since one may doubt there is any such thing, if there were a factor such as Spearman's General Intelligence, it would be unobservable. The phenomena explained in these cases can be viewed sometimes as regularities, sometimes as particular events, or sequences of events, and sometimes as both. The regularity of the superior planets is indeed a regularity, but Copernican theory also explains all of the instances of the phenomenon. Similarly with definite proportions and the motion of Mercury.

The explanation of the deflection of light was, in fact, the explanation of a few events, but there is a corresponding regularity that the theory explains. Spearman's equations neither describe particular events nor express generalizations; I suppose they are best understood as expressing particular features of the population tested. Finally, for each of these explanations, there was available an alternative theory generating competing explanations. The gravamen of the scientific arguments was that the explanations cited were *better* than any others available. Copernican explanations were compared to Ptolemaic explanations, general relativistic explanations to classical ones, and one-factor accounts of correlations were compared to multifactor accounts.

III

Prejudicially, I view one of the chief goals of a philosophical account of aspects of scientific explanation as that of providing a canon or canons for the assessment of scientific theories. To the best of my knowledge, the philosophical literature on scientific explanation provides no clear criteria which can serve to explain why the four cases I have cited might have been viewed as good explanations or, in particular, why they should have been viewed as better than competing explanations. Nonetheless, what I shall have to offer by way of criteria is closely connected with some philosophical articulations of the notion of scientific explanation.

Philosophical theories of scientific explanation can be roughly divided into three types: purely logical theories, which analyze explanation solely in terms of logical relations and truth conditions; theories with extra objective structure, which impose objective conditions on explanation beyond those of truth and logical structure; and theories with extra subjective structure, which impose on explanations psychological conditions of belief, interest, etc. (In the first class fall the theories of Hempel and Oppenheim, Kaplan, Kim, and, more recently, Omer, Thorpe, Cupples, etc. In the second fall theories such as those of Salmon, Brody, Friedman, and Causey. And the third class includes theories such as those of van Fraassen, Skyrms [I believe], Achinstein, and, recently, Putnam.)

The apparent aim of Hempel and Oppenheim (1965) was to provide an account of the logical structure of "explains" in much the way that the logical tradition of Frege, Russell, Whitehead, and Hilbert had provided accounts of the logical structure of "is a proof of." The conditions given by Hempel and by Oppenheim were not intended as an analysis of when it is appropriate to say that someone has explained something to someone; they were intended instead to specify the logical structure which fully explicit, nonstatistical explanations in the natural sciences would typically have if there were any and which actual explanations in the natural sciences typically abbreviate. If Hempel and Oppenheim had been successful in their aim, their criteria would have formulated a critical standard for judging theories and hypotheses. On the basis nearly of logical structure alone, it would be fully determinate what singular sentences a given theory could potentially explain and what sentences it could not possibly explain. Combined with a further account of how to use information about what theories can and cannot explain in order to assess those theories, a logical account such as Hempel and Oppenheim's promised to provide us with an understanding of how explanation is used in deciding what to accept and believe.

Hempel and Oppenheim's account of the logical structure of deductive explanations turned out to be altogether unequal to its task. In effect, given an *arbitrary* true (but not logically true) sentence E and an *arbitrary* universally quantified sentence S which is not a logical truth, there exists a logical consequence of S and a true singular sentence C, such that the logical consequence of S and C together constitute the premises of a potential explanation of E (cf. Eberle, Kaplan, and Montague 1961). In effect, anything explains anything. What was important in Hempel and Oppenheim's work was not the execution but the vision. If one shared the vision, the natural response to the failure of their analysis is to attempt another of the same kind, either in entirely logical terms or perhaps with extra structure.

Several purely logical replacements for Hempel and Oppenheim's analysis have been published. While not all of them are trivial, they are without exception far too weak to provide useful criteria for theory assessment. Perhaps the best attempt is that of David Kaplan (1961), who uses conditions of truth as well as logical structure to constrain explanation. Accounts such as Kaplan's are deficient in tolerating as explanations a range of cases we would dismiss as not explanatory. For any explanation of the form

$$(1) \qquad \frac{\forall x(Dx)}{Da}$$

if P is any property which the individual a also has, then

$$(2) \qquad \frac{\forall x(Px \rightarrow Dx)}{\dfrac{Pa}{Da}}$$

is an explanation meeting Kaplan's criteria and likewise various generalizations of Kaplan's criteria. Again, for any explanation of the form

$$(3) \qquad \frac{\forall x(Fx \rightarrow Gx)}{\dfrac{Fa}{Ga}} \, ,$$

if P is any property which the individual a also has, then

$$(4) \qquad \frac{\forall x((Fx \,\&\, Px) \rightarrow Gx)}{\dfrac{Fa \,\&\, Pa}{Ga}}$$

is likewise an explanation meeting Kaplan's and related criteria. Thus, to use an example of Henry Kyburg's, we can explain the fact that a sample of table salt dissolves in water by citing the (true) claim that the sample of table salt has been hexed and the (true) generalization that all hexed table salt dissolves in water.

One sensible response to these difficulties is that they show nothing more than the incompleteness of the analysis of explanation. Thus, to an account of Kaplan's kind restricting the logical form of explanations, one should add criteria for comparing explanations with the appropriate logical form. For example, when an explanation of form (1) exists, any explanation of form (2) is defeated. Alternatively, one might reconstruct the logical conditions for explanation so that (2) and (4) will not count as explanations at all. Wesley Salmon (1971) proposes to regard deductive explanation as a limiting case of statistical explanation: explaining a single event consists in assigning it to one member of a statistically homogeneous partition of the largest class of events that can be so partitioned. Very cleverly, it follows that, when (1) obtains, (2) does not meet the conditions for explanation (unless P and D are coextensive), and analogously with (3) and (4).

None of this work on explanation provides any criteria which can help to account for the cases mentioned earlier or for others like them. Ptolemaic and Copernican astronomical theories each provided deductions of sentences about the positions of the planets from general lawlike sentences; if that is what is required for potential explanation, then both theories provided it. The atomic theory explained each instance of the law of definite proportions, but, so far as these logical criteria are concerned, equally good explanations were obtained from the view, popular in the nineteenth century, that one ought to abjure atoms and simply make tables of combining weights. Similar remarks hold for the other cases. Salmon's account of explanation, it is fair to say, is such that we simply do not know how to apply it in such contexts.

A simple way to see the limitations of the deductive criteria so far described is to consider what it would be to explain a regularity rather than a single event. One naturally supposes that to explain a regularity is to explain every instance of it, but, if no more is required, then every regularity explains itself. If we further require that what explains a regularity not be logically implied by the regularity itself, we are still no better off, as Hempel and Oppenheim noted, since it will remain the case that by these criteria any regularity can be explained by conjoining that regularity with any other logically independent of it.

I do not know of any writer who has attempted to account for the explanation of regularities in logical terms alone. Inevitably, considerations other than truth and logical form enter into the criteria, and that is quite proper and harmless so long as the further structure does not defeat the very goal of an account of explanation: to help us to understand how explanation can be and is used to assess our hypotheses. Three recent discussions of explanation that appeal to extra structure are especially valuable, not so much for any technical advance they make, but for pointing to features of explanation easily overlooked and for at least suggesting that such features might be formally tractable. One is Michael Friedman's proposal (1974) that what good theoretical explanations do is to somehow reduce the number of hypotheses that must be accepted independently of one another. Another is Robert Causey's careful analysis (1979) of the structure of intertheoretical reductions, with its emphasis on the identity of objects and properties described and postulated by the hypotheses to be explained with complexes of objects and properties postulated by the theory which provides the explanation. A third is Baruch Brody's attempt (1980) to bring Aristotle back to scientific explanation and to take seriously the idea that good scientific explanations show that the event or regularity to be explained is *necessary*, not just a necessary consequence of theoretical assumptions. None of these three accounts suffices to treat the sorts of cases I described earlier, and only Friedman's really constitutes an attempt to do so. Friedman's technical account came a cropper; Causey's analysis has technical difficulties and does not provide applicable comparative criteria; and only in those situations where we know, independently, something about what properties are essential and what properties are accidental can Brody's account of explanation be applied to compare competing claims to explain a phenomenon. Even so, in each of these three cases the motives seem sound in their way, and the scheme that I present below can be viewed as an attempt to unify aspects of each of them.

One of the things that many explanations seem to do, and which appeals to certain intuitions about explanations, is to demonstrate that one phenomenon is really just a variant of, a different manifestation of, another phenomenon. In some sciences the explanations provided in textbooks are very largely of this kind. In classical thermodynamics, for example, the standard exercise repeated throughout textbooks of physical chemistry consists in using thermodynamic principles to derive empirical regularities from one another. The first and second laws are used to derive Joule's law from the ideal gas law or to derive from the gas law the fact that specific heat at constant pressure minus specific heat at constant volume is proportional to the gas constant. Intuitively, an explanation produces a form of understanding if it shows us that the phenomenon we want explained is a manifestation of a different phenomenon we already know about. The understanding consists in a grasp of the unity of pattern or substance behind the apparently disparate phenomena. In physical contexts, the mechanics of such explanations is familiar to anyone who has suffered through problem sets. In the thermodynamic case again, one starts, say, with $C_p - C_v = R$ and explains this relation by relating a subformula of it (e.g., $C_p - C_v$) to thermodynamic quantities, applying the thermodynamic laws to these quantities and deriving the result that $C_p - C_v = PV/nT$.[1]

As a first approximation to a formal account of this feature of explanation, consider theories represented in the predicate calculus. For convenience, let us suppose that all sentences considered are in prenex form. Let a *subformula* of a sentence S be any well-formed subformula occurring in the matrix of the prenex form which is not logically equivalent to the entire sentence. A subformula of a sentence may have several occurrences

1. I am indebted to my student Jeanne Kim for suggesting this example to me.

in a sentence. By a term in S we will mean any term occurring in the matrix of S. Let us say that where T, H, and K are sentences, T *explains* H *as a result of* K if for (nonvalid) subformulas $H_1, \ldots, H_n$ of H, no one of which occurs in H as a subformula of any of the others, and terms $t_1 \ldots t_k$ of H, not occurring in $H_1 \ldots H_n$, there are formulas $K_1, \ldots, K_n$, terms $s_1, \ldots, s_k$, and subtheories $T_1, \ldots, T_{h^\circ\kappa}$, of T such that:

(i) $T_i \vdash H_i \leftrightarrow K_i$.

(ii) $T_n + j \vdash t_j = s_j$.

(iii) $T_i \nvdash H$ for any $i < n + k$.

(iv) If $H (H_i t_j / K_i s_j)$ is the result of simultaneously substituting, for all i and j, K_i for every occurrence of H_i in H, and s_j for every occurrence of t_j in H, then $K \vdash H (H_i t_j / K_i s_j)$.

The third condition above is both natural and necessary to eliminate certain spurious explanations. For example, one could not use any theory including the ideal gas law to explain that very law as a form of the relation $T = T$ (where T is the temperature) by starting with $PV = nRT$, and noting that from the theory it follows that $PV/nR = T$ and that, on substituting T for PV/nR in the ideal gas law, one obtains $T = T$. Such an explanation is excluded by condition (iii). The condition that a term substituted for not occur as part of a subformula substituted for is perhaps unduly restrictive but is intended to avoid complications.

In practice we do not usually consider first-order formulas but rather equations representing theories which are to provide the explanations. In such a context, T explains H as a result of K, where $H(X_1 \ldots X_n) = 0$ and $K(Y_1 \ldots Y_m) = 0$ are equations and T a system of equations, if there are consequences T_i of T, not entailing H, and each T_i entails an equation between a combination of the Xs occuring in H and a combination of Ys, and when the combinations of Ys are substituted in H for the corresponding combination of Xs, one obtains an equation that is satisfied whenever K is.

As I have presented these conditions, they do not satisfy the condition of logical equivalence; that is to say, T can explain H as a result of K, while for logically equivalent formulas T', H', and K', T' fails to explain H' as a result of K'. Of course, what I mean is that T explains H as a result of K if each of these sentences has an equivalent meeting the formal conditions. In fact, the equivalence is generally much wider than logical equivalence, for we count an equation H as explained as a result of K if we can find mathematical equivalents of H and K meeting the conditions. Doubtless, the formal proposal has other defects as well; it is at best a first try. Veterans of the textbook tradition in physics and chemistry will, I hope, be at least tingled by a reminiscence that much of the work in the explanations presented and the problems assigned was to find the right representations of equations, and the right subtheories, in order to derive one thing from another.

A theory may entail each member of a class of regularities without explaining any one of them as a result of any other of them. If, for example, the regularities in question are logically (mathematically) independent, the theory consisting of their conjunction will have this property. Thus, to borrow an example of Friedman's, the theory consisting of the conjunction of Kepler's laws, Boyle's law, Galileo's law of falling bodies and the law of the pendulum will not explain any one of these regularities as the result of any other. Again, the conjunction of the additivity of masses and the law of definite proportions does not explain either of these regularities by the other. By contrast, the atomic theory explains the law of definite proportions as a result of the additivity of masses, and that unification is one of the great virtues of the theory. In some contexts in which theories are being assessed and compared, there is a reasonably well established collection of logically independent empirical regularities pertinent to assessment. Contending theories, such as the atomic theory and the theory of equivalents, may well entail a common class of regularities, but the entire classes of regularities pertinent to the respective theories need not be coextensive: the additivity of masses has nothing to do with the theory of equivalents but a great deal to do with the theory of atoms. In general, we prefer theories that explain entailed regularities as the result of other established regularities to theories which do not. More exactly, I propose the following rather modest principle of comparison:

I. *Ceteris paribus*, if T and Q are theories and for every established pair of regularities, H, K, such that Q explains H as a result of K, T also explains H as a result of K, but there exist established regularities, L, $\mathcal{J}$, such that T explains $\mathcal{J}$ as a result of L but Q does not explain $\mathcal{J}$ as the result of any other established regularity, T is preferable to Q.

The statement is cumbersome, but the idea is almost trivial, and it seems to me to be part of what is correct about Friedman's suggestion that good theories reduce the number of independently acceptable hypotheses. It might be interesting to explore further principles of comparison which would extend to circumstances in which competing theories explain a common set of regularities as the result of other regularities but disagree about which regularities are the result of which others. I leave the matter aside here, except for a special case to be discussed later.

There are two points to emphasize. First, according to principle I, certain theories entailing empirical regularities may be entirely preferable to the conjunction of those regularities alone and may be preferable as well to the conjunction of the regularities entailed with any regularities from which these are claimed to result. The atomic theory illustrates the point. Second, according to principle I, Copernican theory is preferable to its principal rival, Ptolemaic astronomy, and Spearman's explanation of the correlation equations among four measured variables in terms of a single common factor is preferable to many alternative explanations that equally save the phenomena. To see this, it will help to return to Brody's idea.

At least part of Brody's idea is that we may explain a state of affairs by showing it to be necessary, and that such a demonstration is to be given by showing that the state of affairs is a logical consequence of essential properties of the entities concerned, and of laws governing these essential properties. I suggest a different kind of demonstration of necessity: we may show that a regularity has a kind of necessity by explaining it as a result of a necessary truth. In particular, in the schema for "T explains H as a result of K," nothing prohibits that K be a logical or mathematical, and thus a necessary, truth. Logical and mathematical truths, in general, may be considered regularities, and they may certainly be established and used in the assessment of theories. In the case of Copernican positional astronomy, and in the case of Spearman's tetrad equations, regularities or equations of an empirical kind are explained as the result of mathematical truths.

It is a mathematical truth that, if two objects move in closed orbits about a common center, one orbit inside the other, and the inner object moves faster than the outer object, then the number of revolutions of the inner object equals the number of revolutions of the outer object plus the number of times the inner object has overtaken the outer object, whenever all of these numbers are whole. According to Copernican theory, the Earth moves in an orbit interior to the orbit of any superior planet and moves faster in its orbit than does a superior planet. The number of solar years equals the number of revolutions of the Earth in its orbit; the number of oppositions or cycles of anomaly equals the number of times the superior planet is overtaken by the Earth, and (over any long period) the number of revolutions of longitude of a superior planet equals the number of revolutions in its orbit. Thus, the planetary regularity noted by Ptolemy is explained by Copernican theory in terms of a mathematical truth. It is not explained in Ptolemaic theory as the result of any other independently established regularity.

Spearman assumes (tacitly) that the relations between measured variables and unmeasured factors are linear. Assuming that measurement error averages to zero and that errors in measured variables are uncorrelated with one another, it follows that for standardized variables the correlation between two variables is equal to the product of the linear coefficients relating each variable of the pair to the common factor. Thus, with Spearman's causal scheme

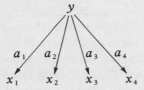

we have that

$$\rho_{ij} = a_i a_j .$$

On substituting the right-hand side of these equations into Spearman's supposed empirical equations,

$$\rho_{12} \, \rho_{34} = \rho_{13} \, \rho_{24}$$
$$\rho_{13} \, \rho_{24} = \rho_{14} \, \rho_{23},$$

we obtain instances of trivial truths of real algebra. Spearman's tetrad equations are thus explained in terms of mathematical truths. Typically, other arrangements of unmeasured factors can be found which are consistent with the tetrad equations but do not explain the tetrad equations as the results of mathematical truths.

One might wonder whether, in the Copernican and psychometrical cases, an empirical regularity has been entirely explained as the result of necessary truths, for the Copernican relations between numbers of orbits and revolutions in longitude and solar years, etc., are at least arguably not necessary truths. Again, psychometric equations relating correlation coefficients to "factor loadings" scarcely seem to be necessary truths. There are cases and contexts, however, in which an empirical relation seems literally to be explained *as* a necessary truth. I have in mind cases in which a regularity is explained as the result of a mathematical truth, and the equations used to connect quantities in the regularity with other quantities (as in [ii] and [iii] of the scheme) are identifications, as Professor Causey requires for microreductions. One case in which there is a tradition—correct or not—providing such an interpretation is that of general relativity. Both the motion of Mercury and the motion of light rays near the limb of the sun provide a special case of the general relativistic law of motion. That law of motion is in turn a consequence of the general relativistic conservation laws, which require that the covariant divergence of the energy-momentum tensor vanish. The field equations of the theory give the energy-momentum tensor as a function of the metric field; as early as 1920, Eddington proposed that the field equations be understood as an identification and not as a contingent relation: these equations, on Eddington's construal, specify what matter-energy *is*. When a function of the metric field which equals the momentum-energy tensor is substituted for the momentum-energy tensor in the conservation laws, the conservation laws are transformed into mathematical truths of the tensor calculus.

I surmise that, other things being equal, we prefer explanations that explain regularities in terms of necessary truths to explanations that explain those same regularities in terms of other empirical generalizations. I suppose, further, that we find particularly virtuous those explanations that explain regularities as the result of necessary truths and do so by means of connections which are themselves necessary. When such explanations succeed, they are complete, and nothing remains in need of explanation. I do not mean to suggest that any or all of these considerations exhaust the criteria by which we judge explanations. That is why there are *ceteris paribus* clauses. It was, of course, true that one of the major reasons for the inferiority of the Newtonian explanations of the perihelion advance and light deflection was that all of these explanations required hypothetical circumstances—hidden masses and concentrations of ether—for which there was no evidence whatsoever.

V

The explanatory principle provides a reason why, in each of the cases described in section 2, the explanation is particularly virtuous and why it is better than were competing explanations. In some of these cases there is even a little historical evidence that these virtues were recognized in something like the form they are given here: Kepler praised Copernican theory in contrast to Ptolemaic astronomy because the former makes the regularities of the planets a "mathematical necessity." In each of the cases considered, the explanatory power is obtained by introducing theoretical connections and structure not explicit to the regularities to be explained, and in each of these cases the connections among empirical regularities could not be made at all without additional theoretical structure. The regularity of the superior planets simply cannot be explained as the result of a mathematical truth unless the quantities occurring in the statement of that regularity are somehow related to other quantities distinct from solar years, oppositions, and revolutions of longitude. The same is true of Spearman's tetrad equations and of the motion of Mercury, and so on. The law of definite proportions cannot be explained as a result of the additivity of mass unless one introduces objects other than the macroscopic samples of elements that were used in the nineteenth-century laboratory. In some of these cases the additional structure which must be introduced in order to obtain the explanatory connections is structure which has turned out to be observable; in other cases that structure has turned out to be not

observable, or at least arguably not observable. But observability or unob-
servability has nothing to do with the explanatory virtue in question.
There seem no grounds, save arbitrary ones, for using a principle of preference to give credence to theoretical claims when the claims turn out to be
about observable features of the world but to withhold credence when the
theoretical claims obtained by the very same principle turn out to be about
unobservable features of the world. The argument for atoms is as good as
the argument for orbits.

I know of three objections to this argument: first, that explanation is
subjective and context-dependent and therefore cannot be an objective
and interpersonal ground for belief; second, that the principle of preference presented in section 4 may be acceptable as a principle governing
what theories we should prefer to accept or display or work with, but it
cannot be acceptable as a principle governing what theories we should prefer to *believe*; and, third, that the introduction of theoretical structure beyond the bounds of observation generates underdetermination without
limit.

Many writers have maintained that explanation is basically a pragmatic
notion. The fundamental explanatory relation is held to be something of
the form "X explains Y for person P." Of course, what will explain Y for
person P will depend on what P already knows about Y, doesn't know
about Y, wants to know about Y, and so on. This view of explanation is
surrounded by examples trading on the ambiguities of why questions and
the context dependence of relevant replies to questions of the kind "What
causes Y?" Undoubtedly, understanding is a pragmatic notion, and, also
undoubtedly, we legitimately call "explanations" those replies to why questions (and questions of other forms) that produce understanding in an audience. None of this has any pertinence to whether or not there are objective unpragmatic relations between theories and statements of empirical
phenomena, relations which when apprehended produce understanding
and give grounds for belief. The claim that explanation is *entirely* interest-
dependent has exactly as much merit as the claim that "proof" is entirely
interest-dependent: based on what one's mathematical audience knows or
believes, one says some things in giving proofs but not other things; there
are arguments which meet the formal conditions for proof but which we do
not call proofs because they are uninteresting (for example, because we
have no reason to believe the premises), as in the "proof" of the Poincaré
conjecture from the premises that (1) Glymour weighs more than 190
pounds and (2) if Glymour weighs more than 190 pounds, then the Poincaré
conjecture is true. An account of "proof" entirely tied to speech acts and,
eschewing characterization of ideal logical form, would be blinding; so
would a like account of explanation.

Bas van Fraassen has argued that preferences of the kind to which I have appealed in principle I cannot be credences or preferences as to what to believe. The reason is as follows: According to principles such as I, Copernican theory, for example, is to be preferred to the conjunction of the empirical laws which it explains. But the conjunction of the empirical laws that Copernican theory explains is a *logical consequence* of Copernican theory. We cannot reasonably give a claim more credence than its logical consequences. Hence, the preferences in principle I cannot be credences or preferences for belief.

Van Fraassen is right, so far as it goes, but the preferences may nonetheless be tied to belief. It can be irrational to believe a collection of empirical laws, to believe that a particular theory entails these regularities, is tested by them, and explains them better than any other possible theory and still not believe the theory. The irrationality is akin to that of believing certain premises, believing that some other sentence is a logical consequence of these premises and refusing nonetheless to believe the further sentence. There is inductive closure quite as much as deductive closure. The preference stated in principle I is a preference for a certain candidate for the inductive closure of the regularities explained. If theory T is preferable to theory Q, according to principle I, that is not to be understood to mean that T is to be believed rather than Q or that there is reason to believe T rather than Q, for Q may be a logical consequence of T. Rather, the preference is to be understood to mean that the inductive closure of the laws explained is (the deductive closure of) theory T rather than (the deductive closure of) theory Q. More simply but less exactly, the preference is for believing T rather than believing *merely* Q when T and Q are consistent or for believing T rather than Q when T and Q are not consistent.

Some years ago, Quine argued that, once we introduce language which does not describe observable features of the world, underdetermination is rampant. "Rampant" means infinite here, for, if only a finite number of theories are consistent with all possible evidence of some kind and satisfy our inductive canons, then there is a preferred conclusion, namely their disjunction. Such sweeping views are unconvincing, for they lack both a convincing account of theoretical equivalence and a plausible characterization of inductive canons. In more restricted cases where a characterization of the class of possible theories is available and clear and applicable criteria for sameness of theories are at hand, familiar arguments for underdetermination fail when inductive canons are applied. For example, Reichenbach's well-known arguments for the underdetermination of space-time geometry fail when the inductive canons I prefer are applied to space-time theories.[2]

It is certainly the case that the principle of explanation that I have de-

2. See my *Theory and Evidence* (Princeton: Princeton University Press, 1980) for a discussion of this case.

scribed, even with elaborations about necessity and identity, is insufficient as a foundation for inductive logic, and, if that principle were the whole of the matter, underdetermination would indeed be rampant. Nonetheless, principle I in conjunction with other available inductive principles can be extremely powerful and can in some contexts virtually eliminate under-determination. Or so I conjecture. Consider the context provided by Spearman's kind of theory. The data consist of correlations among some determinate set of variables. The theories consist of any set of digraphs without cycles, connecting the measured variables with each other or with unmeasured variables, and the associated linear equations. Under the assumptions about error made before the equation, they can be reduced to equations relating the "path coefficients" to one another and to the measured correlations. The correlations may satisfy equations such as Spearman's tetrad equations, or analogous equations involving triples of correlation coefficients, quadruples of correlation coefficients, and so on. Given a set of data satisfying a determinate set of equations of this kind, of the infinity of digraphs consistent with this data, a proper subset will re-duce to mathematical truths all and only the equations of the set satisfied by the data. This set of digraphs will be those preferred by principle I.

In some cases the set of preferred digraphs will, I believe, be finite and small. For example, in addition to the digraph illustrated in connection with Spearman's explanation of his correlation equations, I know of only three kinds of directed graphs that satisfy the principle for the three tetrad equations Spearman assumes hold empirically. They are:

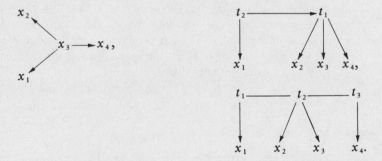

The first digraph, which contains no unmeasured variables, really stands for an entire set of digraphs, since any two variables can be interchanged in the digraph without affecting the derivation of the tetrad equations. Any of the digraphs of this kind can be distinguished from those which introduce any latent variables, since digraphs of this kind imply restric-tions on partial correlations that are not implied by digraphs with latent variables. Various other empirical constraints may distinguish the three al-ternatives with unmeasured variables from one another in the context of a larger theory. Even when other empirical constraints are not available, other methodological constraints may be. In *Theory and Evidence* I described a strategy for testing systems of algebraic equations in real variables with known coefficients. In the context of linear theories of the kind under consideration here, principles founded on bootstrap testing may not themselves lead us to prefer theories with unmeasured or latent variables, but they may lead to determinate preferences among a set of theories all having latent variables. Starting with Spearman's supposed tetrad equations, we have four classes of theoretical digraphs that satisfy principle I. Partial correlations can either establish or eliminate the class of digraphs containing only measured variables. Of the remaining three digraphs, Spearman's original model is the one preferred on bootstrap principles in the absence of other information. That is because the other two digraphs implicitly contain theoretical coefficients which cannot be evaluated from measured correlations even assuming the correctness of the causal relations they postulate. In the econometricians' jargon, they are not identifiable.

This is only an example, and a conjectural example at that. I mean it to illustrate nothing more than that there may well be many surprises in store when issues of underdetermination are examined in more modest and more detailed ways. Certainly, there is at present no particular reason to believe that underdetermination is rampant once we move beyond the observable. Underdetermination is, I expect, indestructible in another way, for logic does not determine a unique set of inductive canons. Antirealists may still have recourse to denying principles of preference or inference such as those I have tried to describe here. My point has been not only to give that description but to argue that the principles I have described are principles we use in our sciences to draw conclusions about the observable as well as about the unobservable. If such principles are abandoned *tout court*, the result will not be a simple scientific antirealism about the unobservable; it will be an unsimple skepticism.

References

Brody, B. 1980. *Identity and Essence*. Princeton: Princeton University Press.
Causey, R. 1979. *Unity of Science*. Dordrecht, Holland: Reidel.
Duhem, P. 1974. *The Aim and Structure of Physical Theory*. New York: Atheneum.
Dumas, J. 1937. *Leçons de Philosophie Chemique*. Paris: Gauthier Villars, 178–79. This edition is a reprinting (with slight alteration of title) of the 1837 edition.
Eberle, R., D. Kaplan, and R. Montague. 1961. "Hempel and Oppenheim on Explanation." *Philosophy of Science* 28:418–28.
Friedman, M. 1974. "Explanation and Scientific Understanding." *Journal of Philosophy* 72:5–19.

Hempel, C. 1965. "The Theoretician's Dilemma." In *Aspects of Scientific Explanation*. Glencoe, Ill.: Free Press.

Hempel, C., and P. Oppenheim. 1965. "Studies in the Logic of Explanation." In *Aspects of Scientific Explanation*. Glencoe, Ill.: Free Press.

Kaplan, D. 1961. "Explanation Revisited." *Philosophy of Science* 28:429–36.

Quine, W. V. O. 1953. "Two Dogmas of Empiricism." In *From a Logical Point of View*. Harvard: Harvard University Press.

Salmon, W. 1971. *Statistical Explanation and Statistical Relevance*. Pittsburgh: University of Pittsburgh Press.

Spearman, C. 1904. "General Intelligence Objectively Determined and Measured." *American Journal of Psychology* 15:201–93.

6   Scientific Realism versus Constructive Empiricism: A Dialogue

Gary Gutting

Note: *The following is a discussion between a scientific realist (**SR**) who has been strongly influenced by the work of Wilfrid Sellars and a constructive empiricist (**CE**) who has been equally influenced by the work of Bas van Fraassen. Indeed, the influence is so great that the interlocutors occasionally lapse into direct quotation of their masters. I do not, however, want anyone to identify the views of my two characters with those of Sellars and van Fraassen. What they say merely represents the dialectic of my own mind as I think through the issues raised by the debate between Sellars and van Fraassen on scientific realism.*

SR.: Realism is encapsulated in the claim that "to have good reason for holding a theory is *ipso facto* to have good reason for holding that the entities postulated by the theory exist."[1] For an appropriate scientific theory (say atomic theory), this claim allows us to argue as follows:

(1) If we have good reason for holding atomic theory, we have good reasons for holding that atoms exist;

(2) We do have good reason for holding atomic theory (it's highly confirmed, fruitful, simple, etc.);

(3) Therefore, we have good reason for holding that atoms exist.

CE: Everything depends on what we mean by "holding a theory." If it means "believing that the theory is true," then premise (1) of your argument is obvious but premise (2) strikes me as false. There's a lot to be said for atomic theory but none of it constitutes a cogent case for its truth. At the most, the evidence shows that atomic theory is empirically adequate, by which I mean that there may be reason to think that all its *observable* consequences are true. But the evidence does not support the existence of the particular unobservable mechanisms and entities the theory postulates. On the other hand, if you take "holding a theory," as I do, to mean

Previously published in *The Monist* 65 (May 1983): 336–49. Reproduced by permission.
1. Wilfrid Sellars, *Science, Perception, and Reality* (New York: Humanities, 1963), 91.

"believing it to be empirically adequate," then there's no problem with premise (2) but, for the reasons I've just been urging, there is no basis for premise (1).

SR: I'm willing to stand by the argument even if we take "holding a theory" to mean "believing it to be empirically adequate," but we need to get clear on just what's involved in empirical adequacy. For example, it won't do to take empirical adequacy in the minimal sense of "accurately describing all the observable phenomena." With this sense of "empirically adequate" the argument will fall to the old problem of the underdetermination of theory by data. Specifically, with this meaning of "empirically adequate," premise (1) says: "If we have good reason for believing that atomic theory accurately describes all the observable phenomena, then we have good reason for believing that atoms exist." But this isn't so since, first, there are an infinity of other sorts of theoretical entities that would produce the same observable phenomena and, second, we could just as well believe only in the phenomena and forget about underlying entities.[2]

CE: Your second point is just my view. I'm not saying theoretical entities don't exist or that talk about them is meaningless. I don't even say there's anything wrong with believing in them if you want to. My point is simply that there's no evidence that makes it irrational to withhold jugment about their existence. I'm defending my right to be an agnostic on the issue. I suspect however that, just like theists who deny the rationality of religious agnosticism, you're going to invoke the explanatory power of your postulations to support their existence.

SR: Of course, though the case for scientific realism can avoid the pitfalls of "theological realism." The point is that the empirical adequacy of a scientific theory needs to be taken broadly enough to include the theory's explanatory power and, specifically, its explanatory superiority to the physical thing language used in the observation framework. Atomic theory and its associated ontology is needed precisely because of the explanatory failures of the observation framework. So, roughly, premise (1) needs to be understood as saying this: If we have good reasons to believe that atomic theory is needed to explain the observable data, then we have good reason to believe that atoms exist.

CE: What I want to question is the move you're trying to make from an explanatory need for a theory to the existence of its entities. Consider, for example, Sellars's fictional case of observationally identical samples of gold that dissolve at different rates in *aqua regia*. I agree that available microtheory might explain the empirical fact of the different dissolution rates. "The microtheory of chemical reactions might admit of a simple modification to the effect that there are two structures of microentities each of

---

2. Cf. Bas van Fraassen, *The Scientific Image* (Oxford: Clarendon Press, 1980), 12.

which 'corresponds' to gold as an observational construct, but such that pure samples of one dissolve, under given conditions of pressure, temperature, etc., at a different rate from samples of the other. Such a modification of the theory would explain the observationally unpredicted variation in the rate of dissolution of gold by saying that samples of observational gold are mixtures of these two theoretical structures in various proportions, and have a rate of dissolution which varies with the proportion."[3] Of course, I'd expect the realist to admit in his turn that it might also be possible to use the correspondence rules of our microtheory to "derive observational criteria for distinguishing between observational golds of differing theoretical compositions."[4] If so, we could formulate two empirical laws (in the observation framework), one for each variety of gold, that would explain the differing dissolution rates.

SR: But remember that it might also happen that a good theory does not allow the formulation of any such empirical generalizations. It might simply itself directly explain the singular observable fact of differing dissolution rates. In such a case, the theory would be necessary to give any explanation of the singular observed facts and so would have to be accepted as true if we were to have any explanation at all of these facts. And, of course, this is precisely my claim regarding postulational scientific theories that have actually been developed. If, for example, we do not accept the existence of atoms, there are numerous singular empirical facts for which we simply have no explanation. This discussion lets me formulate more precisely premise (1) of the argument for realism. We should take it to say: If we have good reason to believe that atomic theory provides the only way of explaining some singular observed facts, then we have good reason to believe in the existence of atoms.

CE: Your more accurate formulation only serves to pinpoint the weakness of your position. Just what is the "explanatory failure" of the observation language? You agree that it can sustain inductive generalizations but insist that it cannot sustain "enough to explain all the singular facts that require explanation. In the case of the gold, one might have achieved a very precise statistical generalization: for each number r, the probability that a random sample of gold dissolves at rate r equal p(r). . . . But faced with the question why a given sample of gold dissolves at rate r, the physical thing language provides us with no property X such that we could say: this is an X-sample, and all X-samples dissolve at that rate."[5] You conclude from this that we need a theory to explain what the observation language cannot. But the claim is far too strong. On your principles, an exactly parallel argument could be made for the existence of hidden variables underlying quan-

3. Sellars, *Science, Perception, and Reality*, cited in note 1, above, pp. 121–22.
4. Ibid., p. 122.
5. Bas van Fraassen, "Wilfrid Sellars and Scientific Realism," *Dialogue* 14 (1975): 611.

tum phenomena. As is well known, the laws of quantum mechanics are irreducibly statistical; that is, they can explain why certain events occur a certain percentage of times over a given period but they cannot explain why one particular event occurs rather than another. For example, quantum mechanics has no explanation of the fact that a particular radium atom decays at a particular time, though it can explain why, over a period of time, a given fraction of the atoms in a sample of radium will decay. Using your principles, a quantum physicist would have to accept the suggestion, made by some physicists, that there are "hidden" entities and processes, not taken account of by quantum theory, that are responsible for the occurrence of the singular facts that theory cannot explain. But this conclusion, generated by an a priori demand for explanation, conflicts with the fact that the irreducibly statistical character of its laws does not, in the mind of the scientific community, constitute a case for the explanatory inadequacy of quantum theory and the need for the acceptance of hidden variables. So, just like the theist with his cosmological argument, you make your case by insisting on the need to explain something that there is no reason to think has to be explained. If the singular facts not explained by quantum theory don't need explanation, neither do the singular facts not explained by the observation framework.

SR: I entirely agree with you about quantum mechanics, and in arguing for realism I do not mean to "demand that all singular matters of fact be capable of explanation."[6] Perhaps the fictional gold example is misleading. The inadequacy of the observation framework does not consist in its inability to explain some singular empirical facts but in its inability to explain without relying on theoretical concepts. "It is not that the 'physical thing framework' doesn't sustain *enough* inductive generalizations, but rather that what inductive generalizations it *does* sustain, it sustains by a covert introduction of the framework of theory into the physical thing framework itself."[7]

CE: I take it, then, that you're revising even your most recent statement of premise (1) in your argument for realism. You now seem to be taking it to mean something like this: If we have reason to believe that all explanations of singular empirical facts (in a given empirical domain) must rely on the concepts of atomic theory, then we have good reason to believe in the existence of atoms.

SR: That's about what I have in mind.

CE: But then we need some clarification of what you mean by an explanation's "relying on the concepts of a theory." You're obviously assuming that explanation is at least partly a matter of subsuming singular empirical facts

6. Wilfrid Sellars, "Is Scientific Realism Tenable?" *PSA 1976*, ed. F. Suppe and P. Asquith (East Lansing, Mich.: Philosophy of Science Association, 1977), 315.
7. Ibid.

under generalizations. So one possible meaning of the claim that all of a set of observation-framework explanations "rely on the concepts of a theory" is that all the generalizations that accurately subsume the singular empirical facts must be expressed in theoretical terms. But taken this way the claim is clearly false. Even when theoretical corrections are provided for empirical generalizations, the results are typically expressible by a new empirical generalization in the observation language. For example, the correction of Boyle's Law by kinetic theory results in van der Waal's Law, which is still entirely observational.

SR: But my claim about the explanatory reliance of the observation framework on theory is not about the *content* of the generalizations used to explain singular empirical facts but about the way in which these generalizations are *inductively justified*. To a theoretical correction of an empirical generalization there will, I agree, typically correspond a revised empirical generalization. Further, this revised generalization will be compatible with the observational evidence. But, I claim, it will not typically be the law that would be accepted "on purely inductive grounds—i.e., in the absence of theoretical considerations."[8] Rather, lacking theoretical guidance, purely inductive reasoning in the observation framework alone would lead us to accept an empirical generalization that is shown by theoretical considerations to be false. So explanations of singular empirical facts rely on theory in the following sense: The empirical generalizations that explain these facts must be justified, ultimately, by theoretical considerations. This can be expressed by putting premise (1) in the following way: If we have good reason to believe that the empirical generalizations that explain singular empirical facts require an appeal to atomic theory for their justification, then we have good reason for believing in the existence of atoms.

CE: I have a number of reservations about your factual claim that the empirical data alone wouldn't lead us to the same law that theoretical considerations yield. We'd have to look at some examples to probe that thesis. But suppose I agree that theory plays the role you say it does, that it's essential for the development of adequate empirical generalizations. I still don't think there's any reason to accept such a theory as a true (or approximately true) description of the world. In other words, I might be prepared to accept the antecedent of your last formulation of premise (1), but I'll still deny the consequent. Here my position is like Duhem's. He shared your conviction that "postulation of unobservable entities is indispensable to science." But since he also held that improved description of observable phenomena is the only basic role for theoretical postulation, he maintained that there is no need for us to accept the existence of postulated entities. "If that is how one sees it, then truth of the postulates becomes quite irrel-

8. Ibid., p. 319.

evant. When a scientific theory plays [its] role well, we shall have reason to use the theory whether we do or do not believe it to be true; and we may do well to reserve judgment on the question of truth. The only thing we need to believe here is that the theory is empirically adequate, which means that in its round-about way it has latched on to actual regularities in the observable phenomena. Acceptance of the theory need involve no further beliefs.''[9]

SR: Don't we at least need an explanation of why the theory is empirically adequate; that is, of why what we observe is just as it would be if the theory were true? And isn't the best explanation just that the theory is in fact true (or anyway near the truth)? Don't we ultimately need to invoke realism to explain the success of science?

CE: It seems to me that you're slipping back into the theological demand for explanation for its own sake. Why are you so sure that we need an explanation for science's success? But let me agree, at least for the moment, that we do need such an explanation. Even so, I think the appropriate explanation is quite different from the one you've proposed. After all, "science is a biological phenomenon, an activity by one kind of organism which facilitates its interaction with its environment."[10] Just as, from a Darwinian viewpoint, the only species that survive are those that are successful in coping with their environment, so too only successful scientific theories have survived. "Any scientific theory is born into a life of fierce competition, a jungle red in tooth and claw. Only the successful theories survive—the ones which in fact latched on to actual regularities in nature."[11] But this process of selection on the basis of epirical success need have nothing to do with the *truth* of the theories selected. Your argument is no different from that of an antievolutionist who holds that the survival of a species can be explained only by some design that has preadapted the species to its environment. But we don't need the hypothesis that theories are successful because truth has preadapted them to the world. We need only the hypothesis that theories that are not empirically successful have not survived.

SR: It seems to me you're ignoring the amazing *rate* at which empirically successful theories have emerged. Perhaps an extended process of trial and error would eventually lead to an empirically adequate scientific description of what we observe. But the use of theoretical postulations has led to success far more quickly than we could reasonably expect from mere trial and error selection. So even from your Darwinian veiwpoint I think we

9. Bas van Fraassen, "On the Radical Incompleteness of the Manifest Image," *PSA 1976*, ed. F. Suppe and P. Asquith (East Lansing, Mich.: Philosophy of Science Association, 1977), 325.

10. Van Fraassen, *The Scientific Image*, cited in note 2, above, p. 39.

11. Ibid., p. 40.

need the realist hypothesis. But this is taking us off the track. The defense of realism that I'm interested in proposing need not be based on putting it forward as an explanation of science's success. Rather I have been arguing from the indispensable role of theoretical postulation in the formulation of empirically adequate laws. Think about it this way: Imagine that we are doing science initially only in the observation framework. We are aware of various singular observational facts and are trying to explain them by subsuming them under empirical generalizations—that is, under inductive generalizations that employ only the concepts of the observation framework.

CE: Excuse me a moment. Just how are you understanding the notion of *observation* when you speak of the observation framework? Some realists, you know, have maintained that anything—even electrons and other postulated submicroscopic entities—are in principle observable.

SR: I sympathize with some of the epistemological motives behind such claims. It is important to reject the idea that the realm of entities and properties we in fact observe functions as an unchangeable given. But in this context we must avoid trivializing the distinction between what is observable and what is not. When someone says that everything is observable, he is envisaging a situation in which concepts from the theoretical framework of science have ingressed into the observation framework. I'm speaking of the observation framework prior to any theoretical ingressions.

CE: All right. So what we observe are the ordinary objects of everyday life and their properties.

SR: Yes, but we need to distinguish two sorts of properties that we attribute to observable things. On the one hand, there are occurrent (non-dispositional) properties that are, strictly speaking, *what we perceive of an object* that we perceive. There is, for example, "the occurrent sensuous redness of the facing side" of a brick.[12] On the other hand, there are dispositional or, more broadly, causal properties that correspond to *what we perceive an object as* (e.g., the brick seen as made of baked clay). Within the observation framework properties of the first sort are a constant factor. They correspond to the way that, for physiological reasons, we must perceive the world. The second sort of properties corresponds to our conceptual resources for classifying objects into kinds with distinctive causal features. These kinds are not constant but can change as "our classification of physical objects . . . becomes more complex and sophisticated."[13] An essential feature of scientific inquiry is its *revision* of the causal concepts of the observation framework in order to arrive at maximally accurate empirical generalizations. In many cases, these revisions take place entirely

---

12. Sellars, "Is Scientific Realism Tenable?" cited in note 6, above, p. 316.
13. Ibid., p. 318.

within the observation framework. In such cases, the causal properties are always built out of concepts expressing the occurrent properties of physical objects. We can imagine that science never required any conceptual revision other than this sort. In that case, the observation framework would be conceptually autonomous; that is, its conceptual resources would suffice for formulating and justifying entirely accurate empirical generalizations about singular observable facts. If this were the case, then the framework of postulational science would be "in principle otiose," [14] and what Sellars calls the "manifest image" would provide a correct ontology for the world. But, as we have learned from the development of science, the observation framework is not autonomous. We cannot do the job of science using only its conceptual resources. Rather, we can arrive at justified empirical generalizations in the observation framework only by appealing to theories that employ concepts that cannot be built out of the conceptual resources of the observation framework.

CE: You seem to be missing the point of my antirealism. I'm willing to admit everything you've said about the *indispensability* of theories. But why should we also have to accept the *truth* of theories? As I see it, the highest virtue we need attribute to a theory needed for the successful practice of science is empirical adequacy. In other words, we need only agree that all of a successful theory's *observable consequences* are true. If we regard a theory as empirically adequate—and of course we are entitled to so regard highly successful theories—then from that alone we have sufficient justification for accepting the empirical generalizations that theory entails. The further assertion of the theory's truth is a gratuitous addition, entirely unnecessary for the fulfillment of science's fundamental aim; namely, an exact account of observable phenomena. The enterprise of science can be entirely successful without ever accepting the truth of its theories. Consider two scientists. One accepts atomic theory in the sense that he thinks it is empirically adequate: he knows that it fits all observations to date and expects that it will continue to fit all further observations. Accordingly, he thinks in terms of atomic theory and uses its conceptual resources to solve relevant scientific problems. However, he remains agnostic on the question of whether atoms really exist. A second scientist shares the first's views abut the empirical adequacy of atomic theory, but he also holds that atoms really do exist. But what sort of work is done by this latter belief? It makes absolutely no difference for what the second scientist expects to observe or for how he proceeds in his scientific work. His expectations and procedures are exactly the same as those of his agnostic colleague. So, while I agree with you that theories are not "otiose in principle," I do maintain that a realistic interpretation of theories is. There is nothing in

---

14. Sellars, *Science, Perception, and Reality*, cited in note 1, above, p. 118.

the aims of scientific inquiry that is in the least affected by the acceptance or the rejection of the existence of theoretical entities.

SR: It seems to me that it's you who are missing the point of my realism. First of all, you misrepresent my view by taking it as a thesis about the aims of science. I am entirely content with the view that science aims only at empirical adequacy. Indeed, I agree that this aim might have (in some other possible world) been attained without the postulation of theoretical entities. My thesis is rather that empirical adequacy in fact requires theories that postulate unobservable entities and that this fact provides good reason for thinking such entities exist. My realism is not a thesis about the aim of science but rather a thesis about the philosophical (specifically, metaphysical) significance of the means that scientists have had to use in fulfilling this aim.

CE: I'm happy to accept this clarification of your views, but it does not affect the central point: that you haven't offered an argument from the indispensability of theories to the existence of the entities they postulate. Even Sellars, who develops the indispensability thesis much more cogently than you do, seems to ignore the need for such an argument. He seems just to assume that, once theories have been shown to be indispensable, the reality of the entities they postulate has been established. But in fact there is a gap that needs bridging. Sellars himself has emphasized this very point in parallel contexts. For example, he agrees that semantic concepts such as *meaning* and moral concepts such as *person* are indispensable, but nonetheless insists this doesn't entail that meanings and persons exist. Similarly, I think you should admit that the indispensability of theoretical language does not entail the existence of theoretical entities. Further, I submit that the only way of bridging the gap between indispensability and existence is by an act of faith that may be permissible for those who want to believe in atoms and similar things but is in no way required by the evidence.

SR: I agree that the gap needs to be bridged and I even agree that it cannot be bridged by a deductive argument. I admit that indispensability does not entail existence; theories could be indispensable and yet theoretical entities not exist. But I insist that there is a good *inductive* case for realism, reasons that make it highly probable that theoretical entities exist.

CE: I take it then that you're about to follow Putnam in presenting realism as a quasi-scientific hypothesis that is the best explanation of the success of science. I've already suggested my criticism of this sort of approach.

SR: On the contrary. I'm not entirely happy myself with this sort of "empirical" approach. Besides the difficulties you raised earlier about possible alternative hypotheses, it seems to me that realism lacks the fruitfulness we require of a good scientific hypothesis. From a purely scientific viewpoint, it looks a lot like an *ad hoc* explanation of one fact with no other explanatory significance. I rather see realism as a *philosophical* thesis based on an analysis of the nature of theoretical explanation in science. I have in mind the following strategy of argument: First find a generally valid type of argumentation from the explanatory power of a hypothesis to the reality of the entities the hypothesis refers to; then show that, in some specific cases, the results of theoretical science enable us to construct an argument of just this type for the existence of theoretical entities. Such a case is inductive because the argument type employed is inductive. But it's philosophical rather than scientific because it is not postulating an explanatory hypothesis but rather pointing out the essential similarity of two ways of arguing.

CE: I need to hear the type of argumentation you have in mind.

SR: The point is really quite simple. There's a standard way of arguing—in both everyday and scientific contexts—for the existence of unobserved entities. The mode of argument is this: from the ability of a hypothesis to (a) subsume all known facts and to (b) predict new and even unexpected facts, we infer the reality of the entities the hypothesis postulates. There's no doubt that we all accept this mode of argument in many cases that involve unobserved though observable entities. For example, this is just the way we proceed in arguing for the past existence of dinosaurs or for the present existence of stars, conceived as huge, tremendously hot, gaseous masses far distant from us. But the very same mode of argumentation used in these noncontroversial cases can be employed to argue for the existence of the unobservable entities postulated by microphysics. Just as we accepted the existence of dinosaurs because the hypothesis of their existence subsumed the known facts and successfully predicted new ones, so too we ought to accept the existence of electrons, neutrinos, etc., for precisely the same sort of reasons. The case for the existence of electrons and neutrinos is logically identical to the case for the existence of dinosaurs and stars. Since you can hardly reject the latter, I submit that you cannot consistently reject the former.

CE: You yourself have mentioned but ignored the crucial point: the non-observability of the entities postulated by microphysics in contrast to the observability of stars and dinosaurs. This difference undermines your claim that the mode of argument for the two sorts of entities is the same. As I see it, the mode of inference at work in the examples you've given is not from a theory's explanatory power to its truth but from its explanatory power to its empirical adequacy. At any rate, the uncontroversial uses of the mode of argumentation—to the existence of stars and dinosaurs—cannot decide between the realist and the antirealist interpretations of it. For these are cases of inference to *observable* entities; and, for such cases, the claim that a postulation is empirically adequate is equivalent to the claim that it is true. For example, 'Stars (as described by modern astronomy) exist' is equivalent to 'All observable phenomena are as if stars exist', since the existence of stars is itself (in principle) observable. So I can accept the inference to stars and reject the inference to electrons, etc., on the grounds that, in both cases, we are only entitled to infer the empirical adequacy of a hypothesis from its explanatory and predictive success. In the case of stars, this is equivalent to inferring their existence; but in the case of electrons, which are not even in principle observable, it is not.

SR: This response keeps your position consistent but at the price of arbitrariness. By maintaining that explanatory and predictive success supports only the empirical adequacy of a theory, you are implicitly committing yourself to a sharp epistemic distinction between the observable and the unobservable. You say the explanatory success of a hypothesis is evidence of its truth only if the hypothesis is about observable entities. But why should observability matter in this context?

CE: The answer depends, of course, on what you mean by 'observable'. As we noted above, in one sense everything is observable: there might be some creature with sense organs appropriate for perceiving it. But, as we agreed when discussing this point, here a more restricted sense of 'observable' is appropriate. Specifically, observability must be taken as a function of certain empirical limitations of human beings. "The human organism is, from the point of view of physics, a certain kind of measuring apparatus. As such it has certain inherent limitations—which will be described in detail in the final physics and biology. It is these limitations to which the 'able' in 'observable' refers—our limitations, *qua* human beings." [15]

SR: I agree with this construal of 'observable'. But my question is, What does observability in this sense have to do with the existence or nonexistence of an entity? You're surely not so much of a positivist as to deny the possibility of the existence of what's unobservable?

CE: You're misconstruing my point. Of course observability in the sense we're taking it has nothing to do with the existence or the nonexistence of an entity. But it has a great deal to do with what we have reason to believe exists. "The question is . . . how much we shall believe when we accept a scientific theory. What is the proper form of acceptance: belief that the theory, as a whole, is true; or something else? To this question, what is observable by us seems eminently relevant." And my answer to the question is this: "To accept a theory is (for us) to believe that it is empirically adequate—that what the theory says *about what is observable* (by us) is true." [16]

SR: Since you refer to what is observable "by us," I take it you make observability relative to an epistemic community, not to individuals?

CE: Of course. The dimmer component of the double star in the Big Dipper's handle is observable because some sharp-eyed people can see it, even if most of us can't.

15. Van Fraassen, *The Scientific Image*, p. 17.
16. Ibid., p. 18.

SR: But then I have a problem. You surely must admit the possibility that our epistemic community might be enlarged, say by the inclusion of animals or extraterrestrials capable of observing things that we now can't observe. For example, we might encounter space travellers who, when we tell them our theories about electrons, say, "Of course, we see them all the time." If this happened, your principles would require that we then, for the first time, accept the existence of electrons. But this seems absurd. Why should the testimony of these aliens be decisive when the overwhelming evidence of our science was not? More generally, isn't it absurd to say that, just because our epistemic community has been enlarged in this way, our beliefs about what there is should change?

CE: Not at all. Your objection has weight only if we believe that "our epistemic policies should give the same results independent of our beliefs about the range of evidence accessible to us."[17] But I see absolutely no reason to believe this. On the contrary, it seems to me that to deny that what evidence is accessible is relevant to what we should believe is to open the door to skepticism or irrationalism.

SR: It seems to me you're equivocating on the expression "evidence accessible to us." Of course such evidence is relevant to our beliefs if it means "the evidence that we are in fact aware of." But this isn't what you mean here. Rather, you're saying that our beliefs ought to depend on the range of evidence that *we might* have even if we don't. Specifically, you're saying that believing that an entity exists ought to depend on whether or not there *could be* direct observations of its existence. Of course, actual observations of an entity are relevant to belief in its existence. And it's also true that, since evidence of actual observations of unobservable entities is not available, it's often harder to produce an adequate case for the existence of such things. But what reason could you have for thinking that the mere question of whether or not an entity could in principle be observed by us is decisive for the question of whether we ought to believe that it exists?

CE: What reason do you have for thinking this isn't the case?

SR: Well, consider an example. Suppose astronomical theory postulates the existence of a far distant star that has not been observed but which we have every reason to think we could observe if we were close enough. It might be, for example, that the star has been postulated as the much smaller double of a known star to explain certain anomalies in its motion. I suppose that if the evidence supporting this postulation is strong enough, you will agree that we have good reason to believe in the existence of this star.

CE: Of course, since it is in principle observable.

SR: All right. But suppose further that, entirely independent of astro-

17. Ibid.

nomical investigations, physiological studies subsequently show that there are previously unknown limits on human powers of observation that make the postulated star unobservable. We may, for example, have assumed that the star was observable because it emitted light in the visible spectrum and so could be seen if we got close enough to it. But physiologists might discover that the visible spectrum is not continuous, that there are small "holes" of invisibility corresponding to specific wave lengths, one of which is that emitted by the star. On your principles, such a physiological result would require us to abandon our conclusion that the star exists, even though all the empirical evidence that led us to postulate it remains the same. But surely such a move would be unreasonable; whether or not the star is observable does not alter the evidence in favor of its existence.

Notice that I'm not saying that observations we in fact have made are not relevant to our beliefs about what exists. But the mere fact that something is observable does not give us any reason to think that it ever has or will in fact be observed. The issue between us is whether mere observability—as distinct from actual observation—is relevant to our beliefs about what exists. I submit that it is not.

Another difficulty for your view derives from the fact that an observable entity may have unobservable properties. The sun, for example, is observable but the temperature of its interior is not. What then is your attitude toward the claim that the temperature at the sun's center is about 20 million degrees Centigrade? It would be odd not to accept it: After all, the claim is very well supported by a calculation based on observed facts (the average temperature of the earth, etc.). If these observed facts were appropriately different, the calculation would yield a temperature of the center of the sun that is observable (e.g., about 10 degrees Centigrade). Since you would accept the truth of the result of the calculation in the latter case, it's hard to see how you could coherently not accept its truth in the former case. But, if you accept the claim that the temperature of the sun's center is about 20 million degrees, then you've implicitly given up your principle that observability is relevant to the justification of existence claims.

CE: Not necessarily. The principle might distinguish between unobservable entities and unobservable properties of observable entities.

SR: Possibly. But then we'd need an explanation of why such a distinction is epistemically relevant.

CE: Of course, but you can see where this would lead. To respond to this objection—and your other one about the star that turns out to be unobservable—in a convincing way would require a very elaborate excursion into the theory of knowledge. "But we cannot settle the major questions of epistemology *en passant* in philosophy of science."[18] I'll just acknowledge

18. Ibid., p. 19.

the relevance of your objections but maintain that more careful epistemological analysis would disarm them. Furthermore, even if I can't answer your objections and your argument stands, remember that the argument is only inductive. This means that, even if I can't directly refute it, I might be able to blunt its force by pointing to overriding considerations that make realism implausible. There is, for example, the fact that, for any theory whose ontology you propose to accept, we can always formulate another theory with a different ontology that is just as well supported by the evidence as the theory you favor. Also, there's the strong historical evidence that scientific postulations are not converging to any single picture of what the unobservable world is like. Until you've dealt with these historical and logical objections to realism, you can't be content with your case for realism.

SR: I agree that the issue isn't fully settled, but your remarks strike me as a strategic retreat that, at least for the present, leaves me in control of the battlefield.

7   Do We See through a Microscope?

Ian Hacking

A couple of years ago I was discussing scientific realism with Dr. Jal Parakh, a biologist from Western Washington University. We had talked about many of the things that philosophers find important. He diffidently added that, from his point of view, a main reason for believing in the existence of entities postulated by theory is that we have evolved better and better ways of actually seeing them. I began to protest against this naive instinct that bypasses the philosophical issues, but I had to stop. Isn't what he says right?

Last fall, during a lecture in Stanford University's "Microscopy for Biologists" course, the professor, Dr. Paul Green, casually remarked that "X-ray diffraction microscopy is now the main interface between atomic structure and the human mind." Dr. Green is a nuts and bolts man, not given to philosophizing. Philosophers of science who discuss realism and anti-realism must needs know a little about the instruments that inspire such eloquence. What follows is a first start, which limits itself to biology and which hardly gets beyond the light microscope. Even that is a marvel of marvels which, I suspect, not many philosophers well understand. Microscopes do not work in the way that most untutored people suppose. But why, it may be asked, should a philosopher care how they work? Because a correct understanding is necessary to elucidate problems of scientific realism as well as answering the question posed by my title. Our philosophical literature is full of intricate accounts of causal theories of perception, yet they have curiously little to do with real life. We have fantastical descriptions of aberrant causal chains which, Gettier-style, call in question this or that conceptual analysis. But the modern microscopist has far more amazing tricks than the most imaginative of armchair students of perception. What we require in philosophy is better awareness of the truths that are stranger than fictions. We ought to have some understanding of those astounding physical systems "by whose augmenting power we now see more / than all the world has ever done before. [1]

Previously published in *Pacific Philosophical Quarterly* 62 (October 1981): 305–22. Reproduced by permission.
1. From a poem, "In Commendation of the Microscope," by Henry Powers, 1664. Quoted in Saville Bradbury, *The Microscope, Past and Present* (Oxford: Pergammon, 1968).

The Great Chain of Being

Philosophers have written dramatically about telescopes. Galileo himself invited philosophizing when he claimed to see the moons of Jupiter, assuming that the laws of vision in the celestial sphere are the same as those on earth. Paul Feyerabend has used that very case to urge that great science proceeds as much by propaganda as by reason: Galileo was a con man, not an experimental reasoner. Pierre Duhem used the telescope to present his famous thesis that no theory need ever be rejected, for phenomena that don't fit can always be accommodated by changing auxiliary hypotheses (if the stars aren't where theory predicts, blame the telescope, not the heavens). By comparison the microscope has played a humble role, seldom used to generate philosophical paradox. Perhaps this is because everyone expected to find worlds within worlds here on earth. Shakespeare is merely an articulate poet of the great chain of being when he writes of Queen Mab and her minute coach "drawn with a team of little atomies . . . her waggoner, a small grey coated gnat not half so big as a round little worm prick'd from the lazy finger of a maid." [2] One expected tinies beneath the scope of human vision. When dioptric glasses were to hand, the laws of direct vision and refraction went unquestioned. That was a mistake. I suppose no one understood how a microscope works before Ernst Abbe (1840–1905). One immediate reaction, by a president of the Royal Microscopical Society, and quoted for years in many editions of the standard American textbook on microscopy, was that we do not, after all, see through a microscope. The theoretical limit of resolution

[A] Becomes explicable by the research of Abbe. It is demonstrated that microscopic vision is *sui generis*. There is and there can be *no* comparison between microscopic and macroscopic vision. The images of minute objects are not delineated microscopically by means of the ordinary laws of refraction; they are not *dioptical* results, but depend entirely on the laws of *diffraction*. [3]

I think that means that we do not see, in any ordinary sense of the word, with a microscope.

Philosophers of the Microscope

Every twenty years or so a philosopher has said something about microscopes. As the spirit of logical positivism came to America, one could read Gustav Bergman telling us that as he used philosophical terminology,

2. W. Shakespeare, *Romeo and Juliet*, 1.4.58, 66–67.
3. William B. Carpenter, *The Microscope and Its Revelations*, 8th ed., revised by W. H.

. . . microscopic objects are not physical things in a literal sense, but merely by courtesy of language and pictorial imagination. . . . When I look through a microscope, all I see is a patch of color which creeps through the field like a shadow over a wall.[4]

In due course Grover Maxwell, denying that there is any fundamental distinction between observational and theoretical entities, urged a continuum of vision: "looking through a window pane, looking through glasses, looking through binoculars, looking through a low power microscope, looking through a high power microscope, etc."[5] Some entities may be invisible at one time and later, thanks to a new trick of technology, they become observable. The distinction between the observable and the merely theoretical is of no interest for ontology.

Grover Maxwell was urging a form of scientific realism. He rejected any anti-realism that holds that we are to believe in the existence of only the observable entities that are entailed by our theories. In his new anti-realist book *The Scientific Image*, Bas van Fraassen strongly disagrees. He calls his philosophy constructive empiricism. He holds that "*Science aims to give us theories which are empirically adequate; and acceptance of a theory involves as belief only that it is empirically`adequate.*"[6] Six pages later he attempts this gloss: "To accept a theory is (for us) to believe that it is empirically adequate—that what the theory says *about what is observable* (by us) is true." Clearly then it is essential for van Fraassen to restore the distinction between observable and unobservable. But it is not essential to him, exactly where we should draw it. He grants the "observable" is a vague term whose extension itself may be determined by our theories. At the same time he wants the line to be drawn in the place which is, for him, most readily defensible, so that even if he should be pushed back a bit in the course of debate, he will still have lots left on the "unobservable" side of the fence. He distrusts Grover Maxwell's continuum and tries to stop the slide from seen to inferred entities as early as possible. He quite rejects the idea of a continuum.

There are, says van Fraassen, two quite distinct kinds of case arising from Grover Maxwell's list. You can open the window and see the fir tree

Dallinger, London and Philadelphia, 1899. Quoted in S. H. Gage, *The Microscope*, 9th ed. (Ithaca: Comstock), 21. Gage contrasts the alternative theory that microscopic vision "is with the unaided eye, the telescope and the photographic camera. This is the original view, and the one which many are favoring at the present day."

4. G. Bergmann, "Outline of an Empiricist Philosophy of Physics," *American Journal of Physics*, 11 (1943):248–58, 335–42. Reprinted in *Readings in the Philosophy of Science*, ed. H. Feigl and M. Brodbeck (New York: Appleton-Century-Crofts, 1953).

5. G. Maxwell, "The Ontological Status of Theoretical Entities," in *Minnesota Studies in the Philosophy of Science*, vol. 3, ed. H. Feigl and G. Maxwell (Minneapolis: University of Minnesota Press, 1962), 3–27.

6. B. C. van Fraassen, *The Scientific Image* (Oxford: Clarendon Press, 1980), 12.

directly. You can walk up to at least some of the objects you see through binoculars, and see them in the round, with the naked eye. (Evidently he is not a bird watcher.) But there is no way to see a blood platelet with the naked eye. The passage from a magnifying glass to even a low powered microscope is the passage from what we might be able to observe with the eye unaided, to what we could not observe except with instruments. Van Fraassen concludes that we do not see through a microscope. Yet we see through some telescopes. We can go to Jupiter and look at the moons, but we cannot shrink to the size of a paramecium and look at it. He also compares the vapor trail made by a jet and the ionization track of an electron in a cloud chamber. Both result from similar physical processes, but you can point ahead of the trail and spot the jet, or at least wait for it to land, but you can never wait for the electron to land and be seen.

Taking van Fraassen's view to the extreme you would say that you have observed or seen something by the use of an optical instrument only if human beings with fairly normal vision could have seen that very thing with the naked eye. The ironist will retort: "What's so great about 20-20 human vision?" It is doubtless of some small interest to know the limits of the naked eye, just as it is a challenge to climb a rock face without pitons or Everest without oxygen. But if you care chiefly to get to the top you will use all the tools that are handy. Observation, in my book of science, is not passive seeing. Observation is a skill. Any skilled artisan cares for new tools. I elsewhere use Caroline Herschel to illustrate the supremely skilled observer.[7] She discovered more comets than anyone, using a rather simple tool, a sky sweeper, and was backed up by the telescopes of her brother William Herschel. Our confidence that she saw comets has, contrary to van Fraassen, nothing to do with a fiction of getting up close and seeing that they are indeed comets—that's still impossible. To understand whether she was seeing, or whether one sees through the microscope, one needs to know quite a lot about the tools.

Don't Just Peer: Interfere

Philosophers tend to look on microscopes as black boxes with a light source at one end and a hole to peer through at the other. There are, as Grover Maxwell puts it, low power and high power microscopes, more and more of the same kind of thing. That's not right, nor are microscopes just for looking through. In fact a philosopher will certainly not see through a microscope until he has learned to use several of them. Asked to draw what

7. I. Hacking, "Spekulation, Berechnung und die Erschaffung von Phänomemen," in *Versuchungen: Aufsätze zur Philosophie Paul Feyerabends*, ed. P. Duerr (Frankfurt: Suhrkamp 1981), 2 Band, 126–58, esp. p. 134.

he sees he may, like James Thurber, draw his own reflected eyeball, or, like Gustav Bergman, see only "a patch of color which creeps through the field like a shadow over a wall." He will certainly not be able to tell a dust particle from a fruit fly's salivary gland until he has started to dissect a fruit fly under a microscope of modest magnification.

That is the first lesson: you learn to see through a microscope by doing, not just by looking. There is a parallel to Berkeley's *New Theory of Vision*, according to which we have three dimensional vision only after learning what it is like to move around in the world and intervene in it. Tactile sense is correlated with our allegedly two dimensional retinal image, and this learned cueing produces three dimensional perception. Likewise a scuba diver learns to see in the new medium of the oceans only by swimming around. Whether or not Berkeley was right about primary vision, new ways of seeing, acquired after infancy, involve learning by doing, not just passive looking. The conviction that a particular part of a cell is there as imaged is, to say the least, reinforced when, using straightforward physical means, you microinject a fluid into just that part of the cell. We see the tiny glass needle—a tool that we have ourselves hand crafted under the microscope—jerk through the cell wall. We see the lipid oozing out of the end of the needle as we gently turn the micrometer screw on a large, thoroughly macroscopic, plunger. Blast! Inept as I am, I have just burst the cell wall, and must try again on another specimen. John Dewey's jeers at the "spectator theory of knowledge" are equally germane for the spectator theory of microscopy.

This is not to say that practical microscopists are free from philosophical perplexity. I quote from the most thorough of available textbooks intended for biologists:

[B] The microscopist can observe a familiar object in a low power microscope and see a slightly enlarged image which is "the same as" the object. Increase of magnification may reveal details in the object which are invisible to the naked eye; it is natural to assume that they, also, are "the same as" the object. (At this stage it is necessary to establish that detail is not a consequence of damage to the specimen during preparation for microscopy.) But what is actually implied by the statement that "the image is the same as the object"?

Obviously the image is a purely optical effect. . . . The "sameness" of object and image in fact implies that the physical interactions with the light beam that render the object visible to the eye (or which would render it visible, if large enough) are identical with those that lead to formation of an image in the microscope. . . .

Suppose however, that the radiation used to form the image is a beam of ultraviolet light, x-rays, or electrons, or that the microscope employs some device which converts differences in phase to changes in intensity. The image then cannot possibly be "the same" as the object, even in the limited

sense just defined! The eye is unable to perceive ultraviolet, x-ray, or electron radiation, or to detect shifts of phase beween light beams. . . .
   This line of thinking reveals that the image must be *a map of interactions
between the specimen and the imaging radiation.*[8]

The author goes on to say that all of the methods she has mentioned, and
more, "can produce 'true' images which are, in some sense, 'like' the
specimen." She also remarks that in a technique like the radioautogram
"one obtains an 'image' of the specimen . . . obtained exclusively from the
point of view of the location of radioactive atoms. This type of 'image' is so
specialized as to be, generally, uninterpretable without the aid of an additional image, the photomicrograph, upon which it is superposed."

   This microscopist is happy to say that we see through a microscope only
when the physical interactions of specimen and light beam are "identical"
for image formation in the microscope and in the eye. Contrast my quotation [A] from an earlier generation, which holds that since the ordinary
light microscope works by diffraction, it is not the same as ordinary vision
but is *sui generis.* Can microscopists [A] and [B] who disagree about the
simplest light microscope possibly be on the right philosophical track
about "seeing"? The scare quotes around "image" and "true" suggest
more ambivalence in [B]. One should be especially wary of the word "image" in microscopy. Sometimes it denotes something at which you can
point, a shape cast on a screen, a micrograph, or whatever. But on other
occasions it denotes as it were the input to the eye itself. The conflation
results from geometrical optics, in which one diagrams the system with a
specimen in focus and an "image" in the other focal plane, where the "image" indicates what you will see if you place your eye there. I do resist one
inference that might be drawn even from quotation [B]. It may seem that
any statement about what is seen with a microscope is theory-loaded:
loaded with the theory of optics or other radiation. I disagree. One needs
theory to make a microscope. You do not need theory to use one. Theory
may help to understand why objects perceived with an interference-
contrast microscope have asymmetric fringes around them, but you can
learn to disregard that effect quite empirically. Hardly any biologists know
enough optics to satisfy a physicist. Practice—and I mean in general doing, not looking—creates the ability to distinguish between visible artefacts of the preparation or the instrument, and the real structure that is
seen with the microscope. This practical ability breeds conviction. The
ability may require some understanding of biology, although one can find
first class technicians who don't even know biology. At any rate physics is
simply irrelevant to the biologist's sense of microscopic reality. His ob-

8. E. M. Slayter, *Optical Methods in Biology* (New York: Wiley, 1970), 261–63.

servations and manipulations seldom bear any load of physical theory at all.

Bad Microscopes

I have encountered the impression that Leeuwenhoek invented the microscope, and that since then people have gone on to make better and better versions of the same kind of thing. I would like to correct that idea.

Leeuwenhoek, hardly the first microscopist, was a technician of genius. His scopes had a single lens, and he made a lens for each specimen to be examined. The object was mounted upon a pin at just the right distance. We don't quite know how he made such marvellously accurate drawings of his specimens. The most representative collection of his lenses-plus-specimen was given to the Royal Society in London, which lost the entire set after a century or so in what are politely referred to as suspicious circumstances. But even by that time the glue for his specimens had lost its strength and the objects had begun to fall off their pins. Almost certainly Leeuwenhoek got his marvelous results thanks to a secret of illumination rather than lens manufacture, and he seems never to have taught the public his technique. Perhaps Leeuwenhoek invented dark field illumination, rather than the microscope. That guess should serve as the first of a long series of possible reminders that many of the chief advances in microscopy have had nothing to do with optics. We have needed microtomes to slice specimens thinner, aniline dyes for staining, pure light sources, and, at more modest levels, the screw micrometer for adjusting focus, fixatives and centrifuges.

Although the first microscopes did create a terrific popular stir by showing worlds within worlds, it is important to note that after Hooke's compound microscope, the technology did not markedly improve. Nor did much new knowledge follow after the excitement of the initial observations. The microscope became a toy for English ladies and gentlemen. The toy would consist of a microscope and a box of mounted specimens from the plant and animal kingdom. Note that a box of mounted slides might well cost more than the purchase of the microscope itself. You did not just put a drop of pond water on a slip of glass and look at it. All but the most expert would require a ready mounted slide to see *anything*. Indeed considering the optical aberrations it is amazing that anyone ever did see anything through a compound microscope, although in fact, as always in experimental science, a really skillful technician can do wonders with awful equipment.

There are about eight chief abberations in bare-bones light microscopy. Two important ones are spherical and chromatic. The former is the result of the fact that you polish a lens by random rubbing. That, as can be proven, gives you a spherical surface. A light ray travelling at a small angle to the axis will not focus at the same point as a ray closer to the axis. For angles i for which $\sin i$ differs at all from i we get no common focus of the light rays, and so a point on the specimen can be seen only as a smear through the microscope. This was well understood by Huygens, who also knew how to correct it in principle, but practical combinations of concave and convex lenses to avoid spherical aberration were a long time in the making.

Chromatic aberrations are caused by differences in wave length between light of different colors. Hence red and blue light emanating from the same point on the specimen will come to focus at different points. A sharp red image is superimposed on a blue smear or vice versa. Although rich people liked to have a microscope about the house for entertainments, it is no wonder that serious science had nothing to do with the instrument. We often regard Bichat as the founder of histology, the study of living tissues. In 1800 he would not allow a microscope in his lab.

> When people observe in conditions of obscurity each sees in his own way and according as he is affected. It is, therefore, observation of the vital properties that must guide us rather than the blurred images provided by the best of microscopes.[9]

No one tried very hard to make achromatic microscopes, because Newton had written that they are physically impossible. They were made possible by the advent of flint glass, with refractive indices different from that of ordinary glass. A doublet of two lenses of different refractive indices can be made to cancel out the aberration perfectly for a given pair of red and blue wave lengths, and although the solution is imperfect over the whole spectrum, it is pretty negligible and can be improved by a triplet of lenses. The first person to get the right ideas was so secretive that he sent the specifications for the lenses of different kinds of glass to two different contractors. They both subcontracted with the same artisan, who then formed a shrewd guess that the lenses were for the same device. In due course, in 1758, the idea was pirated. A court case for the patent rights was decided in favor of the pirate, John Doland. The High Court Judge ruled:

> It was not the person who locked the invention in his scritoire that ought to profit by a patent for such an invention, but he who brought it forth for the benefit of the public.[10]

9. X. Bichat, *Anatomie générale appliquéé á la physiologie et à la médecine* (Paris: Brosson, Gaber et cie, 1801), 51.

10. Quoted in Bradbury (note 1), 130.

The public did not benefit all that much. Even up in to the 1860s there were serious debates as to whether globules seen through a microscope were artefacts of the instrument or genuine elements of living material. (They were artefacts.) Microscopes did get better and aids to microscopy improved at rather a greater rate. If we draw a graph of development we get a first high around 1660, then a slowly ascending plateau until a great leap around 1870; the next great period, which is still with us, commences about 1945. An historian has plotted this graph with great precision, using as a scale the limits of resolution of surviving instruments of different epochs.[11] Making a subjective assessment of great applications of the microscope, we would draw a similar graph, except that the 1870/1660 contrast would be greater. Few truly memorable facts were found out with a microscope until after 1860. The surge of new microscopy is partly due to Abbe, but the most immediate cause of advance was the availability of aniline dyes for staining. Living matter is mostly transparent. The new aniline dyes made it possible for us to see microbes and much else.

Abbe and Diffraction

How do we "normally" see? Mostly we see reflected light. But if we are using a magnifying glass to look at a specimen illumined from behind, then it is transmission, or absorption, that we are "seeing." So we have the following idea: to see something through a light microscope is to see patches of dark and light corresponding to the proportions of light transmitted or absorbed. We see changes in the amplitude of light rays. I think that even Huygens knew there is something wrong with this conception, but not until 1873 could one read in print how a microscope works.[12]

Ernest Abbe provides the happiest example of a rags to riches story. Son of a spinning-mill workman, he yet learned mathematics and was sponsored through the Gymnasium. He became a lecturer in mathematics, physics, and astronomy. His optical work led him to associate. He was taken on by the small firm of Carl Zeiss in Jena, and when Zeiss died he became an owner; he retired to a life of philanthropy. Innumerable mathematical and practical innovations by Abbe turned Carl Zeiss into the greatest of optical firms. Here I consider only one.

Abbe was interested in resolution. Magnification is worthless if it "magnifies" two distinct dots into one big blur. One needs to resolve the dots into two distinct images. G. B. Airy, the English Astronomer Royal, had seen the point already when considering the properties of a telescope needed to

11. S. Bradbury and G. L'E. Turner, eds., *Historical Aspects of Microscopy* (Cambridge: Heffer 1967).

12. E. Abbe, "Beitrage zur Theorie des Mikroscop und der mikroskopische Wahrnemung."

distinguish twin stars. It is a matter of diffraction. The most familiar example of diffraction is the fact that shadows of objects with sharp boundaries are fuzzy. This is a consequence of the wave character of light. When light travels between two narrow slits, some of the beam may go straight through, but some of it will bend off at an angle to the main beam, and some more will bend off at a larger angle: these are the first-order, second-order, etc., diffracted rays.

Abbe took as his problem how to resolve (i.e., visibly distinguish) parallel lines on a diatom. These lines are very close together and of almost uniform separation and width. He was soon able to take advantage of even more regular artificial diffraction gratings. His analysis is an interesting example of the way in which pure science is applied, for he worked out the theory for the pure case of looking at a diffraction grating, and inferred that this represents the infinite complexity of the physics of seeing a heterogeneous object with a microscope.

When light hits a diffraction grating most of it is diffracted rather than transmitted. It is emitted from the grating at the angle of first, second, or third order diffractions, where the angles of the diffracted rays are in part a function of the distances between the lines on the grating. Abbe realized that in order to see the slits on the grating, one must pick up not only the transmitted light, but also at least the first order diffracted ray. What you see, in fact, is best represented as a Fourier synthesis of the transmitted and the diffracted rays. Thus according to Abbe the image of the object is produced by the interference of the light waves emitted by the principle image, and the secondary images of the light source which are the result of diffraction.

Practical applications abound. Evidently you will pick up more diffracted rays by having a wider aperture for the objective lens, but then you obtain vastly more spherical aberration as well. Instead you can change the medium between the specimen and the lens. With something denser than air, as in the oil immersion microscope, you capture more of the diffracted rays within a given aperture and so increase the resolution of the microscope.

Even though the first Abbe-Zeis microscopes were good, the theory was resisted for a number of years, particularly in England and America, who had enjoyed a century of dominating the market. Even by 1910 the very best English microscopes, built on purely empirical experience, although stealing a few ideas from Abbe, could resolve as well or better than the Zeiss equipment. The expensive craftsmen with trial and error skills were doomed. It was not, however, only commercial or national rivalry which made some people hesitate to believe Abbe. In an American textbook of 1916 I find it stated that an alternative (and more "common sense") theory of "ordinary" vision is now once again in the ascendant and

will soon scuttle Abbe![13] Resistance arose partly from surprise at what
Abbe asserted, with the apparent consequence that, as quotation [A] has
it, "there is and can be *no* comparison between microscopic and macroscopic vision."

If you hold (as my more modern quotation [B] still seems to hold), that
what we see is essentially a matter of a certain sort of physical processing in
the eye, then everything else must be more in the domain of optical illusion or at best of mapping. On that account the systems of Leeuwenhoek
and of Hooke do allow you to see. After Abbe even the conventional light
microscope is essentially a Fourier synthesizer of first or even second order
diffractions. Hence you must modify your notion of seeing or hold that
you never see through a serious microscope. Before reaching a conclusion
on this question, we had best examine some more recent instruments.

A Plethora of Scopes

We move on to after World War II. Most of the ideas had been around
during the interwar years, but did not get beyond prototypes until later.
One invention is a good deal older, but it was not properly exploited for a
while.

The first practical problem for the cell biologist is that most living material does not show up under an ordinary light microscope because it is
transparent. To see anything you have to stain the specimen. Aniline dyes
are the world's number one poison, so what you will see is a very dead cell,
which is also quite likely to be a structurally damaged cell, exhibiting
structures that are an artefact of the preparation. However it turns out that
living material varies in its birefringent (polarizing) properties. So let us
incorporate into our scope a polarizer and an analyzer. The polarizer trans-
mits to the specimen only polarized light of certain properties. In the
simplest case, let the analyzer be placed at right angles to the polarizer, so
as to transmit only light of polarization opposite to that of the polarizer.
The result is total darkness. But suppose the specimen is itself birefringent;
it may then change the plane of polarization of the incident light, and so a
visible image may be formed by the analyzer. "Transparent" fibers of

---

13. S. H. Gage (note 3), 11th edition, 1916. The direct quotation from Carpenter and
Dallinger is dropped in the 12th edition of 1917, but the spirit is retained, including the *"sui
generis."* Gage does admit that "Certain very striking experiments have been devised to show
the accuracy of Abbe's hypothesis, but as pointed out by many, the ordinary use of the microscope never involves the conditions realized in these experiments" (page 301). How Imre
Lakatos would have delighted in this degenerating programme of preserving the naive picture
of vision, complete with "monster-barring" of the striking experiments! This passage remained unchanged in essentials even in the 17th edition of 1941.

striated muscle may be observed in this way, without any staining, and relying solely on certain properties of light that we do not normally "see."

Abbe's theory of diffraction, augmented by the polarizing microscope, leads to something of a conceptual revolution. We do not have to see using the "normal" physics of seeing in order to perceive structures in living material. In fact we never do. Even in the standard case we synthesize diffracted rays rather than seeing the specimen by way of "normal" visual physics. Then the polarizing microscope reminds us that there is more to light than refraction, absorption and diffraction. *We could use any property of light that interacts with a specimen in order to study the structure of the specimen.* Indeed we could use any property of *any kind of wave* at all.

Even when we stick to light there is lots to do. Ultraviolet microscopy doubles resolving power, although its chief interest lies in noting the specific ultraviolet absorptions that are typical of certain biologically important substances. In fluorescence microscopy the incident illumination is cancelled out, and one observes only light re-emitted at different wave lengths by natural or induced phosphorescence or fluorescence. This is an invaluable histological technique for certain kinds of living matter. More interesting, however, than using unusual modes of light transmission or emission are the games we can play with light itself: the Zelnicke phase-contrast microscope and the Nomarski interference microscope.

A specimen that is transparent is uniform with respect to light absorption. It may still possess invisible differences in refractive index in various parts of its structure. The phase contrast microscope converts these into visible differences of intensity in the image of the specimen. In an ordinary microscope the image is synthesized from the diffracted waves D and the directly transmitted waves U. In the phase contrast microscope the U and D waves are physically separated in an ingenious although physically simple way, and one or the other kind of wave is then subject to a standard phase delay which has the effect of producing in focus phase contrast corresponding to the differences in refractive index in the specimen.

The interference contrast microscope is perhaps easier to understand. The light source is simply split by a half silvered mirror, and half the light goes through the specimen while half is kept as an unaffected reference wave to be recombined for the output image. Changes in optical path due to different refractive indices within the specimen thus produce interference effects with the reference beam.

The interference microscope is attended by illusory fringes but is particularly valuable because it provides a quantitative determination of refractive indices within the specimen. Naturally once we have such devices in hand, endless variations may be constructed, such as polarizing interference microscopes, multiple beam interference, phase modulated interference, and so forth.

Truth in Microscopy

> The differential interference-contrast technique is distinguished by the follow-
> ing characteristics: Both clearly visible outlines (edges) within the object and
> continuous structures (striations) are imaged in their true profile.

So says a Carl Zeiss sales catalogue to hand. What makes the enthusiastic
sales person suppose that the images produced by these several optical systems are "true"? Of course, the images are "true" only when one has
learned to put aside distortions. There are many grounds for the conviction that a perceived bit of structure is real or true. One of the most natural
is the most important. I shall illustrate it with my own first experience in
the laboratory.[14] Low powered electron microscopy reveals small dots in
red blood cells. These are called dense bodies: that means simply that they
are electron dense, and show up on a transmission electron microscope
without any preparations or staining whatsoever. On the basis of the movements and densities of these bodies in various stages of cell development or
disease, it is guessed that they may have an important part to play in blood
biology. On the other hand they may simply be artefacts of the electron
microscope. One test is obvious: can one see these selfsame bodies using
quite different physical techniques? In this case the problem is fairly readily solved. The low resolution electron microscope is about the same power
as a high resolution light microscope. The dense bodies do not show up
under every technique, but are revealed by fluorescent staining and subsequent observation by the fluorescent microscope.

Slices of red blood cell are fixed upon a microscopic grid. This is literally a grid: when seen through a microscope one sees a grid each of whose
squares is labelled with a capital letter. Electron micrographs are made of
the slices mounted upon such grids. Specimens with particularly striking
configurations of dense bodies are then prepared for fluorescence microscopy. Finally one compares the electron micrographs and the fluorescence
micrographs. One knows that the micrographs show the same bit of the
cell, because this bit is clearly in the square of the grid labelled P, say.
In the fluorescence micrographs there is exactly the same arrangement of
grid, general cell structure, and of the seven "bodies" seen in the electron
micrograph. It is inferred that the bodies are not an artefact of the electron
microscope.

Two physical processes—electron transmission and fluorescent re-
emission —are used to detect the bodies. These processes have virtually
nothing in common between them. They are essentially unrelated chunks
of physics. It would be a preposterous coincidence if, time and again, two

14. I owe a particular debt of gratitude to my friend R. J. Skaer of Peterhouse, Cambridge, who allowed me to spend a good deal of time in his cell biology laboratory in the
Department of Haematological Medicine, Cambridge University.

completely different physical processes produced identical visual configurations which were, however, artefacts of the physical processes rather
than real structures in the cell.

Note that no one actually produces this "argument from coincidence"
in real life. One simply looks at the two (or preferably more) sets of micrographs from different physical systems, and sees that the dense bodies occur in exactly the same place in each pair of micrographs. That settles the
matter in a moment. My mentor, Dr. Richard Skaer, had in fact expected
to prove that dense bodies are artefacts. Five minutes after examining his
completed experimental micrographs he knew he was wrong.

Note also that no one need have any ideas what the dense bodies *are*. All
we know is that there are some structural features of the cell rendered
visible by several techniques. Microscopy itself will never tell all about
these bodies (if indeed there is anything important to tell). Biochemistry
must be called in. Also, instant spectroscopic analysis of the dense body
into constitutent elements is now available, by combining an electron
microscope and a spectroscopic analyzer. This works much like spectroscopic analyses of the stars.

Coincidence and Explanation

Arguments from coincidence have been put to more general use in discussions of scientific realism. In particular J. J. C. Smart notes that good
theories are used to explain diverse phenomena. It would, he says, be a
cosmic coincidence if the theory were false and yet correctly predicted all
the phenomena:

> One would have to suppose that there were unnumerable lucky accidents about
> the behavior mentioned in the observational vocabulary, so that they behaved
> miraculously *as if* they were brought about by the non-existent things ostensi-
> bly talked about in the theoretical vocabulary.[15]

Van Fraassen challenges this and related arguments for realism that deploy
what Gilbert Harman calls "inference to the best explanation," or what
Hans Reichenbach and Wesley Salmon call the "common cause" argument. So it may seem as if my talk of coincidence puts me in the midst of
an ongoing feud. Not so! My argument is much more localized, and com-
mits me to none of the positions of Smart or Salmon.

First of all, we are not concerned with an observational and theoretical
vocabulary. There may well be no theoretical vocabulary for the things
seen under the microscope—"dense body" means nothing else than some-

---

15. J. J. C. Smart, *Between Science and Philosophy* (New York: Random House, 1968), 150.

thing dense, i.e., that shows up under the electron microscope without any staining or other preparation. Secondly we are not concerned with explanation. We see the same constellations of dots whether we use an electron microscope or fluorescent staining, and it is no "explanation" of this to say that some definite kind of thing (whose nature is as yet unknown) is responsible for the persistent arrangement of dots. Thirdly we have no theory which predicts some wide range of phenomena. The fourth and perhaps most important difference is this: we are concerned to distinguish artefacts from real objects. In the metaphysical disputes about realism, the contrast is between "real although unobservable entity" and "not a real entity, but rather a tool of thought." With the microscope we know there are dots on the micrograph. The question is, are they artefacts of the physical system or are they structures present in the specimen itself? My argument from coincidence says simply that it would be a preposterous coincidence if two totally different kinds of physical systems were to produce exactly the same arrangements of dots on micrographs.

The Argument of the Grid

I now venture a philosopher's aside on the topic of scientific realism. Van Fraassen says we can see through a telescope because although we need the telescope to see the moons of Jupiter when we are positioned on earth, we could go out there and look at the moons with the naked eye. Perhaps that fantasy is close to fulfillment, but it is still science fiction. The microscopist avoids fantasy. Instead of flying to Jupiter he shrinks the visible world. Consider the grid that we used for re-identifying dense bodies. The tiny grids are made of metal; they are barely visible to the naked eye. They are made by drawing a very large grid with pen and ink. Letters are neatly inscribed by a draftsman at the corner of each square on the grid. Then the grid is reduced photographically. Using what are now standard techniques, metal is deposited on the resulting micrograph. Grids are sold in packets, or rather tubes, of 100, 250, and 1,000. The procedures for making such grids are entirely well understood, and as reliable as any other high quality mass production system.

In short, rather than disporting ourselves to Jupiter in an imaginary space ship, we are routinely shrinking a grid. Then we look at the tiny disc through almost any kind of microscope and see exactly the same shapes and letters as were drawn in the large by the first draftsman. It is impossible seriously to entertain the thought that the minute disc, which I am holding by a pair of tweezers, does not in fact have the structure of a labelled grid. I know that what I see through the microscope is veridical because we *made* the grid to be just that way. I know that the process of manufacture is reliable, because we can check the results with the microscope. Moreover we can check the results with any kind of microscope, using any of a dozen unrelated physical processes to produce an image. Can we entertain the possibility that, all the same, this is some gigantic coincidence? Is it false that the disc is, in fine, in the shape of a labelled grid? Is it a gigantic conspiracy of 13 totally unrelated physical processes that the large scale grid was shrunk into some non-grid which when viewed using 12 different kinds of microscopes still looks like a grid? To be an anti-realist about that grid you would have to invoke a malign Cartesian demon of the microscope.

The argument of the grid probably requires a healthy recognition of the disunity of science, at least at the phenomenological level. Light microscopes, trivially, all use light, but interference, polarizing, phase contrast, direct transmission, fluorescence, and so forth all exploit essentially unrelated phenomenological aspects of light. If the same structure can be discerned using many of these different aspects of light waves, we cannot seriously suppose that the structure is an artefact of all the different physical systems. Moreover I emphasize that all these physical systems are made by people. We as it were purify some aspect of nature, isolating, say, the phase interference character of light. We design an instrument knowing in principle exactly how it will work, just because optics is so well understood a science. We spend a number of years debugging several prototypes, and finally have an off-the-shelf instrument, through which we discern a particular structure. Several other off-the-shelf instruments, built upon entirely different principles, reveal the same structure. No one short of the Cartesian sceptic can suppose that the structure is made by the instruments rather than inherent in the specimen.

It was once not only possible but perfectly sensible to ban the microscope from the histology lab on the plain grounds that it chiefly revealed artefacts of the optical system rather than the structure of fibers. That is no longer the case. It is always a problem in innovative microscopy to become convinced that what you are seeing is really in the specimen rather than an artefact of the preparation of the optics. But by 1981, as opposed to 1800, we have a vast arsenal of ways of gaining such conviction. I emphasize only the "visual" side. Even there I am simplistic. I say that if you can see the same fundamental features of structure using several different physical systems, you have excellent reason for saying, "that's real" rather than, "that's an artefact." It is not conclusive reason. But the situation is no different from ordinary vision. If black patches on the tarmac road are seen, on a hot day, from a number of different perspectives, but always in the same location, one concludes that one is seeing puddles rather than the familiar illusion. One may still be wrong. One is wrong, from time to time, in microscopy, too. Indeed the sheer similarity of the kinds of mistakes made in macroscopic and microscopic perception may increase the inclination to say, simply, that one sees through a microscope.

I must repeat that just as in large scale vision, the actual "images" or micrographs are only one small part of the confidence in reality. In a recent lecture the molecular biologist G. S. Stent recalled that in the late forties or early fifties *Life* magazine had a full color cover of an electron micrograph, labelled, excitedly, "the first photograph of the gene." [16] Given the theory, or lack of theory, of the gene at that time, said Stent, the title did not make any sense. Only a greater understanding of what a gene is can bring the conviction of what the micrograph shows. We become convinced of the reality of bands and interbands on chromosomes not just because we see them, but because we formulate conceptions of what they do, what they are for. But in this respect, too, microscopic and macroscopic visions are not different: a Laplander in the Congo won't see much in the bizarre new environment until he starts to get some idea what is in the jungle.

Thus I do not advance the argument from coincidence as the sole basis of our conviction that we see true through the microscope. It is one element, a compelling visual element, that combines with more intellectual modes of understanding, and with other kinds of experimental work. Biological microscopy without practical biochemistry is as blind as Kant's intuitions in the absence of concepts.

The Acoustic Microscope

I here avoid the electron microscope. There is no more "the" electron microscope than "the" light microscope: all sorts of different properties of electron beams are used. A simple but comprehensive explanation requires another essay. In case, however, we have in mind too slender a diet of examples based upon the properties of visible light, let us briefly consider the most disparate kind of radiation imaginable: sound. [17]

Radar, invented for aerial warfare, and sonar, invented for war at sea, remind us that longitudinal and transverse wave fronts can be put to the same kinds of purpose. Ultrasound is "sound" of very high frequency. Ultrasound examination of the foetus *in vitro* has recently won well deserved publicity. Over forty years ago Soviet scientists suggested a microscope using sound of frequency 1000 times greater than audible noise.

16. I think Stent must have been referring to LIFE, 17 March 1947, p. 83.
17. C. F. Quate, "The Acoustic Microscope," *Scientific American* 241 (Oct. 1979), 62–69. R. N. Johnston, A. Atalar, J. Heiserman, V. Jipson, and C. F. Quate, "Acoustic Microscopy: Resolution of Subcellular Detail," *Proceedings of the National Academy of Sciences U.S.A.*, 76 (1979): 3325–29.

Technology has only recently caught up to this idea. Useful prototypes are just now in operation.

The acoustic part of the microscope is relatively simple. Electric signals are converted into sound signals and then, after interaction with the specimen, are reconverted into electricity. The subtlety of present instruments lies in the electronics rather than the acoustics. The acoustic microscope is a scanning device. It produces its images by converting the signals into a spatial display on a television screen, a micrograph, or, when studying a large number of cells, a videotape.

As always a new kind of scope is interesting because of the new aspects of a specimen that it may reveal. Changes in refractive index are vastly greater for sound than for light. Moreover sound is transmitted through objects that are completely opaque. Thus one of the first applications of the acoustic microscope is in metallurgy, and also in detecting defects in silicon chips. For the biologist, the prospects are also striking. The acoustic microscope is sensitive to density, viscosity, and flexibility of living matter. Moreover the very short bursts of sound used by the scanner do not immediately damage the cell. Hence one may study the life of a cell in a quite literal way: one will be able to observe changes in viscosity and flexibility as the cell goes about its business.

The rapid development of acoustic microscopy leaves us uncertain where it will lead. A couple of years ago the research reports carefully denied any competition with electron microscopes; they were glad to give resolution at about the level of light scopes. Now, using the properties of sound in supercooled solids, one can emulate the resolution of electron scopes, although that is not much help to the student of living tissue!

Do we see with an acoustic microscope?

Looking with a Microscope

Do we see through a microscope? Let us first do away with the anachronistic word *through*. Looking through a lens was the first step in technology, then came peering through the tube of a compound microscope. The micrograph is more to the point: we study photographs taken with a microscope. Thanks to the enormous focal length of an electron microscope it is natural to view the image on a large flat surface so everyone can stand around and point to what's interesting. Scanning microscopes necessarily constitute the image on a screen or plate. Any image can be digitized and retransmitted on a television display or whatever. Moreover digitization is marvellous for censoring noise and even reconstituting lost information. Do not, however, become awed by technology. In the study of crystal structure, one good way to get rid of noise is to cut up a micrograph in a systematic way, paste it back together, and rephotograph it for inter-ference contrast.

We do not in general see *through* a microscope; we see with one. But do we *see* with a microscope? It would be silly to debate the ordinary use of the word *see*, a word already put to innumerable uses of an entirely intel-lectual sort. "Now I see the point," and kindred employments in mathe-matics. Or consider how the physicist writes of the hypothetical entities. I quote from a lecture listing twelve fermions, or fundamental constituents of matter, including electron neutrinos, deuterons, etc. We are told that "of these fermions, only the t quark is yet unseen. The failure to observe tt' states in e⁺e⁻ anihilation at PETRA remains a puzzle. . . ."[18] Seeing and observing for this high energy physicist are a long way from the eye. (Probably seeing acquired its peculiar association with ocular vision only at the start of the nineteenth century, as is manifested in the twin doctrines called positivism and phenomenology, the philosophies that say seeing is with the eye, not the mind.)

Consider a device for low-flying jet planes, laden with nuclear weapons, skimming a few dozen yards from the surface of the earth in order to evade radar detection. The vertical and horizontal scale are both of interest to the pilot; he needs both to see a few hundred feet down and miles and miles away. So the visual information is digitized, processed, and cast on a head-up display on the windscreen. The distances are condensed and the altitude is expanded. Does the pilot *see* the terrain? I should say so. It would be foolish to put in some unnatural word like *perceive* to indicate that the seeing employs an instrument. Note that this case is not one in which the pilot could have seen the terrain by getting off the plane and taking a good look. There is no way of getting a look at that much land-scape without an instrument.

Consider the electron diffraction microscope with which I produce im-ages either in conventional space or in reciprocal space. Reciprocal space is, roughly speaking, conventional space turned inside out; near is far and far is near. Crystallographers often find it most natural to study their speci-mens in reciprocal space. Do they see them in reciprocal space? They cer-tainly say so, and thereby call in question the Kantian doctrine of the uniqueness of perceptual space.

How far could one push the concept of seeing? Suppose I take an elec-tronic paint brush and paint on a television screen, an accurate picture (I) of a cell that I have previously studied, say, by using a digitized and recon-stituted image (II). Even if I am "looking at the cell" in case (II), in (I) I

---

18. C. Y. Prescott, "Prospects for Polarized Electrons at High Energies," Stanford Lin-ear Accelerator, SLAC-PUB-2630, Oct. 1980: 5.

am only looking at a drawing of the cell. What is the difference? The important feature is that in (II) there is a direct interaction between a wave source, an object, and a series of physical events that end up in an image of the object. To use quotation [B] once again, in case (II) we have a map of interactions between the specimen and the imaging radiation. If the map is a good one, then (II) is seeing with a microscope.

This is doubtless a liberal extension of the notion of seeing. We see with an acoustic microscope. We see with television, of course. We do not say that we saw an attempted assassination *with* the television, but *on* the television. That is mere idiom, inherited from "I heard it on the radio." We distinguish between seeing the television broadcast live or not. We have endless distinctions to be made with various adverbs, adjectives, and even prepositions. I know of no confusion that will result from talk of seeing with a microscope.

Scientific Realism

When an image is a map of interactions between the specimen and the image of radiation, and the map is a good one, then we are seeing with a microscope. What is a good map? After discarding or disregarding aberrations or artefacts, the map should represent some structure in the specimen in essentially the same two- or three-dimensional set of relationships as are actually present in the specimen.

Does this bear on scientific realism? First let us be clear that it can bear in only the modest way. I do not even argue here for the reality of objects and structure that can be discerned only by the electron microscope (that calls for another essay). I have spoken chiefly of light microscopy. Now imagine a reader initially attracted by van Fraassen, and who thought that objects seen only with light microscopes do not count as observable. That reader could change his mind, and admit such objects into the class of observable entities. This would still leave intact all the main philosophical positions of van Fraassen's anti-realism.

But if we conclude that we see with the light microscopes, does it follow that the objects we report seeing are real? No. For I have said only that we should not be stuck in the nineteenth century rut of positivism-cum-phenomenology, and that we should allow ourselves to talk of seeing with a microscope. Such a recommendation implies a strong commitment to realism about microscopy, but it begs the question at issue. This is clear from my quotation from high-energy physics, with its cheerful talk of our having seen electron neutrinos, deuterons, and so forth. The physicist is a realist, too, and he shows this by using the word *see*, but his usage is no *argument* that there are deuterons. Here perhaps is one source of the philosophers' scepticism of Dr. Parakh's suggestion that one can become a convinced realist because of advances in microscopy.

Does microscopy then beg the question of realism? No. On closer inspection Parakh's suggestion is right. We *are* convinced of the structures that we observe using various kinds of microscopes. Our conviction arises partly from our success at systematically removing aberrations and artefacts. In 1800 there was no such success. Bichat banned the microscope from his dissecting rooms, for one did not, then, observe structures that could be confirmed to exist in the specimens. But now we have by and large got rid of aberrations; we have removed many artefacts, disregard others, and are always on the lookout for undetected frauds. We are convinced about the structures we seem to see because we can interfere with them in quite physical ways, say by microinjecting. We are convinced because instruments using entirely different physical principles lead us to observe pretty much the same structures in the same specimen. We are convinced by our clear understanding of most of the physics used to build the instruments that enable us to see, but this theoretical conviction plays a relatively small part. We are more convinced by the admirable intersections with biochemistry, which confirm that the structures that we discern with the microscope are individuated by distinct chemical properties, too. We are convinced not by a high powered deductive theory about the cell—there is none—but because of a large number of interlocking low level generalizations that enable us to control and create phenomena in the microscope. In short, we learn to move around in the microscopic world. Berkeley's *New Theory of Vision* may not be the whole truth about infantile binocular three-dimensional vision, but is surely on the right lines when we enter the new worlds within worlds that the microscope reveals to us.

8   Surface Dazzle, Ghostly Depths: An Exposition and Critical Evaluation of van Fraassen's Vindication of Empiricism against Realism

Clifford A. Hooker

> To be an empiricist is to withhold belief in anything that goes be-
> yond the actual, observable phenomena, and to recognize no objec-
> tive modality in nature. To develop an empiricist account of science
> is to depict it as involving a search for truth only about the em-
> pirical world, about what is actual and observable. Since scientific
> activity is an enormously rich and complex cultural phenomenon,
> this account of science must be accompanied by auxiliary theories
> about scientific explanation, conceptual commitment, modal lan-
> guage, and much else. But it must involve throughout a resolute
> rejection of the demand for an explanation of the regularities in the
> observable course of nature, by means of truths concerning a real-
> ity beyond what is actual and observable, as a demand which plays
> no role in the scientific enterprise. (Van Fraassen 1980, 202–3)[1]

The *surface* of the title is, of course, the empirical, phenomenal world, the
alleged "surface appearance" of things. *Depths* refers to the real entities
and mechanisms "behind" the appearances, causally giving rise to appearances and explaining them. Realists hold that the surface merely dazzles
and should not beguile anyone into mistaking it for reality. Empiricists,
contrariwise, hold that the depths are ghostly, having no independent rationale for their veneration beyond the surface through which they indirectly appear.

I am a persuaded realist. In this essay I shall critically examine recent
arguments for empiricism, with a view to defending realism. The arguments are presented in a recent book, quoted above; I believe it to be
the most important defense of empiricism for more than a decade. Van
Fraassen calls his version of empiricism *constructive empiricism*, hereafter
referred to as CE.

Prior to CE, in my view, realists had a comparatively easy time of it
against empiricists, whether one considered the matter directly at the level
of theory of science or at the less direct, but deeper level of underlying

---

1. All page references in what follows are to this source unless explicitly stated otherwise.

assumptions. At either level the empiricists were forced to make assertions that were so plainly unbelievable for anyone who knew science and took it seriously that the contest was one-sided. For three decades empiricism has been losing adherents to realism; only the simplicity and definiteness of the empiricist program, and the occasional philosopher dazzled by the surface of appearance or drawn by the sirens of certainty, served to maintain some adherents.

Now CE claims to change this position dramatically. According to van Fraassen, access to new and vastly more powerful logical tools plus a shift toward pragmatic accounts of various aspects of science allows the empiricist to say all those things he needs to be able to say to have a plausible position—and more, for it now allows the empiricist to capture everything of value and interest in the realist's account without commitment to the ghostly depths. Van Fraassen would have empiricism left in triumph as the wholly structurally adequate yet more economical, more rational position.

Moreover, I must confess to thinking van Fraassen correct in much of his technical argumentation. Wherefor, I approach this essay with the caution and sobriety appropriate to admitting that the decades of "easy Realism" are over—though realism remains, I believe, the most interesting and exciting position.

First I will briefly review historic empiricism and its traditional objections. (For a good general review, see Suppe 1974; my own analysis of empiricist structure and assumptions is given in Hooker 1975c, from which I draw here. Both sources contain substantial bibliographies.) Then I will outline the basic strategy and technical achievements of CE. The remainder of the essay will be taken up with a critical assessment of CE vis-à-vis realism.

II

Empiricists aim to base knowledge on facts and confine knowledge to the facts. Depending upon the austerity of the doctrine, it is deductive logic or a careful enlargement, inductive logic, which provides the only connection defining the "ascents" from the facts to the full panoply of scientific knowledge. At its crudest and most general level, the following model of science emerges from empiricist studies of recent years:

1. The aim of science is to maximize the number of assertions known to be true.
2. The assertions of science are expressible in the (formal) language employed in making elementary factual claims, and scientific theories are deductively axiomatized classes of elementary factual claims and logical functions thereof.

3. The method of science is to maximize the number of elementary factual claims and to accept only those theories which bear a formal logical relationship (inductive or deductive) to the data base.
4. The history of science is reconstructed as the continuous accumulation of data and the periodic replacement of theories by more accurate and/or more general successors.
5. The choice between competing scientific theories is value-neutral; it is based on objective epistemic criteria alone.
6. All other judgments (e.g., moral, social, political) are either reducible to factual judgments or are noncognitive (because they are nonverifiable or nontestable).

The formal language is the backbone of empiricism; everything else is structured by its expressive resources. Historically, the contemporary empiricist movement first clearly emerged as positivism, which espoused a very austere language, the simple logic of finite truth functions wedded to predication, names, and a class of immediately observationally verifiable descriptive terms. In these terms, very little could be (cognitively) meaningfully said. Later empiricisms successively enlarged the logic but otherwise tried to leave the positivist framework untouched (although some alterations were inevitable). They hoped that richer logical structures would offer richer, hopefully more realistic, descriptions of science while leaving intact the essential positivist philosophical program for interpreting science.

In a little more detail, the positivist position can be regarded as a set of philosophical theses, generally as follows:

D1. The epistemic goal of a rational man is the acquisition of knowledge, understood to be certain truth.
D2. Rationally acceptable argument is restricted to the domain of deductive logic.
D3. Every cognitively meaningful sentence is a finite truth function of cognitively meaningful descriptive terms.
D4. Every cognitively meaningful descriptive term may be logically analyzed into a finite truth function of immediately given or ostensively defined descriptive terms and names.
D5. The cognitively meaningful empirical language comprises the class of cognitively meaningful sentences together with logic.
D6. All analytical knowledge is *a priori* and not empirically significant.
D7. Direct knowledge grounded in sensory experience comprises the class of logical simple sentences issued as immediate serious reports of sensory experience; each such sentence is indubitably true.
D8. All cognitively meaningful knowledge is empirical knowledge and

ultimately founded upon direct knowledge grounded in sensory experience.

To these theses we may add further, less direct, often tacit, claims which appear in typical positivist writings, but these will not be pursued here (see Hooker 1975c).

This general positivist position has a set of supporting, but again usually tacit, metaphilosophical theses. These are theses which would be used to defend the theory from general philosophical criticism. In my study elsewhere I offered the following set:[2]

M1. There is a first philosophy which includes theories of epistemology, rationality, and language. A first philosophy is logically and epistemologically prior to, and hence normative for, all other statements.

M2. All acceptable sentences, whether philosophical or empirical, must be expressible in a logically and semantically clear and precise language. Philosophical theories are to be characterized in terms of logical structures in that language.

M3. The theory of epistemology is the analysis of the certain foundations of human knowledge.

M4. The domain of the theory of rationality is confined to the empirically meaningful; that part of the theory of rationality pertaining to the pursuit of truth is logically independent of any other component of the theory (if any).

M5. The doctrines of philosophy and of metaphilosophy must be logically consistent; in particular, both sets of doctrines are conventions, devoid of empirical content, but in some sense constituting successive logical frameworks for the construction of empirical knowledge.

M6. First philosophy and metaphilosophy are logically and epistemologically prior to, independent of, and normative for the theory of science.

The chief objections to the positivist theory of science are: (1) theories are not definitionally reducible to finitely, observationally verifiable assertions; (2) scientific method is not rationally confinable to entailment from the facts; (3) observation is not a fundamental, transparent category but a complex, anthropomorphic process, itself investigated by science; (4) historical intertheory relations do not fit the accumulative model; (5) accepted

2. See Hooker 1975c, 187–93. I there offered a "Kantian" alternative to M_5:

M_5^*. The doctrines of philosophy and of metaphilosophy are necessarily universally descriptive of intellectual structure.

This alternative may make positivism/empiricism more attractive to some. It seems to have attracted Popper, whom I argue to be basically an empiricist (1975c, 177–231).

observationally based facts do not belong to an eternal, theory-free category but are theory-laden and subject to theoretical criticism. To these I would especially add the criticisms (6, 7) that science is not isolated from the human individual and from society in the manner presupposed by positivism; (8) that method is not rationally universal either across scientists at a time or across history; (9) that logic does not have the privileged status given it by positivism but is itself open to broadly empirical investigation; and (10) that there is not the gulf between the normative and the descriptive presupposed by positivism. (Besides Hooker 1975c; Bjerring and Hooker 1979, 1980/81, and 1981; and Suppe 1974, see Achinstein and Barker 1969; Churchland 1979 and this volume, chap. 2; Eslea 1973; Feyerabend 1965, 1969, and 1975 for these criticisms.)

The first five of these criticisms have overwhelming support and are now broadly agreed to by all sides. They forced an initial retreat to a richer logic, simple quantificational logic, which allowed richer, less direct, and partial definitions of theoretical terms, of the theoretical ontology (hence of the observationally fixed yet unobservable), and of intertheory relations. Ingenius logical constructions, such as Carnap's reduction sentences, were much manipulated. This did something to ease objections 1, 4, and 5, though it never really blunted them; removed objection 2, replacing it with a confirmationist program; but left objection 3 untouched. The confirmationist program soon gave way to an inductivist program proper, requiring a further enrichment of the logical framework, which has provided a rich methodological program. It also added two more traditional objections to replace 2: (2') inductive logic is impossible (inadequate, incomplete) and (2'') it is impossible to distinguish true laws of nature from accidentally true generalizations.

These objections essentially received two kinds of response from those now calling themselves empiricists (= positivists using quantificational logic): first, introduce pragmatic considerations, as empirically noncognitive, to settle questions left unsettled by the resources of quantificational logic; and, second, introduce still richer logical structures, such as Carnap's inductive calculi, modalities, probabilities. The virtue of this twin response was this: as long as the only accretions made could be argued to be either empirically noncognitive or purely logical, empiricists could claim to hang on to the basic positivist philosophical and metaphilosophical position. Van Fraassen, as we shall see, falls squarely in this tradition.

For four decades (1930s–1960s) most philosophers of science focused on the questions made salient by these strategies: Can pragmatics and richer logics ease the objections to empiricism? Is pragmatics really noncognitive (empirically, anyway), and can empiricists really accept the richer logics? Van Fraassen essays positive answers to his versions of both questions.

This rather stark strategic history of empiricist philosophy of science in this century is only slightly complicated by strands of conventionalism and operationalism running through it.[3] It is also only slightly modified by Popper's early attacks (see Hooker 1975c and 1981b for my claim here). It is virtually unaffected by other, largely continental traditions (e.g., phenomenology, Marxism, hermaneuticism), yet it is these that ultimately kept alive the crucially important objections 6–10 above. These latter objections now rival the traditional ones (1–5, 2', 2'') for philosophical attention. We must see what van Fraassen does with them.

III

Recall the main difference between the realist and anti-realist pictures of scientific activity. When a scientist advances a new theory, the realist sees him as asserting the (truth of the) postulates. But the anti-realist sees him as displaying this theory, holding it up to view, as it were, and claiming certain virtues for it.

This theory draws a picture of the world. But science itself designates certain areas in this picture as observable. The scientist, in accepting the theory, is asserting the picture to be accurate in those areas. This is, according to the anti-realist, the only virtue claimed which concerns the relation of theory to world alone. Any other virtues to be claimed will concern either the internal structure of the theory (such as logical consistency) or be pragmatic, that is, relate specifically to human concerns. (P. 57)

This quote and that with which this essay opens define the essence of van Fraassen's position, constructive empiricism (CE). It is focused on the "manifest image" alone, taking that image as a depthless surface without any but observable structure; all else in science is either explained as purely pragmatic or explained away.

But note how much of what had hitherto been considered distinctively realist, or in any case beyond the reach of empiricism, van Fraassen claims to capture for CE:

1. *On observation and theory* he says:

If there are limits to observation, these are a subject for empirical science, and not for philosophical analysis. (P. 57)

. . . observability has nothing to do with existence (is, indeed, too anthropocentric for that). . . . (P. 19)

See also pages 59–64 for detailed examples and discussion.

3. See, e.g., the accounts in Suppe 1974; Hooker 1975c. The account is only slightly complicated also by admitting that logical theory itself had already achieved a sophistication far beyond the theory of finite truth functions or quantification theory at the hands of Frege and Russell at the turn of the century and continued to develop. For the *strategic* development I depict is a history of the uses of logical systems to defend philosophical theses, not a history of logic. (Nonetheless the two interact in complex ways.)

Traditionally, as van Fraassen himself acknowledges, the observation/ theory distinction had been crucial to empiricism; the distinction had been drawn philosophically, as part of a conceptual analysis of the foundation of empirical knowledge. Indeed, in its severest positivist form, observability is deployed as the basis of a semantic that would of necessity reduce all empirically meaningful knowledge to the observable (see part 2 above). Realists, in contrast, have argued for the imperfectness, limitedness, and radical idiosyncrasy of human observation on general scientific grounds and from thence to the conclusion that observability should be no criterion of existence, not even a factual one, let alone a semantic one. But all of this critique van Fraassen is able to grant and still go on to affirm his version of empiricism (see quote opening this part).

2. *On language and theory* van Fraassen agrees that

> [a]ll our language is thoroughly theory-infected. . . . The way we talk, and scientists talk, is guided by the pictures provided by previously accepted theories. This is true also, as Duhem already emphasized, of experimental reports. Hygienic reconstructions of language such as the positivists envisaged are simply not on. (P. 14)

These remarks clearly signal an abandonment not only of positivism, but of the entire twentieth-century empiricist tradition up to and including later Carnap and fellow travelers, for this entire tradition rested on a reconstruction of theoretical language in terms of logical constructions out of observation terms. This rejection of a schizophrenic attitude to the language of science is one of the planks in the realist platform and the semantic counterpart to rejecting a formal observation/theoretical dichotomy.

3. *On truth and theory* he says:

> I would still identify truth of a theory with the condition that there is an exact correspondence between reality and one of its models. . . . And logical relations among theories and propositions continue of course to be defined in terms of truth. . . . (P. 197)

Traditionally, empiricists have wanted to deny meaningfulness to any parts of theory that would create ontic commitments to unobservables or to reduce those unacceptable parts to the acceptable, observable core. At their most liberal, empiricists allowed only commitments to unobservable entities identifiable by definite descriptions using solely observable predicates (e.g., behavioral psychology's intervening variables). Although liberal enough to cause epistemological difficulty for empiricist principles, this latter criterion is still a lot narrower than van Fraassen's position. For, as far as meaning, reference, and truth are concerned, van Fraassen professes to treat theories literally, just as realists insist they should be. Once again the realist critique is deflected as indecisive.

4. *On experiment and theory* he writes:

For theory construction, experimentation has a twofold significance: testing for empirical adequacy of the theory as developed so far, *and* . . . guiding the continuation of the construction, or the completion, of the theory. Likewise, theory has a twofold role in experimentation: formulation of the questions to be answered in a systematic and compendious fashion, *and* as a guiding factor in the design of the experiments to answer those questions. (P. 74, van Fraassen's italics)

Again, the empiricist tradition has been a "bottom up" account in which experimental results were the most important category and inductively determined everything "above" them. (For the positivist it was even deductive determination of theory by experiment.) Van Fraassen notes approvingly Sellars's demolition of this naïve picture (p. 32). By contrast, traditional realism has tended to the opposite extreme, giving all the weight to theory, relegating experiment to theory testing. Only determined realists, trying to take humankind's evolutionary epistemic status seriously, have seriously started to explore the proposition that there are extremely complex two-way interactions and trade-offs between theory and experiment in the risky business of pushing science along (see Munévar 1981; Hooker 1976, 12–14; also Hooker 1975a, 1974). But van Fraassen calmly accepts the insights as his.

Of less theoretical importance, but still of interest, he says of *rational scientific psychology*:

. . . acceptance has a pragmatic dimension: it involves a commitment to confront any phenomena within the conceptual framework of the theory. . . . That is why, to some extent, adherents of a theory must talk just as if they believed it to be true. It is also why breakdown of a long-entrenched, accepted theory is said to precipitate a conceptual breakdown. . . .
    . . . Commitment to a theory involves high stakes. The theories we develop are never complete, so that even if two of them are empirically equivalent, they will be accompanied by research programmes which are generally very different. Vindication of a research programme within the relatively short run may depend more on the theory's conceptual resources and facts about our present circumstances than on the theory's empirical adequacy or even truth. (P. 202)

The overall picture van Fraassen provides of science, then, has remarkable similarities to that promulgated as crucial by the strongest realists, myself included. But we know from the first two quotes in this section that there are deep differences, and it is time to expose these.

First, it is clear that, despite the range of agreement, *there is a deep difference in the theory of rational commitment* espoused by the two theories. Van Fraassen gives as the defining conditions for realism and CE:

Realism: *Science aims to give us, in its theories, a literally true story of what the world is like; and acceptance of a scientific theory involves the belief that it is true.* (P. 8)

CE: *Science aims to give us theories which are empirically adequate; and acceptance of a theory involves as belief only that it is empirically adequate.* (P. 12)

And later he elaborates as follows:

. . . we can distinguish between two epistemic attitudes we can take up toward a theory. We can assert it to be true (i.e. to have a model which is a faithful replica, in all detail, of our world), and call for belief; or we can simply assert its empirical adequacy, calling for acceptance as such. In either case we stick our necks out: empirical adequacy goes far beyond what we can know at any given time. (All the results of measurement are not in; they will never all be in; and in any case, we won't measure everything that can be measured.) Nevertheless there is a difference: the assertion of empirical adequacy is a great deal weaker than the assertion of truth, and the restraint to acceptance delivers us from metaphysics. (Pp. 68–69)

But this cannot be all there is to acceptance for CE, since it would not motivate the CE scientist to take the nonempirical part of theories as seriously as van Fraassen agrees they do. And it isn't; there is a pragmatic dimension, as well.

Theory acceptance has a pragmatic dimension. While the only belief involved in acceptance, as I see it, is the belief that the theory is empirically adequate, *more than belief is involved.* To accept a theory is to make a commitment, a commitment to the further confrontation of new phenomena within the framework of that theory, a commitment to a research programme, and a wager that all relevant phenomena can be accounted for without giving up that theory. . . . Commitments are not true or false; they are vindicated or not vindicated in the course of human history. (P. 88, van Fraassen's italics)

See also the quote from page 202 above; that quote continues:

The depth of commitment is reflected, just as in the case of ideological commitment, in how the person is ready to answer questions *ex cathedra*, using counterfactual conditionals and other modal locutions, and to assume the office of explainer. (P. 202)

But, of course, modal locutions (possibilities, probabilities, counterfactual conditionals, nomic necessities) and explanation are held by CE not to add to the empirical content of theory—the quote concludes with the paragraph quoted at the outset of this paper.

This brings us directly to the second and third major differences between CE and realism, over the *theories of modalities and explanation*, CE relegating them to merely pragmatic status and realism typically taking them to be crucial substantive features of theories.

With respect to explanation, van Fraassen summarizes the realist as involving these claims:

> . . . explanatory power is something quite irreducible, a special feature dif-
> fering in kind from empirical adequacy and strength. . . . [I]t was argued that
> what science is really after is understanding, that this consists in being in a posi-
> tion to explain, hence what science is really after goes well beyond empirical
> adequacy and strength. Finally, since the theory's ability to explain provides a
> clear reason for accepting it, it was argued that explanatory power is evidence
> for the *truth* of the theory, special evidence that goes beyond any evidence we
> may have for the theory's empirical adequacy. (p. 153–54, van Fraassen's italics)

And indeed, in recent realist writing, explanatory capacity is taken, qua
primary aim of science, as a *prima facie* reason to believe in the truth of
the theory, while explanation is linked to understanding via theoretical
ontology.[4]

Van Fraassen sums up his own pragmatic theory of explanation as
follows:

> To call an explanation scientific, is to say nothing about its form or the sort of
> information adduced, but only that the explanation draws on science to get this
> information (at least to some extent) and, more importantly, that the criteria of
> evaluation of how good an explanation it is, are being applied using a scientific
> theory. . . .
>   . . . Being an explanation is essentially relative, for an explanation is an *an-
> swer*. . . . Since an explanation is an answer, it is evaluated *vis-à-vis* a question,
> which is a request for information. . . .
>   Hence there can be no question at all of explanatory power as such (just as it
> would be silly to speak of the 'control power' of a theory, although of course we
> rely on theories to gain control over nature and circumstances). Nor can there
> be any question of explanatory success as providing evidence for the truth of a
> theory that goes beyond any evidence we have for its providing an adequate de-
> scription of the phenomena. For in each case, a success of explanation is a suc-
> cess of adequate and informative description. And while it is true that we seek
> for explanation, the value of this search for science is that the search for expla-
> nation is *ipso facto* a search for empirically adequate, empirically strong theo-
> ries. (Pp. 155–57; van Fraassen's italics)

Here the sense becomes clear in which a pragmatic feature of science adds
nothing to the substantive content of science, nothing to the intrinsic cognitive content and status of science. The same treatment is accorded all
modalities.[5]

---

4. See, for example, the view that inductive inference is in fact inference to the best explanation (Harman 1973) and views that give like weight to explanatory power. And I have, myself, tried to improve on Friedman's worthwhile idea of linking (again) understanding to explanation (see, respectively, Friedman 1974; Hooker 1980a).

5. Since in realist theories of explanation, explanations are normally held to require laws
in their premises expressing natural necessities, the two problems are immediately connected.
Van Fraassen begins with probabilities, which are central to many scientific theories, offering a "modal frequency" interpretation of their occurrence which he informally summarizes thus: "The probability of event A equals the relative frequency with which it would

These, then, are the main differences between van Fraassen's CE and realism. The last question to be asked is how he manages to take over so much of what has hitherto been held distinctive of realism and yet maintain these sharp differences. Specifically, treating, as he does, all terms as theoretically laden, theories as providing a nomic structure, observation as internally specified, and truth conditions as specified by theoretical models, how can he still construct the empirical observable core of a theory, isolated from unobservables and modalities?

The secret lies in van Fraassen's abandonment of syntactic logical constructions and his adoption of model-theoretic semantic constructions.[6]

---

occur, were a suitably designed experiment performed often enough under suitable conditions" (p. 194). He concedes that probabilities are irreducibly central to theories and require reference to a manifold of events only one subsequence of which can be actual. This leaves him defending an apparently inconsistent triad: "*First*: probability is a modality. *Second*: science includes irreducibly probabilistic theories. And *third*: there is no modality in the scientific description of the world" (p. 198, van Fraassen's italics). The second thesis is conceded as a fact; the third expresses the CE commitment; the outstanding issue is how to understand the first so as to reconcile the three. "In my opinion, that solution consists mainly in the correct diagnosis of the problem, which is that modality appears in science only in that the language naturally used once a theory has been accepted, is modal language. This relocates the problem in philosophy of language, for it becomes the problem of explicating the use and structure of modal language" (p. 198).

With regard to this latter theory, van Fraassen suggests that the solution involves two components, (1) further developing the rich, formal model-theoretic semantics for modalities that has exploded on the logical and philosophical scene in the last two decades and (2) allying this to an analysis of pragmatic commitments to show how use of modal language reflects only the latter, not substantive content commitments beyond empirical adequacy. Van Fraassen provides a homely example, the person who is willing to assert that nothing could be both red and green all over. We are to understand use of the modal "could" here by supposing that "this person is guided by his idea of a simple abstract structure, the colour spectrum. We can think of this as a line segment, or an interval of real numbers (the wave lengths). He associates with each colour predicate, such as 'green', a part of that spectrum; he associates disjoint parts with 'red' and 'green'; and when he says that an object is green or red, he is *classifying it*, that is, assigning it a *location* in the spectrum" (pp. 200–201, van Fraassen's italics). The modal sentence, then, says that "no point of the spectrum at all belongs to both parts." "Blatantly modal sentences . . . do occur; but this person evaluates them as true or false by reflection on the structure of the spectrum that guides all his uses of colour terms. His linguistic commitments can be summed up by referring to his use of this spectrum; his theory of colour consists in the family of models each of which is a classification of objects through location in this spectrum" (pp. 200–201). Similarly, all use of modal language is to be explicated, occasion by occasion, in terms of nonmodal features of models for some scientific theory which the speaker pragmatically uses on that occasion for the purpose of exploring the elaboration or testing of that theory, a theory he has accepted only as empirically adequate. (See pp. 201–2).

6. Syntax concerns the actual skeletal structure of a language or sentence, the structured string of logical devices that organize descriptive terms into declarations, questions, and commands (but especially declarations). "And," "if-then," and the like studied in traditional logic are all syntactical operations, while "is a tautology" and "is deducible from" are syntactically definable predicates both within traditional logic and for large, useful fragments of ordinary and scientific language. Semantic features of language are, on the other hand, those that concern its relation to the world; here the main property for statements is truth, and van Fraassen adds empirical adequacy as the other main semantic property for theories (p. 90).

Whereas traditional logic has historically been the recognized vehicle for the study of the syntactic features of formal and natural languages, including theories, the appropriate vehicle for semantics is formal model theory. Then "Theory T is true" can be reexpressed as "Reality exactly matches one model of T" (cf. the quote from p. 197, above). Many other formal semantic features of languages or theories can be defined in model-theoretical terms, e.g., consistency (= has at least one model). The important point is that passage to model-theoretic semantics represents a further strengthening of the logical tools available to the empiricist. Everything that can be syntactically expressed can be expressed model-theoretically, but there are semantic constructions that are inaccessible to the syntactical approach, for example, relative semantic embeddability ("every model of T_1 can be mapped homomorphically into a model of T_2,"):

> This sort of relationship [relative semantic embeddability] which is peculiarly semantic, is clearly very important for the comparison and evaluation of theories, and is not accessible to the syntactic approach.
> The syntactic picture of a theory identifies it with a body of theorems, stated in one particular language chosen for the expression of that theory. This should be contrasted with the alternative of presenting a theory in the first instance by identifying a class of structures as its models. In this second, semantic, approach the language used to express the theory is neither basic nor unique; the same class of structures could well be described in radically different ways, each with its own limitations. The models occupy centre stage. (P. 44)

Van Fraassen goes on to give a devastating critique of the old empiricist approach which relied solely on syntactic methods to define its key structures (e.g., definitions of empirical-observable and theoretical portions of theories; see pp. 53–56). As van Fraassen observes, it is possible to define and identify all manner of unobservable, theoretical entities using only observable empirical predicates and logical devices, so attempting to extract that portion of a theory (defined syntactically via theoremhood) stated in some observational sublanguage will never isolate the empirical-observable content of the theory; rather, it is merely "in a hobbled and hamstrung fashion, the description by T of everything" (p. 55).

Well, then, how does the semantic deployment of model theory work the trick of isolating an empirical core? Like this: The model "models" everything in the theory; it models, therefore, the full theoretical ontology and its complete set of nomicly permissible processes. But it also thereby models that subset of processes within the theory which represent mea-

---

A model is a structured set of entities (abstract or concrete) such that the significant syntactical components of a set of sentences in a language, or a theory, can be mapped onto it in a structure-preserving way and, when they are, the sentence set or theory in question is evaluated true. (A model is a way of "realizing," or satisfying, the sentences or theory in question.) In general, sentence sets or theories will have many different models.

surements or observations and the features of the theoretically specified model which are thereby measured or observed. In so doing, a model has modeled the observable/unobservable distinction, as specified by the theory. The class of all thus modeled observable features forms the empirical core of the theory.

> To present a theory is to specify a family of structures, its *models*; and secondly, to specify certain parts of those models (the *empirical substructures*) as candidates for the direct representation of observable phenomena. The structures which can be described in experimental and measurement reports we can call *appearances*: the theory is empirically adequate if it has some model such that all appearances are isomorphic to empirical substructures of that model. (P. 64, van Fraassen's italics)

Why are these empirical substructures modality-free? Because, if the theory picks out only what is observable in each circumstance, then a catalog of those features will be thoroughly "Humean," modalities won't come out as observable in any one situation, and across situations only correlated bundles of observable features will be listed. Even the modality-inclined realist must admit, if she is realistic, that modalities belong to theory and aren't among the observable features of the world, so no theoretical models of observability will assign them to their observable submodels. Technically the matter is settled by building in these results to the models from the very beginning. Van Fraassen's theory of probability is a good example (see note 5). Theories with irreducible probabilistic features are modeled by families of possible experiments, the relative frequencies of whose outcomes model the probabilities involved. Of course, only individual such experiments can be actual, and only their identifying conditions (apparatus type, etc.) and outcomes will be identified as observable by the theory. Thus, the empirical substructure of the theory is some actual finite collection of these identifying conditions and outcomes; no probabilities or possibilities occur therein.

It is time to take stock of the new situation in the realism/empiricism debate. Van Fraassen has managed to deflect many of the sharpest criticisms from realists by simply incorporating them into his own position. In doing so, he has had to forgo many of the treasured dogmas of his empiricist predecessors, e.g., that there is an epistemologically prior distinction between theory and observation; that theory is a "dependent variable" only, either deductively or inductively determined by experiment; and that there is semantic priority for observational terms. Nonetheless, van Fraassen claims through all of this to hold on to the essential core of empiricist doctrine and thereby to retain the essential disagreements with realism, to wit, commitment only to empirical adequacy, not truth; no modalities within the commitment; explanation not an objective virtue.

How is this liberalized balancing act between concession and defense achieved? Through the wielding of powerful new logical tools to cut the boundary of empiricist doctrine more finely, closer to the core yet consistently defensible, while consigning all of the new leftovers to noncognitive pragmatics. Thus, a recognizable historical pattern is repeated; van Fraassen is following exactly the strategy of his empiricist forebears (recall the discussion in section 2 above).

Have we now, perhaps, reached the ultimate end of this process? Now, after the passage from finite truth functions to quantification, to quantification with definite descriptions, identity, and various abstraction operations, to various inductive calculi; now, as van Fraassen says, after each of these transitions has distracted a generation of scholars; now, with the passage to model theory, may we perhaps suppose there is at last a solid defensible core to empiricism and the issue with realism can be finally joined in less superficial terms than criticisms of logical inadequacy? I frankly suspect not, since I don't think that what divides realists and empiricists ultimately revolves around the empirical adequacy of either of the two philosophies of science.[7] But, being the most recent in a long line of such maneuvers, van Fraassen's empiricist position is certainly the most plausible and powerful version of the doctrine yet, and the logic to which he has access makes it uniquely more powerful than its predecessors; his position deserves to be treated as such.

IV

Central argument

The core of the difference between van Fraassen's CE and realists concerns what kind of epistemic attitude is rational where scientific theories

7. I believe it *should* involve the empirical adequacy of the theories, perhaps more heavily even than most realists would like (see Hooker 1975c). (To be strict, empiricists should not countenance empirical testing of their *philosophy* at all; but it is a historical fact that, when a then-current exposition of their views has failed to win wide acceptance, they proceed to revise the exposition in exactly the manner that any Popperian or realist would do.) Ironically, it is precisely because philosophical theories themselves have a theoretical dimension and, even more importantly, a metatheoretical dimension, as well, that permits them to be clung to even in the face of recalcitrant historical experience. Empiricists may seek ever-richer logics or more powerful theories of pragmatics with which to remedy existing defects. Attempts to dismiss these strategies out of hand usually can be taken to beg the question. See text following note 11.

I tried to illustrate this for the realist in the case of applying Craig's theorem to generate the "theoretician's dilemma" (1968, 161–63). Using a theorem of formal logic, one can separate out a supposed purely observational segment of a theory; empiricists then argued that one could throw away the theoretically infected segment, thus rendering theoretical terms superfluous. One *could* do this, but why would it be *rational* to do so? Reflection reveals that it would not be rational, save on empiricist assumptions. So the argument carries no force against the realist but succeeds only in begging the question.

are concerned. According to van Fraassen, CE says, "Accept theories only as empirically adequate, no more"; realism says, "Believe theories as true on the grounds of their empirical adequacy and explanatory power (and perhaps other virtues)."[8] Consider first the rationality of CE's epistemic attitude.

Van Fraassen says:

> On the view I shall develop, the belief involved in accepting a scientific theory is only that it 'saves the phenomena', that is, correctly describes what is observable. But acceptance is not merely belief. We never have the option of accepting an all-encompassing theory, complete in every detail. So to accept one theory rather than another one involves also a commitment to a research programme, to continuing the dialogue with nature in the framework of one conceptual scheme rather than another. Even if two theories are empirically equivalent, and acceptance of a theory involves as belief only that it is empirically adequate, it may still make a great difference which one is accepted. The difference is pragmatic, and I shall argue that pragmatic virtues do not give us any reason over and above the evidence of the empirical data, for thinking that a theory is true. (P. 4)

I shall reserve for later discussion the question of whether the decision to "continue the dialogue with nature in one conceptual scheme" is pragmatic and, if pragmatic, also nonepistemic. The other question remains, why is it rational to epistemically accept a theory only as empirically adequate? I suppose for this ironical reason: it minimizes risk in the face of the very fact realists have often emphasized, namely, that every actual theory is likely only to be an approximation to the truth, hence strictly speaking false. Faced with this "realistic" assessment of the human epistemic situation, one withdraws to a less risky affirmation, for, even if the theory is overthrown, surely its empirical adequacy *at the time* remains. Thus, the tables are turned on a realist trying to be realistic.[9]

Reply

This important argument invites two major responses: (1) The epistemic strategy recommended is a very poor one in itself, arguably less likely to advance knowledge than alternatives; and (2) there is anyway no reason to suppose that van Fraassen's position *is* less risky than the real-

8. In 1976 I argued for a range of utilities relevant to rational scientific decisions, of which those mentioned were only two. Others I included were expanding technological power, theoretical or explanatory depth, conceptual fecundity, testability, precision. I also argued that there was no effective way to separate out a group of purely cognitive or epistemic utilities from the other moral or pragmatic utilities, a position I regard as strengthened by my arguments in 1982.

9. For example, the tables are apparently turned on myself, for I have argued from evolutionary perspectives to the adoption of certain philosophical positions and strategies (most notably in 1975c and 1976). But we shall see below that the reversal is more apparent than real, for the real force of an evolutionary perspective has not been properly grasped.

ists', hence risk taking cannot be the decisive ground. I shall consider these two replies in order.

1. *Epistemic strategy.* Why is it rational to limit epistemic commitment to observational adequacy out of caution? Popper taught us long ago that caution is a poor policy for creatures born in ignorance and trying to find the truth. Ultimate caution for an empiricist would require sticking just to the evidence as it comes in, risking no theoretical ideas about it at all, but in this case we would learn very little. In fact, we would quickly become mired in dogma, for not only would few (if any) exciting new experiments be tried, but we would be incapable of criticizing our existing experiments and data; our unnoticed biases and errors would quickly accumulate.[10]

Would one, then, always rationally wish to prefer the more empirically adequate scientific theory, i.e., the theory most closely in accord with the available data? I suggest not and for the reasons Popper has already pointed out: it may be explanatorily shallow or even self-confirming, descriptively idiosyncratic, internally fragmented, etc. Indeed, it is typical of scientists to prefer to explore what is judged to be the theoretically deeper theory, even if it has some empirical difficulties, over less theoretically insightful, if more empirically adequate, alternatives. Scientists aim at interesting or valuable truth, not simply truth.[11]

Although this disposes of the naïve empiricist conservative, it doesn't dispose of van Fraassen, for he grants the necessity of bold theoretical conjecture and of theory-directed experimentation and evaluation, and he would be able to find a place for the other virtues of theory just considered—it is just that he insists that all of these factors and considerations are nonepistemic, that they are merely pragmatic. They are to be understood in some way as merely satisfying our individual interests or as mere heuristics in aid of scientific creativity and the like, thereby falling outside of empirical cognition proper.

Again I postpone consideration of the status of alleged pragmatic commitments but wish to note in passing that no argument emerging here can

10. See, for example, the classic discussion by Feyerabend (1965, 1969, 1975)—though I have attempted to back off from the relativist suggestions in his position, ironically on realist grounds, cf. Hooker, 1973b and 1974—and the discussion by Churchland (1979 and this volume, chap. 2).

11. On choosing theoretically interesting programs over mere empirical adequacy, see again the discussions by Feyerabend (1965, 1969, 1975). Discussions by Kuhn (1962) and Lakatos (1970) also contain some valuable examples. The advent of both relativity theories are cases in point, though I do not deny that empirical testing also entered early on as a factor, but hardly the only one. Much contemporary work in quantum field theory is notoriously in the same position.

The insight that what is sought is valuable truth I have from Popper (see, e.g., 1973). It is connected directly to the rational choice model of epistemology I follow, for according to me it is in a utility-uncertainty trade-off structure that a theory of valuable truth is to be spelled out, as are the grounds for trading off theoretical attractions against empirical adequacy.

actually force a realist to adopt the CE view. Consider the situation over the empiricist use of Craig's theorem to attack the relevance of theory. Using Craig's theorem, it is syntactically possible to segregate just those consequences of a theory that can be stated in solely observational vocabulary. Empiricists then argued that this shows that theoretical terms are irrelevant to the assessment and use of a theory and so should be eliminated. Technical complexities aside (see Hooker 1968), this argument presumes that there is an independently justifiable division between observational and theoretical terms, and it presumes that empirical adequacy is the sole locus of epistemic judgment in science. But both of these assumptions are rejected by realists, and so the argument only succeeds in begging the question. Van Fraassen himself concedes the same points in the case (p. 30; cf. an additional example at pp. 21–22). But just so, too, with his own position here.

2. *Risk*. Turning now to the second response to the argument for CE from risk reduction, I shall offer a three-stage argument that van Fraassen's position is no less risky than is that of realism. First, with respect to observation we have already noted van Fraassen's remarking, "Even if observability has nothing to do with existence (is, indeed, too anthropocentric for that), it may still have much to do with the proper epistemic attitude to science" (p. 19). But if observability really *is* too anthropocentric to be a safe guide to existence, why is it rational to base one's epistemic attitude to science solely on it? Isn't this latter attitude therefore a very risky one, relying on a collection of senses known to be limited and idiosyncratic, and sometimes even suspected of deep error?[12]

From an evolutionary point of view, what is risky is cognition generally, whether concerned with observational or nonobservational aspects. Our cognitive capacities include both abilities to perceive and abilities to theorize; both evolved in idiosyncratic ecological conditions from less complex life forms, and both may be supposed to thereby inherit limitations and idiosyncrasies. Why reserve skepticism for just one? Indeed, from an evolutionary point of view, theoretical thought has the impressive achievements of having grasped both mathematics (the simpler, linear part, anyhow) and highly insightful formal theories for dynamics. By contrast, we have merely come to understand the quirks and limitations of human observation better over time without significantly enlarging its nature—why not, then, side with Plato and give theoretical cognition the primary trust? At the least, why not put the two risks on a par and proceed with a unified, realist account in which each risk is traded off against the other?

12. Indeed, deep suspicion of human senses on just this sort of ground has been a constant philosophical theme; cf., for example, Platonism and Zen Buddhism. (See also Churchland 1979 and this volume, chap. 2.)

Second, there is a further risk with CE commitment, for empirical adequacy refers to a theory's fitting the observable phenomena *for all time*, and this can never be guaranteed by the evidence available now (see p. 69, quoted above; also p. 13). Thus, if it is the mere underdetermination of a theory by its observational core that makes realism unacceptably risky, then CE ought also to be rejected, since any empirical evidence available to us will in general radically underdetermine a theory's empirical adequacy. If it is the volatility of theoretical ideas and principles under criticism that favors CE, then observational terms and data have shown volatility as well (e.g., the starry dish above).

Third and finally, it is not as though accepting a theory à la CE is risk-free. There is the risk of committing oneself to confront nature in its terms and within its resources. Considering the very strong role which van Fraassen has theories play in the scientific process, these commitments are every bit as risky as are the corresponding realist commitments.

In sum, van Fraassen's CE risks are in fact exactly the same as are those of the realist. This seems to me to dispose of the argument for CE from risk. If there is to be an advantage for CE here, it must hinge on the pragmatic status CE accords most of these risks. So let us postpone this issue no longer but consider it directly.

Pragmatic-Cognitive

Consider van Fraassen's account of the role of theory in experiment (see quote from p. 202, above). If theories are to structure the questions to be asked and to inform the experimental designs needed to pose them properly to nature, surely theories must be taken seriously epistemologically. How can scientists rely on them to this extent and remain committed only to their empirical adequacy?

Van Fraassen's answer is, by carefully distinguishing pragmatic use of the models of a full theory from the empirical core of that theory. Thus, if, in testing theory T_1 in some situation S, theory T_2 suggests a factor F in S which either T_1 might not account for or which must be controlled for if a parameter crucial to T_1 is to be measured, then what is really happening is no more than this: the models of T_1 are extended in certain ways, viz., to include models of F drawn from models of T_2, and the empirical cores of the extended models are then compared with those of T_1 alone and with experimental outcomes (p. 80). Everything can be done with models and empirical cores, and the focus is on the adequacy of those cores, so a realist account is unnecessary. But, like the older use of Craig's theorem, this reply misses the point; why rely on these theoretical models so heavily if they have no cognitive significance? If it is unreasonable to believe that these models capture something of the structure of things beyond the observable, why the meticulous reliance on their detail?

Van Fraassen can consistently answer, as he does, that the only *cognitive* reason for reliance is empirical adequacy, that the remaining reliance is purely pragmatic and present only because we have no option but to take risks in the epistemic game. So let us revive again the two claims I suppressed at the outset of this discussion: Is the reliance on theory pragmatic and is the pragmatic noncognitive? Van Fraassen adopts affirmative answers to both questions, but what arguments does he offer for them?

So far as I can see, none at all, except for the following brief remark:

> . . . the requirement that our epistemic policies should give the same results independent of our beliefs about the range of evidence accessible to us. That requirement seems to me in no way rationally compelling; it could be honoured, I should think, only through a thorough-going scepticism or through a commitment to wholesale leaps of faith. (Pp. 18–19)

Here van Fraassen is facing the objection that "observable" really means "observable to us," but, since other species may count our unobservables among their observables, observability is a poor criterion of cognitive commitment. Van Fraassen doesn't deny the basis for the objection, he simply argues that the alternative is worse: "thorough-going scepticism" or "wholesale leaps of faith." If realist epistemology could be spiked on the horns of that dilemma it would argue in favor of van Fraassen's agnosticism. But it cannot, because we have just seen that a good CE scientist will risk practical reliance on theory no less than his realist counterpart. The risks are as great and, since a CE scientist suspends belief in theory but is still forced to rely on it, whatever the dangers involved, the degree of skepticism involved must be at least as great. Beyond this I can find no argument in the book.

And van Fraassen himself says that ultimately semantics, which contains the notions of truth and empirical adequacy, is only an abstraction from pragmatics (p. 89); in that case, why claim that pragmatics is noncognitive while semantics is cognitive? No direct answer is given, but the answer van Fraassen's comments suggest is that pragmatics concerns relations of language or theories to users of the language, while semantics concerns relations of language or theories to the world (p. 89; cf. p. 100). It is not clear what this distinction really comes to; van Fraassen himself agrees that there is no neat way to demarcate semantics within pragmatics (p. 89).

In any case, why does this distinction render the pragmatic noncognitive? Even supposing explanation were a relation between a theory and its human users, as van Fraassen claims, surely it is a significant feature of a theory that humans can use it to explain and predict. Indeed, whatever the limitations and idiosyncrasies of humans, it would be an objective feature of a theory that humans could use it to map the world. Is this not a significant cognitive feature of the world? More generally, it would be a significant cognitive feature of any species/theory pair (S, T) that S's could explain and predict successfully using T. Martian scientists might incorporate just such an account of our activities in their science, and we of them. More relevantly, we are rapidly building a science of ourselves which is now beginning to be able to reflect on our own species use of theories for various purposes. Theoretical reflection on human perception, the basis of empirical adequacy, and the subsequent theory-guided modification and extension of human perception, is an excellent case in point.[13]

A unified evolutionary account of intellectual development undermines any attempt to dismiss the relating of science to human purposes as merely pragmatic; at bottom, all human purposes are understood to be at least partly cognitive (survival and enhancement require information), and those connected with scientific activity (theory construction and testing, yes, but also technology development, development of epistemic institutions, metatheories of explanations, methodology, etc.) deeply so.

I must conclude that van Fraassen's position is just that, a merely convenient position on the matter. This is reminiscent of an earlier empiricist, Carnap, who simply claimed that there were two kinds of questions, internal and external, and that the latter were pragmatic and so noncognitive, also without giving any argument.[14]

Unmatched Advantages to Realism

Are we, then, at a standoff between CE and realism, each with a consistent account of practice and no deciding arguments either way? I think not. To begin with, the realist is better able to explain why he didn't always choose to rely on the most empirically adequate theory. But there are other arguments to be marshaled.

From an evolutionary, naturalist point of view, there is a general argument from a unitary view of mind. As remarked, the human cognitive apparatus developed with both theoretical and practical abilities; without all of these abilities, and without their intimate interaction with each other,

13. In the specific case of perception, we have used our sense organs to acquire the empirical base to build science, including a psychology of perception and sciences of instrumental data acquisition, all of which allow us to criticize our sensory perception and, indeed, to virtually abandon it altogether for the purposes of doing science. Similar transformations are beginning elsewhere; for example, the psychology and economics of organizations allied to policy theory and philosophy of science as applied to science itself are the beginnings of the ability to criticize and eventually transform the elementary and largely unconscious social bases on which science has hitherto been organized. Similar remarks apply to sociology and biology applied to the genetic selection of scientists, and they apply as well to psychology and education as applied to scientific training. (Taken together, these developments amount to the reshaping of our cognitive neotony.) And so on. (Cf. Hooker 1980b.)

14. See Carnap 1956. Presumably, the motive was to protect the *a priori* yet noncognitive status for empiricism's first philosophy (cf. my discussion of empiricist metaphilosophy in 1975c). In any case, it is clear that Quine's reply (1966), though technically telling, does not strike the heart of the position, which is metaphilosophical. (Cf. also Goldstick 1979.)

our cognitive capacity would be much less. What theoretical reason in cognition is there, then, to distinguish among these capacities, giving special status to some and not to others? There is, e.g., a strong case for the role of theoretical capacities in perception itself.[15] In the absence of counter-argument, the evidence favors treating all cognitive abilities on a par, within a single framework, and, as suggested above, refusing to draw any basic cognitive/pragmatic distinction.[16]

An important structural reflection in science of this deeply interacting cognitive unity is the now rapidly increasing interaction between scientific theory and normative metatheory. One of the most striking examples is that of physics and logic, where physics has contributed a revolution in logic (whether or not one holds with quantum logic in physics) and logic has expanded the possibilities for physics.[17] But, according to empiricism, logic is supposed to be normative, *a priori*, arrived at by intellectual rigor and proof, and physics is supposed to be descriptive, *a posteriori*, arrived at by observation and experiment. How, then, can they learn from each other? On the other hand, if one is learning from radical evolutionary ignorance so that both logic and physics have to be explored in a critical and experimental manner, then the current situation is perhaps to be expected. Similar interactions are occurring between rationality theory and economics and sociology, between epistemology and psychology and neurophysiology, and between metaphysics and physics. Indeed, such intense internal critical awareness and interaction is one of the hallmarks of the present stage of the development of science. Unhappily, empiricsm really cannot make any sense of this crucial development.[18]

And there is a further general argument for realism from the "scientific image" itself. The fact is, science has confirmed the evolutionary, naturalist picture of humankind as mammals with complex nervous systems slowly building up their individual and collective cognitive representations of the world using all of the complex and idiosyncratic array of devices to hand: sensory perception, behavior, language, technology, institutions, culture, etc. So realism in philosophy is confirmed by realism within science. For CE, however, empiricism in philosophy is in tension with the picture of *Homo sapiens sapiens* presented in the content of science. The CE scientist is forced continually to play a schizoid game of pretend, to pretend that the theories in which he is immersed are informative and, be-

15. A state of seeing a scene has, all research shows, many levels of hypothesislike operation involved in it. It is, for example, a function of past knowledge and of perceptual expectations. See nearly any modern work on perception, for example the volume in which my essay (1978) is included.

16. This conclusion is enhanced by Churchland's elegant thought experiment (1979, chap. 2).

17. The development is captured in Hooker, 1975b, 1979c. (See also Hooker 1979d, 1973a.)

18. Cf. Hooker 1980b and 1982 for other discussions of this issue.

cause the information they convey contradicts his philosophical self-view, at the same time to be reminding himself to suspend belief, to treat science as, after all, only a game. It is possible to thus play games; our very theoretical or imaginative capacities make it so: fiction and drama would be impossible otherwise. The question is, is it *rational*? I suggest that it is not rational unless there are specific reasons for doing so (as there are in drama), but here we have none.

Finally, there are the continuity and conjunction arguments for realism, based on scientific methodology and the history of science. The continuity argument is this: Theoretical ideas and theoretical structures crop up in various contexts throughout the history of science, successively refined, modified, or conditionalized in the light of further experience. If theories are merely empirically adequate, there is no reason to expect this, unless it can be shown that such ideas are merely of pragmatic convenience. But it is not plausible to regard the correction of Newtonian physics in relativity theory or of Lavoisier's oxygen chemistry by electrochemical valence and that by molecular quantum chemistry, etc., in this way. Hence, the history of science demands that theoretical structures be treated seriously as grasping, albeit imperfectly, the unobserved structure of reality.

The conjunction argument is this: Scientists regularly, and surely rationally, conjoin two theories to derive consequences unavailable in either. This practice is understandable if the theories are taken to be true, since, if A is true and B is true, then $(A \& B)$ is true. But the conjunction of empirically adequate theories need not be empirically adequate, or even consistent. Hence, the justification of this practice as rational requires a realistic understanding of theories.

Van Fraassen is aware of the conjunction objection, devoting a special section to its discussion (pp. 83–87). His reply consists of three claims: (1) There are enough observable phenomena overlapping the domains of two or more theories to provide a sufficient empiricist case to explain the scientific desire to conjoin theories (pp. 86–87); (2) much conjoining turns out to be correction of one or both theories and so falls outside the argument (pp. 83–84, 87); and (3) the real and anyway methodologically sufficient motive for conjunction is to test the empirical adequacy of the conjoined theories (p. 85).

I believe van Fraassen wrong on all three counts. With respect to the first claim, it is granted that explanation of empirical phenomena often requires conjoining theories; indeed, conjoining them far more powerfully than van Fraassen suggests, for laboratory phenomena typically require the conjunction of four or more theories to properly understand them, the passage from theory to observatism is typically *theoretically* complex.[19] Even so, it seems clear that theory conjunctions have not always been

19. See my discussion in 1975a and the example worked out in the appendix there.

motivated by appropriate observable phenomena, at least not in the direct sense van Fraassen has in mind. Relativisitic quantum field theory of interior nuclear structure has no directly observable consequences, in the way that van Fraassen's man on the moon requires conjoining terrestrial physiology theory and celestial gravitation theory. The relativistic theory was developed historically first and foremost to "see how it goes," to gain *theoretical* insight; it would still be developed on this basis today. Such observational results as have emerged are quite indirect, requiring the mediation of several other theories to trace their causal claims "up" to the observable surface. Much the same can be said for the development of the kinetic theory underpinnings and corrections to thermodynamics (it still cannot explain such obvious thermodynamic phenomena as phase transitions, but eventually doing so is expected to be a source of *theoretical* insight), of quantized general relativity theory, and so on. Of course, van Fraassen has himself covered; he need only transform theoretical insight into structural knowledge of the models of joint theories and claim that enlarged empirical adequacy is the real long-run goal (cf. p. 93) and he can claim to accommodate the attitude. What is at issue is not the logical consistency of the CE doctrine but its intellectual and empirical plausibility.

Second, it is mistaken to appeal to theory correction at conjunction as rendering that conjunction irrelevant to the issue. The necessity for correction arises from another theorem concerning preservation of truth, $(A \rightarrow \sim B) \rightarrow \sim (A \cdot B)$, if A and B are not consistent their conjunction cannot be true. So a main motive for correction must be truth. (We may follow scientific practice and set aside trivial entailments, e.g., through some notion of relevant entailment, hence holding that with respect to empirical adequacy the conjunction of two inconsistent theories might well be empirically adequate, for their nontrivial inconsistencies might be confined to their theoretical components.)[20] In any case, the original form of the conjunction argument must apply to the corrected theories. Again, van Fraassen might consistently reply by appealing to enlarged empirical adequacy as the real long-run goal and to consistent model structure as the intermediary. Thus, whether conjunction involves correction or not makes no difference to the argument.

The third claim begs the question. It is not denied that testing joint empirical adequacy is a primary motivation of conjunction. But enough has already been said in discussion of the first claim to indicate that basic theo-

---

20. In terms of truth functions

$$(\sim A \lor \sim B) \supset \sim (A \cdot B)$$

Notice that if the negation sign is interpreted as "is not empirically adequate," that is, as "no model has a submodel adequate to the data," then the relationship holds; but, if one attempts to argue from the falsity of either A or B to the failure of empirical adequacy of $A \cdot B$, then the argument fails.

retical understanding in the pursuit of truth can plausibly be looked upon as the independent (though interacting) factor, an equally important reason along with empirical adequacy for examining joint theories. Any attempt to rule out this latter motive would simply beg the question. Van Fraassen might agree, arguing instead that the theoretical motives were all pragmatic, hence not cognitively significant. This would only return us to an earlier argument, one where I believe the realist has a distinctive advantage.

There is no need to rehearse the continuity objection to CE further; it is clear that the argument would run a similar course to that of the conjunction argument.

In sum, neither objection is decisive, but both add weight to the argument that, to support itself, CE must advocate a strained and fragmented picture of science.

The point might alternately be put in this way: Recalling the dependence of observability and hence empirical adequacy on the epistemic community concerned (sections 3 and 4, above), we may say that science itself theorizes its own epistemic community and on that basis modifies it.[21] In addition, humans metatheorize (philosophize) science. This sort of complex internal and meta-self-reflection can itself be understood from the point of view of an emerging scientific theory of humankind, the evoluionary theory of self-organizing cognitive systems.[22] The realist picture fits well together as a single, if complex, pattern. Again, it is possible to ignore this picture, gather up the apparatus of models, empirical cores, and agnosticism, and give an account in terms of joint models of theory and theorizer and of metamodels of model making, etc. But unless there is a definite reason for it, why would it be reasonable to do so when the realist alternative gives a more unified, more deeply insightful theory of what is happening?

V

In all of this discussion the matter of metaphilosophical commitment has been set aside. The reader will recall my summary of empiricist meta-

21. See the discussion in note 18. In 1980b I asked about the fate of objectivity in science if cognitive demands were matched by selective breeding, resulting in a human community split into the "understands" and the "understand-nots." The same might apply to surgical intervention to enlarge perceptual range or to modify thought processes. Empiricism sometimes smacks of an old-fashioned politics, holding to an almost mystical equality among people, no matter how diverse their personalities and abilities, nor how radical the impact of our interventions upon them. Are we to be forever stuck with Everyman's perceptions? (And not because any other way is risky; *every* way is risky.)

22. For an introduction to this rapidly expanding area, see for example Jantsch 1980; Maturana and Varela 1980; Prigogine 1980; Varela 1979; Yovits, et al. 1962.

philosophical commitments in section 2, above. It is now time to inquire of the corresponding commitments behind CE.

Van Fraassen never explicitly mentions metaphilosophy and, with his general aversion to speculation and commitment, might well eschew it altogether, perhaps along the conventionalist lines of M_5 for empiricism (section 2, above).[23] Certainly occasional comments reveal a lack of awareness of the often intimate relations between philosophical and metaphilosophical commitments; e.g., van Fraassen remarks that interpretation of science and methodology of science are two independent doctrines (p. 93), but this is not even strictly true of empiricist metaphilosophy, for choice of empiricist methodology is heavily constrained by empiricist interpretation of science (cf. the bearing of M_2, M_3 on M_4), and it is strongly false of the realist metaphilosophy I shall shortly describe. (Nevertheless, the lack of interest in metaphilosophical matters and the assumption of interpretation/method independence are characteristic of empiricism). However, and as with earlier empiricists, *we* can ask what system of metaphilosophical commitments makes most sense of CE, what would be needed to defend CE, whether or not these commitments occur explicitly in the text.

I think that two of the empiricist commitments very clearly stand behind CE as well, viz., M_2 and M_4. M_2, because van Fraassen's primary effort is devoted, as we have seen, to explicating key philosophical concepts and principles in terms of logical structure in a newly enriched formal language, that of model-theoretic semantics. M_4, because the thrust of the distinction between cognitively directed action and pragmatic action is in line with demarcating the rational pursuit of cognitive ends from all else. Moreover, within cognitive action the sole focus of rational choice concerns the obtaining of empirical adequacy; other factors in theory choice are also pragmatic.

To my mind, CE also embodies the spirit of M_3 and M_6. M_3 because the insistence on agnosticism about all beyond empirical adequacy suggests an attachment to perceptual experience as carrying an epistemological surety that the rest of science does not; correspondingly, empirical adequacy is appealed to as the ultimate ground of knowledge claims in science. However, the requirement of certainty must likely be relaxed; and there is no commitment to the last clause except insofar as the mind must serve as the intuitive source of doctrine satisfying M_6 and M_1. As for M_6, van Fraassen's book is written on the tacit assumption that philosophers can provide a philosophy of science from within the resources of philosophy alone (which they can if M_2 holds universally of philosophy) and that this doctrine should dictate the approach to science. This assumption sur-

---

23. There is a "Kantian" version of M_5, M_5^*; see note 2. I do not think it sits at all well with the spirit of empiricism, but it has its own attractions.

faces only when it is contrasted with realism's strong denial of it (see below).

That leaves us with "classical" empiricism's M_1 and M_5/M_5^*.[23] On both of these I remain agnostic as to CE's commitment. As remarked earlier, I tend to favor a form of M_5 as most fitting CE and it is some version of M_1 that sits best with the other doctrines, but there is too little evidence to argue the matter.

All told, CE shows the distinctive pattern of metaphilosophical commitments characteristic of empiricism. This is not surprising, since van Fraassen sees himself as squarely in the empiricist tradition. What it does do, however, is open up a new front for the evaluation of CE; are its metaphilosophical commitments convincing? Indeed, it is this level of debate which I have long argued to be of crucial importance to the realist/empiricist issue (see, for example, 1974, 1975c).

Stated and enlarged upon elsewhere (1975c), the metaphilosophical commitments of the evolutionary naturalistic realism I advocate can be briefly summarized as follows:

MNR$_1$. There is no first philosophy; no group of statements is to be held immune from criticism in the light of other groups of accepted statements.

MNR$_2$. The normativeness of philosophy consists in its being a general theory of the world, able to criticize belief, being, and practice in each area of life and to be criticized for its inadequacy to the subject matter in turn (cf. the theory/experiment relation). [Metaphilosophy has a similar status vis-à-vis philosophy.]

MNR$_3$. The aim of epistemology is to explain how science, and cognition generally, is possible and what are its limitations. [I would now add: and what are the relations of epistemology to personal and social development.]

MNR$_4$. The philosophical theory of language should be founded on a scientific theory of information use and exchange generally, and in humans in particular.

MNR$_5$. The aim of rationality is to construct a theory of the well-functioning mind, where "well-functioning" is informed by scientific theory. [I would now add: Judgment, at once evaluative and factual, and risk, are the basic categories of rationality theory, and rationality is the central concept in the philosophy of human functioning.]

MNR$_6$. Philosophy of science is a theory of science.

These principles are constructed so as to reflect the evolutionary naturalistic scientific image of humankind. As promised, they express a radically different view from CE of the nature and practice of philosophy and of its relations to science in particular.

I hold that this realist view provides a much greater degree of coherence with the scientific image than does the metaphilosophy of CE and that it thereby itself acquires a greater claim on our commitment. If we are, as science suggests, cognitively organized mammals, exploring the world from an original position of ignorance, why would one suppose that philosophic distinctions and capacities were as empiricist metaphilosophy suggests?

Surely logic and rationality are, e.g., subjects we are still exploring, still theorizing about. Don't we have here as our main guides theoretical virtues, e.g., consistency, unity, and adequacy to the phenomena, i.e., to our natures and to our historical practices and needs?

This is a moderately radical proposal, but philosophers regularly underestimate the radicalness of an evolutionary perspective. As already noted, I believe this perspective also to underwrite a model of the individual in which judgment and risk are the basic factors, rationality principles specifying the dynamics. Human beings are self-organizing systems, and this suggests a framework in which fact and value are sophisticated structural differentiations of a common underlying judgmental activity; risk taking in the light of developmental goals shifts to the fore in epistemology, personality theory, and institutional organization theory alike. A study of our kinds of evolutionary ignorances, their generation and impact, assumes a special importance.[24] The evolutionary perspective, then, leads to a profound revision of much of philosophy, as these few brief examples suggest.

Still on the theme of taking the evolutionary metaphilosophy seriously, consider the theory of intelligence, a subject which modern cognitive and computer studies have revolutionized in the past decade; can we confidently assert that epistemology has nothing to learn from it? The reverse is already the case, since cognitive science has been one of the importance factors in the emergence of the class of "economic" models of knowledge in which knowledge claims are viewed as context-dependent, risky, or uncertain commitments. This represents a profound break with traditional epistemology. Our cognitive resources are being invested increasingly in extrasomatic machines and institutions, all designed in line with our present theories but also thereby helping to transform those very theories. As this happens, who can confidently say what the future of cognition is?

Van Fraassen speaks easily of a "completed human biology" (p. 17), but it is not clear that there ever will be such a thing. As science gives us greater scope to intervene in the genetic and physiochemical functioning

24. On the role of judgments underlying fact and value, see, for example, my discussions in 1979b and 1982 and the writings of Sir Geoffrey Vickers (1968, 1970, 1980), from whom I have learned. As to the role of risk taking, modern psychoanalysis has emphasized the extent to which personality formation is a risky organizational enterprise (cf. the moving account in Becker 1973), and it is a commonplace of the economics of the firm and of organizational theory that the initiation and development of any enterprise or organization is a risk-ridden affair. I have only sought to apply these ideas to science (to epistemic institutions) and to cultures generally.

of our species, not to mention the intrusion of extrasomatic devices, the whole idea of a species biology becomes blurred and the future possibilities for *Homo sapiens sapiens* gape wide open—*in vitro* fertilization (achieved), storage and genetic intervention (in process of achievement), and multicellular cloning (achieved) are but the beginning. Only the beguilement of the mind by the empiricist idea of a spectator model of science (in which we ideally simply record truth as it is presented to our senses) could lead to this ignoring of what is nowadays daily news. And if not a completable biology, then not a completable perceptual structure, and hence not a completable notion of observable or empirical—in everything, even van Fraassen's empiricist touchstone, we rely on spinning out a survivable order via our theorizing. Nor will there likely be a completable theory of cognitive science, hence not a completable theory of rational cognitive strategy, and hence not a completable philosophy of science. I cannot help but think that, measured against the real excitement and danger of the human adventure, empiricism has a meanness of vision, no matter how elegantly expressed; that it is rather like Blake's Newton gazing at the ground when overhead, unseen, the cosmos is ablaze with light.

But I recognize that this last sentiment is unfair, to van Fraassen anyway, for he aims to capture much of the rich complexity of science, and his antirealism is promulgated in the interests of advocating the merits of caution and of protecting us from the excesses of speculation. This is, then, the time to say candidly that realism has had its excesses and has had to undergo modification to survive. The balance of the essay will be restored if these matters are frankly addressed, which I do in the next section.

VI

Realism in science has no doubt often been fostered by a belief that in science we really do come to know the world once and for all with a degree of surety that warrants the claim to knowledge. This naïve realism has a sister theory of perception, naïve direct realism, according to which we directly perceive the world exactly as it is. Both naïve realisms have had to give way before our historical experience. We have today strong inductive grounds for believing that all of our extant theories of any given time will eventually be overthrown, possibly in favor of theories with substantially different structures which are more than mere refinements of them. And we have good scientific reason to believe that perception is interpretation-laden and replete with idiosyncratic features, all of which give rise to characteristic illusions and limitations. Thus, naïve realisms have to become critical or sophisticated if they are to remain plausible. All this was established a century or more ago.

Recently discussions of realism have especially focused on the writings of Putnam, Boyd, and others who have committed their version of realism to the following group of ideas and doctrines:

A. There is a general progression of science toward increasingly accurate and general theories; where theories have been successful in the past, their successes have been, rightly, incorporated into succeeding theories, usually as limiting cases of the successor.

B. Successful theories are so because their theoretical terms really do succeed in referring to what there is, and the theories assert true, or approximately true, relations among those entities.

C. B explains the historical achievement A asserts and also why the historical practice described in A is the right one and B offers the only plausible explanation of the success and history of science, which would otherwise be a mystery.

Unfortunately for this version of realism, many of these claims, or those that support them, are open to serious question, and van Fraassen (chap. 1) and Laudan (1981) between them have leveled a very serious critique of them. In considering theories T_i from the history of science, Laudan, for example, makes the following points: (1) The mere fact that a T_i's terms succeed in referring doesn't guarantee or explain its empirical success and, vice versa, T_i's empirical success doesn't guarantee or explain referential success for T_i's terms, the history of science being replete with theories successful at a given time and now judged to be nonreferring; (2) T_i's being approximately true doesn't guarantee that T_i's terms refer, and vice-versa; moreover, even were it given that no T_i whose terms failed to refer could be approximately true, approximate truth would not be guaranteed by empirical success; and, finally, no realist has furnished a workable definition of approximate truth; (3) historical intertheory relations in general do not exhibit the subsumption-as-limiting-case relation; furthe:, that relation, even if it held, would guarantee neither reference for the subsumed T_i's terms nor even approximate truth for the subsumed theory, and no other plausible realist intertheory relation holds historically. Van Fraassen also levels some of these criticisms indirectly, adding directly that it is equally possible to explain the success of theories as simply the elimination of empirically unsuccessful by empirically more successful theories, in quasi-Darwinian fashion, avoiding any reference to theoretical terms or more global approximate truth.[25]

25. For the latter, see p. 40. The argument is superficially enticing but a slippery one for van Fraassen, because it does suggest looking for a *theory* of the selecting environment and of the internal dynamics of environmental-species interactions *and* for a *theory* of the internal organizations of the evolving species, all of which together would explain the ensuing dynamics. But the demand for the same theories in the present case is natural for my own position and anathema for van Fraassen.

Many of these points are quite valid and by no means confined to critics of realism. I have myself, for example, emphasized the complexity of historical intertheory relations, that retention-as-a-limiting-case does *not* guarantee any substantive reduction relation, that empirical success does *not* guarantee either reference or many versions of approximate truth.[26] At the same time I have argued at length for realism.

But, before turning to other doctrines of realism, let us briefly consider where these criticisms leave the Putnam, et al., realism. First, though it leaves this version of realism in a greatly weakened position, it is not quite so weak a position as Laudan makes out. Partly by prejudicial formulation,[27] partly by overstatement,[28] Laudan makes out a stronger-seeming case than his argument sustains. Second, it must be admitted that these criticisms do undermine important parts of the doctrine, e.g., at least the second part of A, all of B as it stands, and, on any acceptable version of B, C clauses two and three. However, third, there is a version of B which I think defensible: B'. Successful theories T_i are so because their basic (presumably theoretical) terms either (1) really do refer or (2), while not referring as intended, there is a true theory in which the T_i theoretical descriptions can be approximated (the T_i entities can be "aped"), and (3) T_i asserts true or approximately true relations among them. Moreover, in my view, B' is connected to the structure of understanding in explanation, and it is suited to the use of Sellars's version of intertheory relations, which Laudan has done nothing to refute.[29] And this complex of doctrines *does*, I

26. See most explicitly 1979a and 1981c, but for earlier comments see also 1974.

27. For example, by his dropping that part of my claim B (sec. 6) from "and" onward; see his "S" theses at pp. 23–24. Truncating my B in this manner yields a weak thesis which is too easily attacked.

28. For example, the only account of approximate truth examined in any detail is Popper's, but it is quite idiosyncratic. Surely a more natural place to look for a definition of approximate truth is to model theoretic notions. And when Laudan says that "within a successor T_2, any genuine realist must insist that T_1's underlying ontology is preserved in T_2's, *for it is that ontology above all which he alleges to be approximately true*" (Laudan's italics), he is simply uttering a nonsequitur, as field theory—atomic theory transitions demonstrate (cf. Hooker 1973c), thereby again setting up a straw man.

Similarly, Laudan is unjust to Sellars's suggestion that a successor theory T' should explain why its predecessor T was as successful as it was. Laudan recognizes that this is an improvement over the rigid demand that T' explain T but argues that this accomplishment is irrelevant to the epistemic appraisal of T' (or T). But I understand Sellars to be suggesting that an adequate successor T' to T should be able to model T adequately within its resources, where this latter involves modeling the conditions under which T is true or approximately true and those under which it is not in such a way that all of T's true or approximately true consequences become true or approximately true consequences of T' (though not necessarily under T''s ontology) while none of the other of T's consequences do. But this is no more than to demand that T' preserve the empirical evidence and reflect the structural reasons for that evidence (albeit in a critical manner, as represented in T'), without which T' would be disconfirmed. And this *is* relevant to appraising T' (and T).

29. On explanation and understanding see Hooker 1980a, where I link understanding to deeper ontologies and hence to approximate modeling and approximate truth for shallower ones. (Cf. Munèvar 1981; Hooker 1976, 12–14; also Hooker 1975a, 1974.) For Laudan on Sellars, see note 28.

still think, explain the success and history of science, though, as van Fraassen's alternative makes clear, not uniquely so.[30] Finally, fourth, this leaves the realist with a substantial framework within which to address the dynamics of science but not with the argument for realism which A→C was intended to provide. So be it.

If other realists run afoul of these criticisms, then it should be recognized that by no means all versions of realism are involved. The historical record which Laudan cites to such effect is damning of simple structural models of the dynamics of science, but I have long argued that this only represents one more naïveté to be surpassed in the passage to an adequate realism.

However, there are in addition specific features of contemporary science which seem, or have seemed, especially recalcitrant to realist analysis, but reflection suggests that the evidence is quite ambiguous. Earlier this century, relativity theory was often hailed as a triumph for the empiricist-operationalist method; Eisenstein himself occasionally dressed it up that way. Recent analyses have shown that, to the contrary, a realist analysis is both possible and plausible, indeed that the drift in the evolution of the theory is toward an objective absolutism (of a relativistic kind).[31] Of course, van Fraassen can sidestep all this as I have already indicated, but the move to remove the presence of a substantive space-time from the models of the theory has faltered and will likely fail.

Another case often cited in this same connection is modern quantum theory. The theory is strikingly empirically adequate, well understood mathematically but poorly understood conceptually and ontologically. If CE was the rational position to hold, then quantum mechanics should be a paradigm of a successful, satisfying scientific theory. Some scientists do take the empirical success, combined with the interpretational obscurity, of the theory as proof that empiricism is the correct attitude, but this has not been widespread. In fact, no theory has drawn more interpretational discussion in the history of science, and not just (or even mainly) by philosophers but primarily by scientists themselves seeking theoretical understanding. There are also many scientists, it must be admitted, who point to the empirical adequacy and say, "Enough, get on with experiment," but these scientists have really dismissed the interpretational issue rather than decided it, and they are seen to have their pragmatic motivations. These scientists will be satisfied by crude empiricist renditions like that of the physicist Ballantine,[32] and they ought to be satisfied by the elegantly worked

30. Here I repeat that van Fraassen's explanatory tack may not bear deeper reflection; see note 25.

31. For literature arguing in this way, see, e.g., Nerlich 1976 (cf. my review [1981a]) and Hooker 1971.

32. Ballantine 1970. It is my view that this doctrine pretty clearly represents a strategy of emasculating by fiat the semantic content of quantum mechanics until it says only what is compatible with empiricism—always a *possible* approach.

out but ontologically noncommittal models van Fraassen himself has provided (chap. 6 and 1972). The irony is that, for all its elegance, van Fraassen's interpretation is being, and is bound to be, largely ignored by scientists just because it doesn't treat the theoretical problems seriously in the sense required.[33] With few (any?) exceptions, all of the outstanding scientists of this century have worried the problem, seeking theoretical insight.

It must be admitted, however, that relativity and quantum theory still pose difficult unresolved problems for realism. Added to the retreat from naïve forms of the doctrine documented above, it makes it pertinent to ask for the surviving distinguishing core of the position.

In an earlier essay (1974), I cited as the distinctively realist claim the thesis that "the intended and proper sense of the theories of science is as literal descriptions of the physical world." This should be compared with van Fraassen's version (p. 8, quoted earlier). It is now clear that the semantical thesis by itself may be necessary but is not sufficient, for, as already noticed, van Fraassen, too, adopts a position in which the full theoretical content determines the truth conditions (p. 197, quoted in section 3). For van Fraassen the conflict with the realist comes over rational commitment and, correlatively, over the (rational) aim of science.

I think it is correct to locate one locus of the dispute at this level, and here the distinctive tenets of realism are surely these: that there is a reality independent of our intellect and sentience with which we interact and represent to ourselves in theories, that observable and unobservable features of a theory are on an equal footing ontologically and epistemologically, and hence that it is the overall value of a theory, of which empirical adequacy is but one component, which should determine our commitment to it and that our most valuable theories are our most acceptable guides to the nature of that reality.

These are propositions which are not relinquished in the retreat from naïveté and which CE must deny. (One might add, CE is committed to this denial however much of the resulting realist account of science CE claims to mirror within itself. Indeed, part of the burden of argument in section 4 is that CE gives recognition to all too much of scientific practice to remain plausibly empiricist.)

In spelling out realism, however, it is also necessary not to be naïve about realist rational commitment. As van Fraassen remarks of his own

33. That is, it doesn't yield an insightful physical ontology (cf. note 29). On quantum theory and ontology, see Hooker 1972, 1973c. For a review of the literature, see for example Jammer 1966, 1974. A good example of a physicist's physical intuition at work is given by Prigogine (1980). In my view, even so-called realist versions of the quantum logical approach have suffered from a tendency to be divorced from physical understanding and an appreciation of the intuitive problems to be solved (cf. my discussions in 1972, 1973c), but they have nonetheless rejected van Fraassen's maneuvers in an effort to do more justice to the theoretical structures (see, for example, Hooker 1973a, 1975b, 1979c).

formulation, "It does not imply that anyone is ever rationally warranted in forming such a belief" (p. 9). Van Fraassen had in mind someone who only ever attached probabilities less than one to scientific theories, but earlier (1976, 13–15) I had suggested a context-dependent formulation which I still find more fitting to our actual circumstances: there is a complex of epistemic contexts; rational commitment is always commitment relative to one of these.

Relevant contexts are a "commonsense" specified practical setting (e.g., carpentry), a theory specified practical setting (e.g., laboratory experiment), a theory-field (or paradigm?) specified setting (e.g., "normal science" in any discipline), an epistemic institution specified setting (e.g., arguing competing gravitational theories within the codes of argument currently specified as scientific), and a cultural or perhaps transcultural setting (e.g., debate about the limitations of science as a way to knowledge, debate about the idiosyncrasies of human intelligence). For some purposes it may be useful to draw finer distinctions, and these are probably not the only sorts of context involved; but they suffice to indicate what is intended. Commitment to a proposition P is relative to acceptance of a context S. But it would be wrong to adopt a simple hierarchical structure to contexts to express the structure of context acceptance. Contexts exhibit complex cognitive interrelations for, as indicated in the realist metaphilosophy (section 4), commitments in one context can influence commitments within, and acceptance of, other contexts. (E.g., commitment to a theory of human intelligence may well affect acceptance of science, in some theoretical characterizations, as the best route to knowledge.) The distinction among contexts is like Carnap's internal/external distinction pluralized, except that the realist holds all acceptance and commitment decisions to be cognitive and mutually influencing. Or, since Quine demolished the formal significance of Carnap's distinction, like epistemic propositional attitudes partitioned across Quine's web of belief.[34]

With respect to realist attitudes to commitment/acceptance, then, the position I have been sketching earlier suggests the following structure: In all theory-specified contexts, the theory is accepted as literally true and the commitments, whether to belief or action, are as determined by the theory; (but) in all theory-field specified contexts the general theoretical propositions underpinning the research program are accepted as true and theory commitment is only to what follows from these, while the details of subprograms for particular problems may be practical commitments only

34. On Carnap and Quine, see note 14. The web metaphor is spelled out in Quine and Ullian 1970. The web analogy is suggestive, its chief defect from my point of view being its tendency to obscure the substitution of rational commitment criteria for some combination of purely logical structure and noncognitive pragmatic choice. Cf. notes 5 and 9 and accompanying text.

(though context acceptance might dictate commitment as true); (but) in epistemic institution specified contexts the acceptance is of scientific procedure, and commitments in this context are thus constrained but are compatible with agnosticism or skepticism about the successfulness of any one theory or research program; and so on. In sum, commitment at one level is compatible with skepticism at another, but not necessarily "higher," level; one may be committed, for example, to a particular theory of human intelligence and hence be relatively more skeptical of science as theorized in a certain way. The whole interacting "web" of acceptances, agnosticisms, skepticisms, and commitments is then juggled in the light of the evolving totality of theoretical insight and practical experience. I have no good theory of precisely how this is ultimately done, and there may be none.[35]

As remarked, none of this is essentially new. But with respect to realism it allows two important points to be made: (1) This debate on the structure of acceptance/commitment does not touch the core propositions of realism; hence (2) realists are not forced into the naïve position of not being able to be critical of science, even radically criticial, just because they are realists. To try to impose some sort of hierarchical subsumption strucure on acceptances is just to ignore the introduction of risk and trade-off to epistemology, to deny an "economic" theory of knowledge and insist instead that logic suffices to structure epistemic attitudes.[36]

We have now touched on the final locus of realist/empiricist difference, the one not addressed by van Fraassen: the conflict over metaphilosophical commitment. It is clear that realism as conceived here differs fundamentally with empiricism over the aims of epistemology and rationality theory. The two also conflict over the significance and role of logic in philosophical theory and normally conflict over the status of philosophical theory (though in the case of van Fraassen's CE it's hard to tell). But behind these differences lies a more general and more fundamental difference: realism insists on informing the basic content of philosophical theory with the scientific image; empiricism maintains the separation of philosophy and science. This is the basic difference because it leads to the general differences over the aims of philosophical theory just mentioned and, via them, to the other conflicts over theory of science already discussed.

35. There may be none because understanding cannot itself be conceptually understood or because the human person more generally transcends its own realizations (cf. Hooker 1980b) or because at bottom the world is radically nonconceptual.

36. I owe the introduction to a rational commitment approach to epistemology to Leach (1977). The only technically sophisticated development of the idea for a philosophy of science is as yet by Levi (1967), but this is modeled on Popper's position, which I regard as defective (cf. 1981b, 1975c), and it also has its own difficulties (cf. Bogdan 1976; Levi 1979). Though in its infancy, the general idea seems to me to be extremely promising. Cf. also notes 5 and 9 and accompanying text.

There are, then, three loci of difference between empiricists and realists (viz. theory of science, theory of rational commitment, and metaphilosophy). It is idle to hope that the differences between them will fade as empiricism enriches its logic and realism sheds its naïveté, although some differences in their specific descriptions of science may be removed in this manner. It is idle to hope that either of these positions will be broken simply by poking holes in their respective theories of science, although doing this may force change upon them and indirectly weaken their claims on our commitment. Because it fosters the foregoing two hopes, it is misleading to focus the issue solely at the level of their theories of science. Both have survived the criticism flung at them and evolved into more vigorous forms under its selective pressure, but the one with the greater claim on our commitment is, I still think, realism.

I shall now test this conclusion once more by expanding the context beyond the "interior" of science to include the scientific process in its wider social setting.

VII

The basic divergences between realism and CE have been delineated, but this does not exhaust the potential areas of mutual opposition. A last area of disputes remains, the relations between philosophy of science and the larger role of science in the human evolutionary process of the planet, i.e., disputes concerning the social, political, and cultural place and significance of science. I have substantial enough commitments in these areas, as will shortly become clear, and regard the issues as very important. However, there is presently so much controversy about the intellectual significance of the connections I shall discuss that I have deliberately segregated the discussion so that criticisms of CE leveled here can clearly be distinguished from the assessment concluded with section 6.

Science is concerned with creative cognitive construction in the face of evolutionary ignorance. The philosophy of science is itself an example of this process—hence the relevance of metaphilosophical considerations to the assessment of realism (section 5). But science is also the leading edge of the intrusion of cognitive organization into every area of human life, and philosophy of science, broadly construed, has a central part to play in this process—hence the relevance of asking whether CE forms an adequate basis for understanding these social processes.

As to the process itself, I can summarize my own view in this way (see Bjerring and Hooker 1980/81; Hooker 1979b, 1982): The human species has been undergoing a slow but fundamental reorganization around human cognitive organization, a process which has evidently accelerated this century. This process can be conveniently, if artificially, divided into two dimensions, science as remaking the world at large and science remaking itself. The net effect is that, not only is more and more coming within the object of scientific inquiry, but those objects are being increasingly transformed into human artifacts. Artifacts, however, are only partially natural objects, for they exhibit human designs which embed human values, reflecting human choices from among what is possible. Correlatively, the focus of theory shifts from a description of what is, to a theory of what is possible. I shall argue that these fundamental changes in the nature, scope, and human significance of science do not sit well with CE but that they do find a natural framework for their understanding within the kind of realism I have been defending.

The fact that science is remaking the world around it hardly needs emphasizing. From our impact on the global climate, forests, ocean ecologies, hydrology, and so on to the explosion of urban megalopolises, manmade agricultural species and electronic communication networks, and in a hundred other like ways, science-based and -organized technological development has been transforming the biophysical world in quite dramatic ways into a human artifact in new patterns. These patterns have not been universally approved, far from it, but they are human designs nevertheless. At the same time, the social and personal conditions of human life are being similarly transformed. At the personal level we now have neurosurgical intervention, deep-drug therapies and psychotherapies, behavior modification, and so on, all of which make the very boundaries of the self problematical, though in addition there are role model and social learning theories and the like which make the social conception of the individual problematical. And medical intervention now encompasses *in vitro* fertilization, genetic "engineering," and cloning, the ingredients of transforming any species into an artifact. At the social level, to take but one example, new developments in electronic communication and information processing are transforming public and private media, work-role distributions, work/leisure relations, and the basic concepts of money/wealth and of the political process. New planetary institutions have emerged on a scale undreamed of a century ago, made possible, and necessary, by modern science and technology, designed and managed, at least in part, through the use of scientific theories.

It is perhaps less obvious that science has been transforming *itself* in a fundamental way, but this is so. First, there is a growing internal scientific self-consciousness which is transforming the conduct of science. There are several aspects to this, of which philosophy of science is obviously one, for it raises the self-consciousness of scientific methodology, including theory construction and testing, intertheoretical relations, and so on. Another important dimension has been the growing sensitivity to the history of science with its perspectives on long-term developments in science, the history of scientific concepts, changing relations of science to mathematics and technology, and so on. And, finally, there has been a growing mathematical sophistication concerned with the general foundations of scientific description which has for the first time led to the formulation of major alternatives to basic scientific assumptions.[37]

Second, there is a rapidly developing scientific interest in the institutional nature of science and in its social relations. Historical studies already called attention to the social context for science, but there has been a recent explosion of work in the sociology of knowledge, economics of research and development, science and technology policy and related areas.[38] At the same time science itself, and, of course, technology development, has been institutionalized on a scale that would have been scarcely comprehensible a century ago. Research strategy has a global structure, journal articles number in the millions (even in a single discipline), specialization has reached the point where not even sub-subdisciplines may be able to communicate. The cognitive characteristics of science are increasingly influenced by its institutional structure. Social studies of science and technology have become increasingly aware of this development. Finally, science itself has given birth to a collection of new theories, distinctive of the twentieth century and directed at the management of complex systems (including science itself and its immediate social and environmental relations): systems theory, decision theory, operations research, cybernetics, artificial intelligence, irreversible and network thermodynamics, control theory, and so on. The net effect of these three synergistic processes is that science is increasingly redesigning itself in its own image, even while it is transforming the world around it.

Taking both the external and internal transformations of science together allows one to see the depth of the transformation that has occurred over the past three centuries in the role which science plays in human affairs. In the seventeenth century science might reasonably have been regarded as an intellectual tool of individuals for the objective (because non-interfering) description of the antonomous workings of a given natural object, the world. Beyond that, if science could be applied in humankind's service, it could be understood as a tool like any other, subject to the prior interests of policy, economy, community, and individuals. Today the image of science as a simple tool, intellectual or utilitarian, must be replaced by the concept of a powerful self-organizing and environment-organizing sys-

37. It is hardly necessary to review the post–World War II explosion in philosophy of science. As to the history of science and its role here, see for example the references in Kuhn 1962; Lakatos 1970; Lauden 1981, as well as those in Feyerabend 1965, 1969, 1975. For the mathematical revolution, see, e.g., Hooker 1975b, 1979c, 1979d, and references therein.

38. See, e.g., Blume 1974; Eslea 1973; Gordon and Raffensperger 1969; Knorr, et al. 1975, 1978, 1980; Merton 1973; Pusey and Young 1980; Spiegel-Rösing and de Solla Price 1978; Tisdell 1981; van Melsen 1961; Bjerring and Hooker 1979.

tem. The noninterfering description has been replaced by redesign on a massive scale, the individual by something rapidly approaching a species-wide organization, the autonomous object by a human artifact, and the various prior interests are now increasingly molded by the available images and designs. That this process of species "encephalization" enhances the scope for disaster and the delicacy of survival conditions argues not against either its existence or importance.

What, then, are we able to make of it from within CE? What approaches does CE offer for its understanding and assessment? I believe CE serves only to obscure both the process and what is at stake for humanity in it. CE retains the empiricist view that science is at bottom value-neutral and separate from social processes as such, and it emphasizes fact and logic as the sole cognitive determinants (though not perhaps the sole determinants) of science. These characteristics introduce, in my terms, two profoundly distorting effects.

1. They deflect attention away from the object of science as increasingly an artifact, obscuring the understanding of the nature-to-artifact transformation. Artifacts are not value-neutral. CE must relegate treatment of their evaluative components to some extrascientific realm, say politics. But is this the most helpful way to understand the process? Correlatively, CE encourages an image of pure science as separate from applied science or technology, insofar as the latter is bound up with the pursuit of human interests not determined by fact and logic. (The pure/applied distinction is but an application of the fact/value distinction.) However, the pure/applied distinction cannot be sustained when the object of theorizing is itself an artifact and when technology plays so fundamental a role in experimentation and even conceptual evolution.[39] The point here is not simply to repeat internal criticisms of CE but to draw attention to ways in which the inherent structure of CE represents a barrier to really coming to grips with the actual character of human cognitive processes. In particular, CE discourages any value-based critique of "pure" science, defending its neutrality with entrenched distinctions. In specific cases it might well be appropriate to defend science in this way, e.g., against wanton political intervention.[40] But, in a world in which increasingly the objects of scientific study are themselves human artifacts, it is important to be able to subject

39. See my discussion in Hooker 1982 at notes 2 and 13.
40. Historically, empiricism played an important intellectual role in the resistance to Nazi interference in science. Empiricism supported a vision of a rational, valuable future based on free inquiry and objective improvement. If the hope of disentangling science and value, and both from politics, proved naïve, still the vision has its valuable insights, and it served an important purpose. Marxism can also be so interpreted as to yield unreasonable political interference in scientific processes (and has been historically). It must be similarly resisted, though one now knows that the grounds for doing so must be much more sophisticated than those empiricisms offered if they are to be convincing.

the general directions of all science to evaluative scrutiny. For example, it is at least conjecturable that humans might well be so structured as to make some scientific models of them self-fulfilling prophecies if universally applied. (One is tempted to think of sufficiently crude behavioral models or models of political governance here.) Such a circumstance would require a carefully integrated normative-factual evaluation. It is also important to be able to subject the details of science to the same scrutiny as one now more easily applies to technologies.[41] The really critical questions for our century are (1) how to decide when science is to be defended and when it is to be criticized and (2) how to institutionalize these processes compatibly with proper cognitive and political freedoms. These are difficult and urgent questions in our cognitive and political circumstances. CE's absolute dichotomies reduce their allowable answers to disastrously simpleminded proportions.

2. Historically, empiricism has been associated with the relegation of extrascientific thought and behavior to a cognitively inferior status. In its most extreme form, only causality, not cognition, operated outside science; all extrascientific belief and behavior was to be explained causally, e.g., via Skinnerian reinforcement schedules, but not to be expected to exhibit an intrinsic cognitive aim and rationale. Hobbes at the opening of the modern liberal-empiricist tradition and the twentieth-century positivists at its apogee all held some version of this doctrine. But this view leaves no distinctively ethical-spiritual, social, or rational structure to extrascientific life, after all the major substance of life. The only operative terms that can be rescued for theory construction, besides cause, are power (wealth/force), persuasion, desire satisfaction and their cognates sociopolitically, and efficiency and cognate terms economically. There is a correlative theory of rationality holding that outside of logic there either is no rationality or rationality is purely economic, restricted to choosing efficient means to satisfy desires.

My view that this schizoid conception of cognition and of rationality is internally unsatisfactory has already been explained. But I also hold the view that widespread adoption of this position would be a historical sociopolitical mistake for humans, and not simply an intellectual one. It would be a mistake because it would reinforce a self-interested and cynical conception of social and political life at a time when we desperately need a more humane vision of its possibilities and a more humane practice for our present tenuous civilizing institutions. It would help to demean human culture, giving an unwarranted legitimacy to fashion, sophistry, and propaganda in the hands of the merely powerful. It would aggrandize narrowly

41. See in this connection Martin's stimulating study (1979). Cf. other less direct literature, e.g., Churchland, this volume, chap. 2; Livingstone and Mason 1978; Rose and Rose 1969; Sahlins 1976.

economic assessments of courses of action, where efficiency is ultimately dominated by the wealthy and otherwise powerful, when humans badly need a more ecologically and socially sane process of social choice. The cultures of Western nations already suffer enough from these distortions of an extreme liberalism reinforced by positivist social sciences, without masking them with a plausible-seeming theory.

Van Fraassen might well reply that these criticisms reflect a self-righteous, moralizing approach to sociopolitical life and culture, an approach that exhibits an inevitable tendency to authoritarianism and ultimately to totalitarianism. However, he would certainly add (cf. p. 14), it fails to reflect our real and basic capacity to be skeptical, to enter life-styles and discourses experimentally or even pragmatically, to change our commitments. And I would concede that authoritarianism is a danger of the position I have been outlining, though not an inevitable outcome of it. I would also agree that we do have the capacities just mentioned, at least within limits. And I would join van Fraassen in opposing authoritarianism and in insisting on recognition of basic human capacities. But, though I have no utopian formula for the culture well balanced between individualist antiauthoritarianism and social value commitment, I am equally sure that it lies to the commitment side of van Fraassen's position. And, though I have no insightful formulas to capture the proper balance between skepticism and commitment, I am sure that it lies well to the commitment side of always playing at accepting theories and practices. (There is a tendency in van Fraassen's approach to encourage a nihilist theory of life-style.)

Van Fraassen might well reply that these areas of dispute lie well beyond the philosophy of science and that it is unfair to assess his view in a perspective much vaster than the one to which he confined himself. He might go further and claim that it was a straight intellectual blunder to attempt this kind of assessment. I am sure that the latter claim would only be begging the question and fairly confidently conjecture that it is false, and hence conclude the first claim unwarranted. But I have no neat account to offer of how we intelligently juggle narrowly epistemic and broadly evaluative commitments; I am only sure that we do and that it seems to be of the essence of our circumstances and natures that we do so. For myself, I am inclined to regard the evaluative dimensions as also cognitive, properly understood, and to attempt to see the process as some kind of unified cognitive exploration strategy.[42] But we are now well into controversial and speculative waters. I rest content with the insistence that a more complex, more integrated theory than CE's dichotomies will permit is required if the character of evolutionary strategies is to be understood and the challenges of our present circumstances are to be met.

42. See the model tentatively explored in Hooker 1979b and cf. Hooker 1978, 1980b.

References

Achinstein, P., and S. Barker. 1969. *The Legacy of Logical Positivism*. Baltimore: Johns Hopkins Press.

Ballantine, L. E. 1970. "The Statistical Interpretation of Quantum Mechanics." *Reviews of Modern Physics* 42:358–81.

Becker, E. 1973. *The Denial of Death*. New York: The Free Press.

Bjerring, A. K., and C. A. Hooker, 1979. "Process and Progress: The Nature of Systematic Inquiry." In *Perspectives in Metascience*, edited by J. Barmark, Berlings Sweden: Lund.

———. 1980/81. "The Implications of Philosophy of Science for Science Policy." In *The Human Context for Science and Technology*, edited by C. A. Hooker and T. Schrecker. 2 vols. Ottawa: Social Sciences and Humanities Research Council of Canada.

———. 1981. "Lehrer, Consensus and Science: The Empiricist Watershed." In *Profiles: Keith Lehrer*, edited by R. J. Bogdan. Dordrecht, Holland: Reidel.

Blume, S. 1974. *Toward a Political Sociology of Science*. London: Collier-Macmillan.

Bogdan, R. J. 1976. *Local Induction*. Dordrecht, Holland: Reidel.

Carnap, R. 1956. "Empiricism, Semantics and Ontology." *Revue Internationale de Philosophie* 4 (1950). Reprinted in *Meaning and Necessity*. 2d ed. Chicago: University of Chicago Press.

Churchland, P. M. 1979. *Scientific Realism and the Plasticity of Mind*. Cambridge: Cambridge University Press.

———. 1985. This volume, chap. 2.

Eslea, B. 1973. *Liberation and the Aims of Science*. London: Chatto and Windus (for Sussex University Press).

Feyerabend, P. K. 1965. "Problems of Empiricism." In *Pittsburgh Studies in the Philosophy of Science*, vol. 2, edited by R. Colodny. Englewood Cliffs, N.J.: Prentice-Hall.

———. 1969. "Problems of Empiricism II." In *Pittsburgh Studies in the Philosophy of Science*, vol. 4, edited by R. Colodny. Pittsburgh: University of Pittsburgh Press.

———. 1975. *Against Method*. London: New Left Books.

Friedman, M. 1974. "Explanation and Scientific Understanding." *The Journal of Philosophy* 49:250–61.

Goldstick, D. 1979. "The Tolerance of Rudolph Carnap." *Australasian Journal of Philosophy* 49:250–61.

Gordon, T. J., and M. J. Raffensperger. 1969. "A Strategy for Planning Basic Research." *Philosophy of Science* 36:205–18.

Harman, G. 1973. *Thought*. Princeton: Princeton University Press.

Hooker, C. A. 1968. "Craigian Transcriptionism." *American Philosophical Quarterly* 5:152–63.

———. 1971. "The Relational Doctrines of Space and Time." *The British Journal for the Philosophy of Science* 22:97–130.

———. 1972. "The Nature of Quantum Mechanical Reality: Einstein versus Bohr." In *Paradoxes and Paradigms*, edited by R. Colodny. Vol. 5 of *Pittsburgh Studies in the Philosophy of Science*. Pittsburgh: University of Pittsburgh Press.

————, ed. 1973a. *Contemporary Research in the Foundations and Philosophy of Quantum Theory.* Dordrecht, Holland: Reidel.

————. 1973b. "Empiricism, Perception and Conceptual Change." *Canadian Journal of Philosophy* 3:59–75.

————. 1973c. "Physics and Metaphysics: A Prolegomnon for the Riddles of Quantum Theory." In *Contemporary Research in the Foundations and Philosophy of Quantum Theory,* edited by C. A. Hooker. Dordrecht, Holland: Reidel.

————. 1974. "Systematic Realism." *Synthese* 26:409–97.

————. 1975a. "Global Theories." *Philosophy of Science* 42:153–79.

————. 1975b. *The Logico-Algebraic Approach to Quantum Mechanics,* vol. 1. Dordrecht, Holland: Reidel.

————. 1975c. "Systematic Philosophy and Meta-Philosophy of Science: Empiricism, Popperianism and Realism." *Synthese* 32:177–231.

————. 1976. "Methodology and Systematic Philosophy." In *Proceedings, 5th International Congress on Logic, Methodology and Philosophy of Science,* vol. 3, edited by R. E. Butts and J. R. Hintikka. Dordrecht, Holland: Reidel.

————. 1978. "An Evolutionary Naturalist Realist Doctrine of Perception and Secondary Qualities." In *Minnesota Studies in the Philosophy of Science,* vol. 9, edited by C. W. Savage. Minneapolis: University of Minnesota Press, 1978.

————. 1979a. Critical Notice of *Reduction in the Physical Sciences,* by Yoshida. *Dialogue* 18, no. 1:81–99.

————. 1979b. "Explanation and Culture." *Humanities in Society* 7:223–44.

————. 1979c. *The Logico-Algebraic Approach to Quantum Mechanics,* vol. 2. Dordrecht, Holland: Reidel.

————. 1979d. *Physical Theory as Logico-Operational Structure.* Dordrecht, Holland: Reidel.

————. 1980a. "Explanation, Generality and Understanding." *Australasian Journal of Philosophy* 58:284–90.

————. 1980b. "Science as a Human Activity, Human Activity as a . . ." *Contact* 12.

————. 1981a. Critical Notice of *The Shape of Space,* by G. C. Nerlich. *Dialogue* 20:783–98.

————. 1981b. "Formalist Rationality: The Limitations of Popper's Theory of Reason." *Metaphilosophy* 12:247–66.

————. 1981c. "Towards a General Theory of Reduction, Part I: Historical Framework." *Dialogue* 20:38–59.

————. 1982. "Scientific Neutrality versus Normative Learning: The Theoretician's and the Politician's Dilemma." In *Science and Ethics,* edited by D. Oldroyd. Kensington, N.S.W.: NSW University Press, 1982.

Hooker, C. A., with J. Leach and E. McLennan. 1977. *Foundations and Applications of Decision Theory.* 2 vols. Dordrecht, Holland: Reidel.

Jammer, M. 1966. *The Conceptual Development of Quantum Mechanics.* New York: McGraw-Hill.

————. 1974. *The Philosophy of Quantum Mechanics.* New York: Wiley.

Jantsch, E. 1980. *The Self-organizing Universe.* New York: Pergamon.

Knorr, K. D., R. Krohn, and R. Whitley, eds. 1980. *The Social Process of Scientific Investigation.* Dordrecht, Holland: Reidel.

Knorr, K. D., H. Strasser, and H. G. Zillian, eds. 1975. *Determinants and Controls of Scientific Development*. Dordrecht, Holland: Reidel.

Krohn, W., E. T. Layton, Jr., and P. Weingart, eds. 1978. *The Dynamics of Science and Technology*. Dordrecht, Holland: Reidel.

Kuhn, T. S. 1962. *The Structure of Scientific Revolutions*. Chicago: University of Chicago Press.

Lakatos, I. 1970. "Falsification and the Methodology of Scientific Research Programmes." In *Criticism and the Growth of Knowledge*, edited by I. Lakatos and A. Musgrave. Cambridge: Cambridge University Press.

Lauden, L. 1981. "A Confutation of Convergent Realism." *Philosophy of Science* 48:19–48.

Leach, J. J. 1977. "The Dual Function of Rationality." In *Foundational Problems in the Special Sciences*, edited by R. E. Butts and J. Hintikka. *Proceedings, Fifth International Congress of Logic, Methodology and Philosophy of Science*. Dordrecht, Holland: Reidel.

Levi, I. 1967. *Gambling with Truth*. New York: Knopf.

———. 1979. "Inductive Appraisal." In *Current Research in Philosophy of Science*, edited by P. D. Asquity and H. E. Kyburg, Jr. East Lansing, Mich.: Philosophy of Science Association.

Livingstone, D. W., and R. V. Mason. 1978. "Ecological Crisis and the Autonomy of Science in Capitalist Society: A Canadian Case-Study." *Alternatives* 8:3–10.

Martin, B. 1979. *The Bias of Science*. Canberra: Society for Social Responsibility in Science.

Maturana, N. R., and F. Varela. 1980. *Autopoiesis and Cognition: The Realization of the Living*. Dordrecht, Holland: Reidel.

Merton, R. K. 1973. *The Sociology of Science*. Chicago: University of Chicago Press.

Munèvar, A. 1981. *Radical Knowledge*. Hackett.

Nerlich, G. 1976. *The Shape of Space*. Cambridge: Cambridge University Press.

Popper, K. R. 1973. *Objective Knowledge*. Oxford: Oxford University Press.

Prigogine, I. 1980. *From Being to Becoming*. San Francisco: W. H. Freeman.

Pusey, M., and R. Young. 1980. *Knowledge and Control*. Canberra: Australian National University Press.

Quine, W. V. O. 1966. "On Carnap's Views on Ontology." In *The Ways of Paradox*, by W. V. O. Quine. New York: Random House.

Quine, W. V. O., and J. A. Ullian. 1970. *The Web of Belief*. New York: Random House.

Rose, H., and S. Rose. 1969. *Science and Society*. Harmondsworth, England: Penguin.

Sahlins, M. 1976. *The Use and Abuse of Biology*. Ann Arbor, Mich.: University of Michigan Press.

Spiegel-Rösing, I., and D. deSolla Price, eds. 1978. *Science, Technology and Society: A Cross-disciplinary Pespective*. London: Sage.

Suppe, F., ed. 1974. *The Structure of Scientific Theories*. Urbana, Ill.: University of Illinois Press.

Tisdell, C. A. 1981. *Science and Technology Policy*. London: Chapman and Hall.

Van Fraassen, B. C. 1972. "A Formal Approach to the Philosophy of Science." In *Paradigms and Paradoxes*, edited by R. G. Colodny. Pittsburgh: University of Pittsburgh Press.

——. 1980. *The Scientific Image*. Oxford: Clarendon Press.

Van Melsen, A. G. 1961. *Science and Technology*. Pittsburgh: Duquesne University Press.

Varela, F. 1979. *Principles of Biological Autonomy*. New York: Elsevier.

Vickers, G. 1968. *Value Systems and Social Process*. London: Penguin.

——. 1970. *Freedom in a Rocking Boat*. London: Penguin.

——. 1980. *Responsibility—Its Sources and Limits*. Seaside, Calif.: Intersystems.

Yovits, M. C., et al. 1962. *Self-organizing Systems*. Washington: Spartan Books.

9 Realism Versus Constructive Empiricism

Alan Musgrave

The demise of logical positivism has been followed by a rising tide of scientific realism. Bas van Fraassen is to be congratulated for swimming against that tide. But we must also ask whether he manages to make much headway. I shall argue that he does not. My first section explores van Fraassen's rather attenuated antirealism and the distinction between truth and empirical adequacy on which it depends. My second section argues that van Fraassen succeeds no better than his predecessors in answering a major objection to antirealism. My third section examines the link between realism and explanation and van Fraassen's attempt to sever that link.

I. Truth, Empirical Adequacy, Empirical Equivalence

Scientific realism is an old issue, and over the years both realism and antirealism have taken various forms. Van Fraassen defines realism thus: "*Science aims to give us, in its theories, a literally true story of what the world is like; and acceptance of a scientific theory involves the belief that it is true.*" He says that this is a minimal formulation which "can be agreed to by anyone who considers himself a scientific realist."[1] Later, however, van Fraassen extends this minimal formulation by adding to it a realist 'demand for explanation.' As we will see, his antirealism stems in large part from criticisms of this demand. As we will also see, his version of the demand is an absurdly strong one.

What is the nature of van Fraassen's antirealism? The most radical opponents of realism (the instrumentalists) deny that scientific theories have truth-values at all. Van Fraassen's antirealism is not of this radical kind. He accepts a "literal construal of the language of science" whereby "the ap-

Previously published in a shorter form as "Constructive Empiricism versus Scientific Realism," *Philosophical Quarterly* 32 (July 1982): 262–71. Reproduced by permission. I am grateful to Cliff Hooker for his helpful suggestions about what I might focus attention upon in the paper and to Greg Currie, Bob Durrant, and Martin Fricke for comments on earlier versions.

1. *The Scientific Image* (Oxford: Clarendon Press, 1980), 8. Henceforth, all page numbers in the text refer to this book.

parent statements of science really are statements, *capable of* being true or false" (p. 10). In the same vein, he rejects positivist interpretations of scientific language, whereby the 'real meaning' of theoretical assertions is somehow cashed out in terms of the observable:

> Most specifically, if a theory says that something exists, then a literal construal may elaborate on what that something is, but will not remove the implication of existence. . . . If the theory's statements include "There are electrons," then the theory says that there are electrons. If in addition they include "Electrons are not planets," then the theory says, in part, that there are entities other than planets. (P. 11)

Thus, contrary to the positivists, two theories may say exactly the same thing about the observable yet remain distinct and perhaps incompatible theories.

All this puts van Fraassen firmly in the realist camp as far as the *interpretation* of scientific theories is concerned.[2] His antirealism proceeds entirely on the *epistemological or methodological* level. (The same can be said of the antirealism espoused by Larry Laudan in *Progress and Its Problems*.) He thinks that, although scientific theories are capable of literal truth, they "need not be true to be good" (p. 10). Accordingly, it is not the aim of science to provide true theories, and to accept a theory is not to believe it to be true. What matters in science is that theories are correct so far as the observations and experiments go. Hence, constructive empiricism: "*Science aims to give us theories which are empirically adequate; and acceptance of a theory involves as belief only that it is empirically adequate*" (p. 12). A theory is empirically adequate "exactly if what it says about the observable things and events in this world, is true—exactly if it 'saves the phenomena'." (p. 12).

The distinction between truth and empirical adequacy, and hence between realism and constructive empiricism, is a subtle one. For theories about the observable, truth and empirical adequacy coincide (p. 21). For theories about the unobservable, truth entails empirical adequacy but not

2. Here and in what follows I ignore, through lack of space, a central feature of van Fraassen's position, his preference for a semantic approach to scientific theories whereby they emerge as sets of models rather than as sets of (true or false) sentences. I have two excuses for this. First, in much of his own discussion van Fraassen ignores it too, and talks as though theories consisted of true or false sentences. Second, and more important, I think that there is little to choose between the two approaches from a logical point of view. As van Fraassen himself once wrote, "There are natural interrelations between the two approaches: an axiomatic theory may be characterized by the class of interpretations which satisfy it, and an interpretation may be characterized by the set of sentences which it satisfies. . . . These interrelations make implausible any claim of superiority for either approach" ("On the Extension of Beth's Semantics of Physical Theories," *Philosophy of Science* 37 [Sept. 1970]: 325–39; cf. p. 326). I am indebted, both for the general point and for the reference to van Fraassen's endorsement of it, to John Worrall's review article of *The Scientific Image*, which will appear in the *British Journal for the Philosophy of Science*.

vice versa: such a theory may be empirically adequate yet false. Accordingly, to believe that a theory about the unobservable is true is more risky than to believe that it is empirically adequate. Not that the latter is without risk: empirical adequacy "goes far beyond what we can know at any given time" since it requires that the theory save all the phenomena in its field, past, present, and future, not merely all actually observed phenomena (p. 69). Now, the chief difficulty for realism has always been skeptical arguments to the effect that we can never know a scientific theory to be true nor ever be rationally warranted in accepting, however tentatively, a theory as true. This is as much a difficulty for constructive empiricism. The same skeptical arguments might be used to show that we can never know a scientific theory to be empirically adequate nor ever be rationally warranted in accepting, however tentatively, a theory as empirically adequate. Van Fraassen insists, however, that the positions are different:

> There does remain the fact that . . . in accepting any theory as empirically adequate, I am sticking my neck out. There is no argument there for belief in the truth of the accepted theories, since it is not an epistemological principle that one might as well hang for a sheep as for a lamb. (P. 72)

Epistemological or not, the principle that one might as well hang for a sheep as for a lamb is a pretty sensible one. Given two criminal acts A and B whose risks of detection and subsequent penalties are the same, but where A yields a greater gain than B, the sensible criminal will do A. But are the risks and penalties of realism and constructive empiricism the same? And does realism bring with it gains that constructive empiricism does not? Van Fraassen addresses these questions; to evaluate the cogency of his position we must address them too.

Suppose the realist tentatively accepts a theory as true, while the constructive empiricist tentatively accepts it as empirically adequate. The realist does take a greater risk. But he takes no greater risk of being detected in error *on empirical grounds*. So, given strict empiricism (the principle that only evidence should determine theory choice), it seems that we might as well be hung for the realist sheep as for the constructive empiricist lamb.

The trouble is, van Fraassen argues, that realism and strict empiricism do not mix and that realism must pay the penalty of rejecting strict empiricism. He makes the point by considering the case of empirically equivalent yet incompatible theories. This is not, of course, the humdrum case where the *available* evidence fails to discriminate between two incompatible theories. This case need not trouble the realist, who may always hope to show that the two theories are not empirically equivalent and then press for an experimental decision between them. Rather, it is the esoteric case where such hopes are unfounded, where two incompatible theories say exactly the same things about *all* matters observational. The constructive empiricist could accept both theories (believe both to be empirically adequate); the realist cannot on pain of contradiction believe both to be true. But how is the realist to choose between them? In the nature of the case, empirical evidence cannot guide his choice, which must therefore be made on nonevidential grounds. Realism runs counter to strict empiricism and allows nonevidential or 'metaphysical' considerations to intrude into matters of theory choice.

How might the realist respond to this? Presumably, as a realist, he will have no truck with the positivist idea that empirically equivalent theories are really the same theory and not incompatible after all. Nor, as a realist, will he have any truck with the related idea (perhaps it is the same idea in new dress) that there are no 'verification-transcendent truth-conditions' and therefore no truth of the matter for the two theories to disagree about. These ideas, after all, seem to entail that Berkeley's immaterialism is really the same theory as the commonsense belief in independently existing material objects or that there is no truth of the matter for Berkeley and commonsense to disagree about. And these conclusions are anathema to the commonsense realist, let alone the scientific one.[3]

Taking a cue from this example, one might wonder whether the problem is philosophical or metaphysical rather than scientific, in which case metaphysical considerations would not be an intrusion after all. How often have empirically equivalent but incompatible theories occurred in real science? Van Fraassen gives one example, and it is a notorious one. Newton hypothesized that the center of gravity of the solar system is at rest in absolute space. He also pointed out that the appearances would be no different if that center were moving through absolute space at any constant velocity v. So all of the theories $TN(v)$—Newton's theories of mechanics and gravitation plus the postulate that the center of gravity of the solar system has constant absolute velocity v for any v—were claimed by Newton to be empirically equivalent (p. 46).

Van Fraassen's account can be disputed: Newton only claimed the empirical equivalence of the theories $TN(v)$ *so far as appearances within the solar system are concerned*. Hypothesize that some other star is at rest in absolute space, for example, and the empirical equivalence vanishes: if the solar system has any nonzero velocity, then it will approach or recede from

---

3. Here I assume that Berkeley's immaterialism is empirically equivalent to commonsense realism. I am not sure that this is so. Immaterialism can be formulated so as to be consistent with all possible experience and so irrefutable by it. But empirical adequacy should require more than mere consistency with the evidence; it should require (at least) that the theory in question entail the evidence. It might be argued that commonsense realism entails consequences about the stability and reobservability of physical objects which Berkeley's immaterialism does not. Berkeley does invoke God's benevolence to "explain" *post factum* the stability of tables and trees. But Berkeley cannot predict it because of his admission that God might always make an exception to his "laws of nature" and work a miracle instead.

that star and, given sufficient time, the effects of this will become apparent.

Here I resort to a realist ploy whose efficacy van Frassen considers. This is to say that, where equivalent theories occur, by *extending* these theories (that is, embedding them in wider theories) their equivalence will disappear (that is, the wider theories will not be empirically equivalent). In the trivial example just cited, the wider theories are formed simply by adding the statement that some star is at absolute rest to each of the existing theories. The example is trivial because we can, by the Newtonian principle of relativity, construct empirically equivalent theories to each of these wider theories (including the only empirically adequate one): simply consider theories attributing an absolute velocity v to the center of mass of the extended system consisting of the solar system and the star. The process can be continued (assuming the number of masses is finite) until they are all taken into account. And then, again by the Newtonian principle of relativity, we will have an infinite family of empirically equivalent theories each of which consists of Newton's laws plus the hypothesis that the center of mass *of the universe* has velocity v for any value of v.

Van Fraassen considers a more interesting extension or embedding of his Newtonian example, the attempt to combine it with Maxwell's electromagnetism where forces depend upon velocities and not upon accelerations as in Newton. This feature made it possible to devise experiments to detect absolute velocities. The null results of such experiments were an important factor in leading scientists to abandon the Newtonian doctrines of absolute space and time in favor of relativistic ones. And this was to abandon *all* of the empirically equivalent Newtonian theories. Van Fraassen asks us to imagine, however, that null results had not been obtained, that, on the contrary, an absolute velocity for the center of mass of the solar system had been measured. Here it might seem that one of the empirically equivalent Newtonian theories had been confirmed and the rest refuted, and hence that they were not empirically equivalent after all. Van Fraassen finds this reasoning spurious (p. 49). But I find his reasoning, if not spurious, at least hard to follow. Operating within this piece of science fiction (or, rather, history-of-science fiction), he says that we could make compensating adjustments in electromagnetic theory so as to retain whichever of the empirically equivalent Newtonian theories we like. In other words, had the history of science been different, we could construct a new family of empirically equivalent combinations of mechanics and electromagnetism.

But, first, van Fraassen has done nothing to impugn the fact that his empirically equivalent Newtonian theories, when combined with Maxwell's electromagnetism, ceased to be empirically equivalent. Second, could the Newtonian readily have accepted that electromagnetic forces depend upon absolute velocities rather than absolute accelerations? Van

Fraassen concedes that, had his piece of history-of-science fiction occurred, it would have "upset even Newton's deepest convictions about the relativity of motion" (p. 48). But did not these convictions follow from Newton's laws of mechanics and the doctrine of absolute space? Last, and perhaps most important, all of this is a piece of history-of-science fiction: the historical facts are that in this notorious real example of empirical equivalence, the only good example known to me, the actual development of science removed the problem.

Van Fraassen has a further retort to the idea that empirical equivalence can be removed by extension or embedding. He can say that it is the empirical adequacy of the extended theories which counts and that one should accept the victor only as empirically adequate, never as true. And, to offset the scarcity of empirically equivalent theories in real science, he can point to the fact that we can artificially concoct empirically equivalent alternatives to any theory by resorting to notorious logical tricks. The simplest such trick is to conjoin any theory with "The Absolute is lazy" to form a new theory empirically equivalent with the original.

The standard response to such tricks is to eliminate the concocted theory on the ground of its reduced simplicity or unity. Van Fraassen does not object to the appeal to simplicity, but he insists that simplicity is a *pragmatic virtue* of a theory which has nothing to do with that theory's truth or likelihood of being true. The realist, for whom accepting a theory is believing it true, must forge a link between simplicity and truth if he is to appeal to the former. And the link can be forged only by a metaphysical principle:

> Simplicity . . . is obviously a criterion in theory choice, or at least a term in theory appraisal. For that reason, some . . . suggest that simple theories are more likely to be true. But it is surely absurd to think that the world is more likely to be simple than complicated (unless one has certain metaphysical or theological views not usually accepted as legitimate factors in scientific inference). The point is that the virtue, or patchwork of virtues, indicated by the term is a factor in theory appraisal, but does not . . make a theory more likely to be true (or empirically adequate). (P. 90)

So the argument seems to be this: the realist can solve the problem of empirical equivalence only by appealing to simplicity; but he can appeal to simplicity only if he assumes a metaphysical principle ("Nature is simple" or some such); realism therefore involves an illegitimate intrusion of metaphysics into science and the abandonment of strict empiricism.

Is the constructive empiricist in any better position? Presumably he, too, will prefer a respectable theory to an artificially concocted empirically equivalent alternative. He, too, will appeal to simplicity and abandon strict empiricism. But he, apparently, can do this in good conscience, cheerfully admitting that pragmatic virtues such as simplicity have nothing to do with the real aim of science, empirical adequacy. Indeed, how could simplicity have anything to do with that aim? To say that the simpler of two empirically equivalent theories is more likely to be empirically adequate is to contradict oneself.

Returning to the realist, there are several ways he might respond to van Fraassen's argument. The first is simply to admit that there is nothing to choose between empirically equivalent theories. This is hardly satisfactory in view of the ubiquity of the logical tricks. The second is to spice scientific realism with a dash of pragmatism, admitting that there is nothing to choose on realist grounds between empirically equivalent theories but preferring some on the pragmatic ground of simplicity. Despite van Fraassen's argument, I see no reason why the realist cannot appeal to pragmatic virtues just as the constructive empiricist does. The third response is to say that simplicity is not, after all, a merely pragmatic virtue. Realists and constructive empiricists alike value empirical strength; they value it for different reasons, but both connect it with the central aim of science. Is it not a sufficient reason to eliminate concocted alternatives to existing theories that they are not empirically stronger than the theories from which they are concocted?

The realist has a problem here, however. Insofar as simplicity and strength go together (and they do not always), simplicity is not merely a pragmatic virtue. But insofar as simplicity and strength go together, simplicity and truth cannot: the stronger theory is, in some intuitive sense at least, less likely to be true. And here lies the problem for any realist seeking to forge a link between simplicity and truth. Yet the problem may not be completely intractable. "Nature is simple" is a metaphysical principle and a hopelessly vague one to boot. But scientists have made various attempts to say more precisely what it means and to construct theories which conform to it.[4] This transforms it into a metaphysical principle which can, at first remove so to speak, be empirically assessed: roughly speaking, it is acceptable metaphysics if theories constructed under its aegis are empirically successful, while theories which violate it are not.[5] In our post-

4. Einstein was always appealing to simplicity or unity. For an analysis of how these vague appeals were articulated into quite powerful principles of theory construction, see E. Zahar, "Why Did Einstein's Programme Supersede Lorentz's (II)?" *British Journal for the Philosophy of Science* 24 (1973): 223–62.

Philosophers of science still, of course, lack a precise and general account of what simplicity is; perhaps there is none to be had and simplicity is, as van Fraassen says, a patchwork of virtues, some pragmatic and some not. Popper's identification of simplicity and strength works nicely sometimes ("All swans are white" is simpler and stronger than "All non-Australasian swans are white") and badly at other times ("All swans are white and ferocious" is less simple and stronger than "All swans are white").

5. For more detail of how certain metaphysical principles can be rationally assessed in this way, see Watkins, "Confirmable and Influential Metaphysics," *Mind* 67 (July 1958): 344–65, especially pp. 363–65. Watkins does not apply these ideas to principles of simplicity.

positivistic age, we should not regard the intrusion of this kind of meta-physical principle into science as illegitimate. If vague appeals to simplicity can be transformed into precise principles of theory construction and if such principles are acceptable (in the sense roughly defined), then the vir-tue they indicate is not merely pragmatic. It may not be absurd to think that Nature is simple (in some carefully specified sense or senses), if we can point to the empirical success of science in vindication of our belief.

I do not know whether this third response, which I have merely sketched, will work in the end. Perhaps it could be shown (though it would be a far from trivial task) that, for any precise and acceptable sense of the term *simple*, one could concoct empirically equivalent and equally simple theories. I would not see this as the demise of scientific realism, for (and this is the second response again) I cannot see why the realist is barred from invoking a pragmatic virtue to deal with the problem of em-pirical equivalence just as the constructive empiricist does.

II. Theory and Observation

Antirealists need to draw a dichotomy between theory and observation. Van Fraassen is no exception: after all, he could not even distinguish truth from empirical adequacy without it. An old and powerful objection to antirealist views is that no such dichotomy exists. How does van Fraassen deal with this objection?

He first agrees that no such dichotomy can be drawn in scientific language, agreeing with the realist that "All our language is thoroughly theory-infected" (p. 14) and pointing out against the positivist that highly theoretical assertions can be made using only so-called 'observational vo-cabulary' (pp. 54–55). (Here I was reminded of how Popper formulated "There exists an omnipotent, omnipresent, and omniscient personal spirit" in a physicalistic observation language.)[6]

Van Fraassen does insist, however, that some objects and/or events are observable and some not. He concedes the familiar realist point that there is a continuous spectrum between 'directly observing' an object and 'indi-rectly detecting' it using instruments. This only shows that *observable* is a vague predicate. But a vague predicate is perfectly usable provided it has clear cases and clear noncases—and this one has:

> A look through a telescope at the moons of Jupiter seems to me a clear case of observation, since astronauts will no doubt be able to see them as well from close up. But the purported observation of micro-particles in a cloud chamber

6. See K. R. Popper, *Conjectures and Refutations* (London: Routledge and Kegan Paul, 1963), 274–76.

seems to me a clearly different case—if our theory about what happens there is right. . . . while the particle is detected by means of the cloud chamber, and the detection is based on observation, it is clearly not a case of the particle's being observed. (Pp. 16–17)

What if we had microscopic or electron-microscopic eyes? (Actually, we do, only they are not built into our heads!) Could we not then observe things which at present we can only detect, showing that they were not unobservable in principle? Lockean speculations like this merely change the subject:

The human organism is, from the point of view of physics, a certain kind of measuring apparatus. As such it has certain inherent limitations—which will be described in detail in the final physics and biology. It is these limitations to which the "able" in "observable" refers—our limitations, *qua* human beings. (P. 17)

But current physics and biology tell us that what is observable by humans varies (some of us are color-blind) and depends on our particular evolutionary history (other organisms can observe things we cannot). So, even if we can draw a rough and species-specific distinction between what is observable by humans and what is not, should any philosophical significance be attached to it?

Van Fraassen agrees with the realists against the idealists that it has no *ontological* significance: things that humans do not happen to be able to observe may nonetheless exist (p. 18). (Actually, I will argue later, there are problems about van Fraassen's making this concession.) But he wants to give the distinction an *epistemological* significance: humans should never believe to be true a theory about what they cannot observe; instead, they should believe such theories only to be empirically adequate, to tell the truth about what they can observe (p. 18).

Can a distinction which is admitted to be rough-and-ready, species-specific, and of no ontological significance really bear such an epistemological burden? Van Fraassen gives an example of a so-called inference to the best explanation:

I hear scratching in the wall, the patter of little feet at midnight, my cheese disappears—and I infer that a mouse has come to live with me. Not merely that these apparent signs of mousely presence will continue, not merely that all the observable phenomena will be as if there is a mouse; but that there really is a mouse. (Pp. 19–20)[7]

7. Here, incidentally, there is a curious prejudice in favor of vision. True, I have not *seen* the mouse, but have I not *heard* it, and is this not a way of observing it? There is a curious tension in the view that, though we can see (and touch) things, we never hear (or taste or smell) things but only the *noises* they make (or the tastes and smells they emit). Notoriously, one way to resolve the tension is to say that we never really see things, either, but only visions

Will not the same style of argument lead us to the conclusion that there really are electrons (or whatever)? Van Fraassen thinks not. He accepts "inference to the best explanation" but puts his own gloss upon it: such inferences should (and do) only lead us to accept the best explanation as empirically adequate (p. 20). If the best explanation is a theory about the observable, then empirical adequacy and truth coincide and we can (and do) conclude that there really is a mouse (or whatever). But if the best explanation is a theory about the unobservable, empirical adequacy and truth do not coincide and we cannot (and do not) conclude that there really are electrons (or whatever).

There is an empirical claim here (about what scientists actually do infer) and also a methodological claim (about what they ought to infer). I find the methodological claim quite unreasonable. On any plausible theory of evidential support, one would have to admit that there could be far better evidence for an explanation couched in terms of unobservables than for an explanation couched in terms of observables. Is the evidence for the existence of electrons better or worse than the evidence for the existence of the yeti or of the mouse in van Fraassen's wainscoting? It is a curious sort of empiricism which sets aside the weight of *available* evidence on the ground that a casual observer might one day see his mouse or yeti, while the scientist can never see (but can only detect) his electrons.

Van Fraassen's factual claim (that scientists do infer only the empirical adequacy of theories about the unobservable but never their truth) is even harder to swallow. Admittedly, I have not done a sociological survey to settle the matter. And, even if such a survey were to reveal, as I believe it would, that realism is the instinctive philosophy of working scientists, this would not of course settle the methodological question. But to indicate how difficult it is to avoid realist ways of thinking and talking, let us see how van Fraassen thinks and talks. He talks of *detecting* an electron in a cloud chamber. Can one say truly that one has detected an object without also believing it to be true that the object really exists? Later he describes how Millikan *measured* the charge of the electron (pp. 75–77). Did not Millikan think it true, and does not anyone who accepts Millikan's results

---

(visual sense-data) caused by them (and the same will have to go, even more implausibly, for touch). That way leads to idealism. Realists resolve the tension by saying that we can hear, taste, and smell things as well as see and touch them. Van Fraassen, for all his talk about hearing "an apparent sign of mousely presence" rather than a mouse, is once again with the realists. He says that "sense-data, I am sure, do not exist" (p. 72). And he has to be with the realists if truth and empirical adequacy are to be the same so far as observable things are concerned. If all observable phenomena were only apparent signs of mousely presence (mousely visions, mousely noises, mousely smells, etc.), then "All observable phenomena are as if there is a mouse in the wainscoting" would not entail "There is a mouse in the wainscoting," contrary to what van Fraassen says on p. 21.

think it true, that electrons exist and carry a certain charge? Can one say truly that one has measured some feature of an object without also believing that the object really exists?

I shall quote at some length what I *think* is van Fraassen's answer to very obvious questions like these:

> The working scientist is totally immersed in the scientific world-picture. And not only he—to varying degrees, so are we all. . . . But immersion in the theoretical world-picture does not preclude "bracketing" its ontological implications. . . . To someone immersed in that world-picture, the distinction between *electron* and *flying horse* is as clear as between *racehorse* and *flying horse*: the first corresponds to something in the actual world, and the other does not. While immersed in the theory, *and* addressing oneself solely to problems in the domain of the theory, this objectivity of *electron* is not and cannot be qualified. *But this is so whether or not one is committed to the truth of the theory*. It is so not only for someone who believes, full stop, that the theory is true, but also for . . . someone who . . . holds commitment to the truth of the theory in abeyance. For to say that someone is immersed in theory . . . is not to describe his epistemic commitment. . . . it is possible even after total immersion in the world of science . . . to limit one's epistemic commitment while remaining a functioning member of the scientific community. (Pp. 80–83)

This is, I fear, nothing but a sleight-of-hand and an endorsement of philosophical schizophrenia. The sleight-of-hand converts belief in the reality of electrons (belief in the objectivity of *electron*, belief that the term *electron* corresponds to something in the actual world) into belief in, belief full stop in, and finally commitment to something called "the theory of electrons." But there have been several theories about electrons, and no scientist believes them all to be true. As for the most up-to-date theory about electrons, sensible scientists would do well not to believe it to be wholly true either, for details of it are quite likely to be further refined. All this is quite consistent with a pretty firm belief in the reality of electrons, with a refusal to "bracket" this particular ontological implication of science. The philosophical schizophrenia stems from talk of immersion (even total immersion) in the "scientific world-picture" or the "world of science". These metaphors are meant to suggest, if I understand them rightly, that scientists should believe in electrons or whatever while immersed in their scientific work, but should become agnostic about everything they cannot observe once they leave their laboratories. I suppose that split-minded scientists like this are possible, but I wonder whether they are desirable.

Finally in this section, I want to argue that van Fraassen's treatment of the observable/unobservable distinction verges on the incoherent. He insists that what is observable by humans is a "function of facts about us *qua* organisms in the world," so that it is for *science* to tell us what is observable and what is not (pp. 57–58).[8] Now, suppose some theory T does distinguish "the observable which it postulates from the whole it postulates" (p. 59). T might even be van Fraassen's "final physics and biology," if such a theory is possible. T will say, among other things, that A is observable by humans, while B is not. Of course, if we are to use T to delineate the observable, we must *accept* it. But van Fraassen cannot have us accept it as true, since it concerns in part the unobservable B. The constructive empiricist can accept T only as empirically adequate, that is, believe to be true only what T says about the observable. But "B is not observable by humans" cannot, on pain of contradiction, be a statement about something observable by humans. And, in general, the consistent constructive empiricist cannot believe it to be true that *anything* is unobservable by humans. And, if this is so, the consistent constructive empiricist cannot draw a workable observable/unobservable dichotomy at all.

It might be objected that

(1) B is not observable by humans

is logically equivalent with

(2) Everything observable by humans is distinct from B

since (2) is a statement about the observable, so is the logically equivalent (1). But even accepting that there is a sense in which (2) is "about" the observable, it is *also* about the unobservable B and therefore cannot be accepted as true by the constructive empiricist.

Nor does it help if we say that "observable by humans" is an "observational predicate," that we humans can tell from observation that a thing is observable by us. For one thing, this marks a retreat from van Fraassen's insistence that there is no observable/unobservable dichotomy in scientific *language*. For another thing, "observable by humans" will be a nonstandard observational predicate whose negation is not also observational, a predicate akin, for example, to the predicate "is an inscription of finite length." For we cannot observe that anything has a property without also observing that thing. Anyone who claims to have *observed* that something is unobservable contradicts himself. But if "unobservable by humans" is *not* an observational predicate, our conclusion stands. We can grant that "observable by humans" is an observational predicate so that the constructive empiricist can accept as true, on the basis of observation, statements of the form "A is observable by humans." But the consistent con-

8. In case anyone is reminded here of the talk of "observables" in quantum mechanics, we should remind ourselves that the so-called "observables" of quantum mechanics are in the present context remotely calculable theoretical quantities. If electrons are not observable, neither is their charge, momentum, or spin.

structive empiricist cannot accept as true, on the basis of observation or anything else, a statement of the form "*B* is not observable by humans." Constructive empiricism requires a dichotomy which it cannot consistently draw.

III. Realism and Explanation

Realism and explanation are doubly linked. Realists think science explains facts about the world, and they think realist philosophy of science explains facts about science. I will consider the latter claim first.

The claim is that only a realist philosophy of science can explain the fact that science has had a great deal of predictive success. If the unobservables postulated by (successful) science really exist and if what (successful) science says about them is true or nearly so, then we can explain predictive success. Otherwise, such success is just a lucky accident. As Putnam famously remarks, realism is "the only philosophy that doesn't make the success of science a miracle" (cited on p. 39).

Van Fraassen gives short shrift to this Ultimate Argument for realism:

> The explanation provided is a very traditional one—*adequatio ad rem*, the "adequacy" of the theory to its objects, a kind of mirroring of the structure of things by the structure of ideas—Aquinas would have felt quite at home with it.
> . . . Will this realist explanation with the Scholastic look be a scientifically acceptable answer? I would like to point out that science is a biological phenomenon, an activity by one kind of organism which facilitates its interaction with the environment. And this makes me think that a very different kind of scientific explanation is required.
> I can best make the point by contrasting two accounts of the mouse who runs from its enemy, the cat. St. Augustine . . . provided an intensional explanation: the mouse *perceives that* the cat is an enemy, hence the mouse runs. What is postulated here is the "adequacy" of the mouse's thought to the order of nature: the relation of enmity is correctly reflected in his mind. But the Darwinist says: Do not ask why the *mouse* runs from its enemy. Species which did not cope with their natural enemies no longer exist. That is why there are only ones who do.
> In just the same way, I claim that the success of current scientific theories is no miracle. It is not even surprising to the scientific (Darwinist) mind. For any scientific theory is born into a life of fierce competition, a jungle red in tooth and claw. Only the successful theories survive—the ones which *in fact* latched on to actual regularities in nature. (Pp. 39–40)

Amusing though this is, it does no more than play cat-and-mouse with the argument. The scientist does ask why the mouse runs from the cat and answers in roughly the terms made fun of here: the mouse perceives the cat, perceives the cat as an enemy, and runs. This does not commit the scientist to ascribing thoughts, adequate or otherwise, to the mouse: his response might be quite instinctive. But with this proviso, there is nothing unscientific or un-Darwinian about this kind of explanation. Of course, the Darwinian question is not "Why does the mouse run away from the cat?" but, rather, "How did this piece of mouse behavior evolve?" The Darwinian answers this question roughly in the terms suggested by van Fraassen: given an environment full of mouse-hunting cats, cat-fleeing mice are more likely to survive, reproduce, and pass their cat-fleeing behavior on to future generations. But the Darwinian explanation is not a substitute for the "intensional" one, for they are addressed to quite different questions. The Darwinian explains what the "intensionalist" postulates: that the mouse's perceiving the cat as an enemy (or, better, the mouse's genetically programmed behavioral response to cats) is adequate to the order of nature.

Just as with cats and mice, so also with scientific success. It is one thing to explain why some theory is successful and quite another to explain why only successful theories survive. Van Fraassen's Darwinian explanation of the latter can be accepted by realist and antirealist alike. But to say that only successful theories are allowed to survive is not to explain why any particular theory is successful.

Not that a realist explanation of this in terms of the theory's *adequatio ad rem* will do as it stands. The Ultimate Argument is actually very old, and a brief look at an old example of it should give us pause. Christopher Clavius (in his *Commentary on the Sphere of Sacrobosco* of 1581) said it was incredible to suppose that Ptolemaic astronomy could correctly predict eclipses even though its eccentrics and epicycles were mere figments. But the eccentrics and epicycles were figments, and it was no miracle at all that a geometrical model expressly devised to yield some phenomenal regularity (periodic eclipses) should be successful in doing so. It is different, however, if a theory devised to accommodate some phenomenal regularities should turn out to predict *new* regularities. The realist has a ready explanation: the entities postulated by the theory really exist, and what the theory says about them is true (or nearly so). The antirealist seems forced to say that figments dreamed up for one purpose have turned out, miraculously, to be well adapted for a quite different purpose. So it was that thoughtful realists such as William Whewell distinguished two kinds of predictive success (predicting known effects and predicting new ones) and argued that the antirealist cannot explain the latter. So it was that a thoughtful antirealist such as Duhem, seeing the force of the argument, came to spice his instrumentalism with a whiff of realism: a theory is able to make successful *novel* predictions because it is not "a purely artificial system" but, rather, "a natural classification [whose] principles express profound and real relations among things."[9]

As this brief historical excursion shows, the only form of the Ultimate Argument which *might* work is that which focuses on *novel* predictive success. Yet this focus is lacking in recent discussions, both from prominent defenders of the argument (such as Putnam) and from prominent critics of it (such as Laudan).[10] Difficulties remain, of course, not least that of making precise the intuitive distinction between known effects and novel predictions. These difficulties notwithstanding, van Fraassen has done nothing to impugn the Ultimate Argument in its refined form.

The Ultimate Argument proceeds on the metalevel: epistemology is to be naturalized, and philosophy of science is to explain facts about science. But there is a more direct argument, which proceeds from the assumption that science should explain facts about the world. The connection between the demand for explanation and the realist demand for true theories is apparently very obvious. An explanation is not adequate unless what does the explaining is true. So, given that theories figure in explanations, adequate explanations require true theories. It is worth noting that Duhem found the argument cogent and confessed that since in his view science did not aim at true theories, it could not explain anything either. Others adopt the curious view that science does aim truly to describe the world but cannot really explain any features of it. Van Fraassen's response to our simple argument is twofold. He attacks the realist demand for explanation; and he argues that explanation, where it can be had, does not require true theories, that explanatory power is a pragmatic virtue for which an empirically adequate theory will do just as well as a true one.

He softens us up with a linguistic point. We still speak of explanations even when we think the explanatory principles false:

> I say that Newton could explain the tides, that he had an explanation of the tides, that he did explain the tides. In the same breath I can add that this theory is, after all, not correct. Hence I would be inconsistent if by the former I meant that Newton had a true theory which explained the tides. (P. 99)

9. P. Duhem, *The Aim and Structure of Physical Theory* (Princeton, N.J.: Princeton University Press, 1954), 28 (see also pp. 297ff).

10. For example, Laudan presents as historical counterexamples successful theories which did not genuinely refer and were not true or nearly true ("A Confutation of Convergent Realism," *Philosophy of Science* 48, no. 1 [March 1981], 19–49). But few, arguably none, of the theories cited had any *novel* predictive success. Laudan also saddles the realist with the principle that successful reference alone breeds success. I do not know if any realist has thought this, but no realist should think it. For, as Laudan shows, we can construct referring theories which will be quite unsuccessful: take a successful theory containing the term *t* and negate it. Successful reference is a necessary condition for truth (or near truth) but not a sufficient one. And when it comes to success, it is the truth (or near truth) of what a theory says about its theoretical entities which counts, not whether those entities exist.

Quite so. We can speak without contradiction of a false explanation because truth is not a *defining* condition of explanation but an *adequacy* condition upon it. (In a similar way, we can say that Bode's Law is false without contradicting ourselves.) Van Fraassen says that we may agree that a theory is false without at all undermining our previous assertion that it explained many phenomena (p. 98). But could we say that a theory *adequately* explained many phenomena even though it is false? Realists think not. Scientists appear to agree: no modern astronomical text cites the vortex theory of planetary motion as the explanation of why the planets all go around the sun in the same direction, though some may have no other explanation of this fact. If this is wrong, and truth is not required for adequate explanation, then it will take more than a linguistic point to show it—van Fraassen gives us more.

First, he attacks the realist demand for explanation. He claims that in science this demand is severely limited, that explanation is not a "preeminent" or "rock-bottom" scientific virtue:

> If explanation of the facts were required in the way consistency with the facts is, then every theory would have to explain every fact in its domain. Newton would have had to add an explanation of gravity to his celestial mechanics before presenting it at all. (P. 94)

But Newton did present his theory. He "declined to explain," admitting famously that he had "not been able to discover the cause of . . . gravity" (P. 94). Thus:

> Newton's theory of gravitation . . . did not (in the opinion of Newton or his contemporaries) contain an explanation of gravitational phenomena, but only a description. (P. 112)

It is the same in modern physics, where the "unlimited demand for explanation leads to a demand for hidden variables, which runs counter to at least one major school of thought" (p. 23). And, quite generally, to demand that regularities be explained and shown to be more than cosmic coincidences is self-defeating. For what of the regularities postulated to do the explaining?

Something is obviously wrong somewhere in all this. On the one hand, Newton explained the tides (p. 99) and, on the other hand, Newton's theory did not explain gravitational phenomena at all (p. 112). What has gone wrong is a tacit conflation of realism with essentialism, of the demand for explanation with the demand for ultimate explanation. This will take a bit of explaining.[11]

11. Further detail can be found in my "Explanation, Description and Scientific Realism," *Scientia* 112 (1977), 727–55.

Suppose we explain the phenomenal regularity that sticks look bent when half-immersed in water by postulating, among other things, that unobservable light rays refract when passing through media of different densities. This is not, of course, to explain the refraction of light, though we might then try to do so. But at any point in our explanatory endeavors there will be things for which we have no explanations, namely the deepest explanatory principles we have reached at that point. One realist response to this situation is to demand that these principles require explanation in their turn. Another realist response, quite the antithesis of the first, is to demand that our deepest explanatory principles should require no explanation, that they should somehow be ultimate or self-explanatory. This second response is central to the tradition of Aristotelian essentialism, the tradition which holds that the only genuine explanations are ultimate or self-explanatory explanations.

The essentialist tradition is buttressed by rhetorical questions like these. Do we *really* explain why sticks look bent in water by postulating some mysterious and unexplained law of refraction? Did Newton *really* explain the tides by postulating his mysterious and unexplained force of gravity? The intuition is that a *real* explanation does not remove one mystery by postulating another one. And behind this intuition lies another one: the intuition that a real explanation should serve the *pragmatic* function of removing puzzlement, setting the curiosity of the inquirer at rest. It is because nonultimate scientific explanations do not serve this pragmatic function, do not remove puzzlement so much as relocate and enhance it, that they are said not to be real explanations at all.

I use the term *pragmatic* here advisedly and roughly as van Fraassen uses it (sometimes). Whether an explanation removes puzzlement depends very much on the person we are considering. What relieves one man's puzzlement may enhance the next woman's. I dare say that some of the most efficacious puzzlement relievers in the history of thought have been explanations which are not scientific at all and which are, from a scientific point of view, quite inadequate (what about "God moves in mysterious ways," said in explanation of anything whatever which is puzzling?). I dare say that on occasions the incurious have had their puzzlement removed by a scientific explanation—but they should not have. For if it is feelings of puzzlement we want to get rid of, we should turn not to science but to the whiskey bottle!

I think that essentialism and the intuitions which lie behind it are to be rejected. And I think it one of Newton's chief claims to methodological fame that he was among the first to see this. Newton admitted that he could not explain gravity *and* that gravity was a perfectly proper thing to try to explain (since it was not an "essential property" of matter). Yet in the same breath he insisted that his theory of gravity did explain celestial motions and the tides:

> Hitherto we have *explained* the phenomena of the heavens and of our sea by the power of gravity, but we have not yet assigned the cause of this power . . . hitherto I have not been able to discover the cause of . . . gravity from phenomena, and I frame no hypotheses . . . to us it is enough that gravity does really exist, and act according to the laws we have explained, and abundantly serves to *account for* all the motions of the celestial bodies, and of our sea.[12]

Elsewhere Newton contrasted his procedure with that of his opponents, the Cartesian essentialists: he gives precise deductive explanations; they mutter about essences and can explain nothing:

> To tell us that every Species of Things is endow'd with an occult specifick Quality by which it acts and produces manifest Effects, is to tell us nothing: But to derive two or three general Principles of Motion from Phaenomena, and afterwards to tell us how the Properties and Actions of all corporeal Things follow from these manifest Principles, would be a very great step in Philosophy, though the Causes of those Principles were not yet discover'd: And therefore I scruple not to propose the Principles of Motion above mention'd, they being of very general Extent, and leave their Causes to be found out.[13]

Van Fraassen quotes from the first of these famous passages (p. 94), but he misunderstands Newton (as Duhem and others have misunderstood him) in a way that can only stem from a tacit conflation of realism with essentialism. He says that Newton "decline[d] to explain" (p. 94) and that Newton's opinion was that his theory "did not contain an explanation of gravitational phenomena, but only a description" (p. 112). But Newton explicitly claimed to have explained or accounted for gravitational phenomena such as the tides *by* describing precisely how gravity works. The antithesis between explanation and description is quite illusory: we explain one thing *by* describing another. Newton did decline to explain gravity.

---

12. *Principia*, Book 3, General Scholium; Motte's translation, revised by Cajori, vol. 2 (Berkeley and Los Angeles: University of California Press, 1962), 546–47. Italics mine. Also important is the famous passage where Newton says he will treat forces "not physically but mathematically" (*Principia*, Book 1, Definition 8; Cajori, vol. 1, 5–6). Newton is saying that he will describe how gravity works in precise mathematical terms, rather than try to explain it physically. He is often misinterpreted as saying that gravity does not really exist.

13. *Opticks*, Book 3, Query 31; Dover edition (New York, 1952), 401–2. The most extended defense of Newton and attack on Cartesian essentialism is, of course, Roger Cotes's preface to the second edition of the *Principia*. Yet Cotes is sometimes interpreted as defending the essentialist view that gravity is, after all that Newton had said to the contrary, essential to matter (for example, by Popper in *Conjectures and Refutations*, 106). It is unlikely that Newton would have allowed Cotes to defend a view he himself had specifically denied. Cotes himself specifically denied that this was the view he was defending, in a letter to Samuel Clarke, who had questioned him on the point (see Cajori's appendix, n. 6; vol. 2, 634–35).

But to take this as a confession that nothing can be explained by the law of gravity is to father upon Newton a view that was not his.

It is, however, the view which van Fraassen calls the "realist demand for explanation" and attacks. He formulates the demand innocently enough as "every theory should explain every fact in its domain" and then takes "every fact" to include *the theory itself* (p. 94, quoted earlier). Only a theory which was somehow self-explanatory could meet this demand. But one can demand explanation without also demanding ultimate or self-explanatory explanation, as Newton tried to teach us. Van Fraassen's rejection of the latter demand leaves the former quite intact.

The tacit conflation of explanation with ultimate explanation also emerges from the delightful joke which forms the last chapter of van Fraassen's book. There, taking his cue from the remark that "everyone believed in the existence of God until the Boyle lectures proved it" (p. 229), he modifies Aquinas's Five Ways into proofs of scientific realism. What gets proved is actually Aristotelian essentialism, which is perhaps not surprising considering the provenance of the arguments. The First Way gives the flavor of the whole:

> *So I argue*: Everything that is to be explained, is to be explained by something else. That some things are to be explained is evident, for the regularities in natural phenomena are obvious to the senses and surprising to the intellect. So we must either proceed to infinity, or arrive at something which explains, but is not itself, a regularity in the natural phenomena. However, in this we cannot proceed to infinity. (P. 205–6)

Now, what gets "proved" here is that science can achieve something which explains but is not itself a regularity in the natural phenomena, something, in other words, which is not itself to be explained. What gets proved is the essentialist view that science can achieve ultimate explanation. But the "proof" has a missing premise (required to obtain the statement beginning with "So"): the essentialist principle that A does not really explain B if A also requires explanation and has not received it. Van Fraassen hopes that readers skeptical of ultimate explanations will reject the conclusion, miss the missing premise, and infer by *modus tollens* that science cannot explain anything at all. But I am spoiling a clever joke.

I cannot leave this topic without saying a word about van Fraassen's favorite example of how science has transcended realist demands for explanation, quantum mechanics. Hidden variable explanations of quantum mechanics are said to "run counter to at least one modern school of thought" (p. 23). A philosophy of science is not refuted by pointing out that it runs counter to a scientific school of thought, not even a dominant school of thought. But this school can point to *proofs* that hidden variable theories are impossible, and this should give the realist pause. Van Fraassen's comments on these proofs (which he seems to approve of) reveal some interesting things. One proof apparently assumes that "if we cannot point to some possible differences in empirical predictions, then there is no real difference at all" between two theories (p. 34). In endorsing this proof, van Fraassen's earlier resolute antipositivism has wavered, for this assumption requires a positivist reinterpretation of scientific language to show that empirically equivalent theories are really the same theory. Not that realists would be too happy with an explanation of quantum mechanics which was demonstrably empirically equivalent with it: such an explanation could have no *independent* evidence in its favor. Now, if we assume the truth and completeness of quantum mechanics (or even its empirical adequacy and completeness), we will be able to prove that no explanation of it could have independent evidence in its favor. And if we assume that quantum mechanics is not only true and complete but also ultimately so, we will be able to prove that no explanation of it (independently confirmable or not) will be adequate. These various assumptions are no part of quantum mechanics; rather, they are philosophical assertions about quantum mechanics. Insofar as the various proofs rest upon assumptions like these (I do not know whether they do), they are not so much proofs as philosophical arguments, and pretty questionable ones to boot. Finally, the issue of determinism, important though it is in other contexts, is something of a red herring in this context. It is true that some hidden-variable theorists wanted a deterministic explanation of quantum mechanics. But there is no *a priori* reason why a deeper explanation of quantum mechanics has to be deterministic.

Van Fraassen says nothing to impugn a modest realist demand for non-ultimate explanation. And realists can defend such a demand by pointing to cases where the attempt to explain, even to explain theories regarded as empirically adequate, has paid off handsomely. Van Fraassen is not impressed with the argument:

> Paid off handsomely, how? Paid off in new theories we have more reason to believe empirically adequate. But in that case even the anti-realist, when asked questions about *methodology* will *ex cathedra* counsel the search for explanation! We might even suggest a loyalty oath for scientists, if realism is so efficacious. (P. 93)

Realists might retort that explanation has a payoff in terms of *understanding* the world—but that is unlikely to impress van Fraassen. And realists who are also empiricists are impaled on the horns of a dilemma here (as the case of the hidden-variable theories suggested). Realists who are also empiricists will want any proposed explanation to yield empirical regularities other than those it was devised to explain; otherwise there could be no independent evidence for the truth of the explanation. If an explanation does yield them, then the constructive empiricist can value it, too, but for its predictive rather than its explanatory power. If an explanation does not yield them, then it should be rejected as mere "metaphysical baggage." Heads constructive empiricism wins, tails realism loses:

> I think we must conclude that science, in contrast to scientific realism, does not place an overriding value on explanation in the absence of any gain for empirical results. . . . the point is that the true demand on science is not for explanation *as such*, but for imaginative pictures which have a hope of suggesting new statements of observable regularities and of correcting old ones. (P. 34)

This true demand is not, it seems, to be vouchsafed to scientists themselves. They are to take an oath of loyalty to realism, the desire to understand the world, and the search for explanatory truths. Realism is the constructive empiricist's Noble Lie, propounded *ex cathedra* in case scientists should find the true aim (enhancing the empirical adequacy of "imaginative pictures") uninspiring! More seriously, is van Fraassen right to say that "the interpretation of science, and the correct view of its methodology, are two separate topics" (p. 93)? I think it preferable to have an interpretation of science which harmonizes with methodological pronouncements.

At any rate, this is part of what van Fraassen means when he calls explanation a *pragmatic* virtue. The search for explanation *works* because theories which are good explainers will *ipso facto* be good savers of phenomena and the real "name of the game is saving the phenomena" (p. 93). But this is not all that he means. The rest is meant to undercut the simple realist idea that adequate explanations must contain theories that are true (or nearly so). Van Fraassen defends a "pragmatic" analysis of explanation according to which theories do not figure in explanations at all but somehow lie behind or underpin them. And he says that good explanations can be underpinned by empirically adequate theories just as well as true ones. At least, I think that is what is going on.

Charles Morris divided the study of language into syntax, semantics, and pragmatics. The last was meant to deal, among other things, with *context dependence*, as when the truth or falsity of "I'm hungry" depends upon the context of utterance, upon who says it and when. In philosophical circles, "pragmatic" (or better "merely pragmatic") has also come to mean "useful but not true." There is no obvious connection between these two philosophical usages. Utterances can express both truths and falsehoods in virtue of contextual factors. And an utterance can be useful but not true whether or not contextual factors enter into it. ("He went thataway" may be useful for diverting the pursuer though it is false; it is also heavily context-dependent. "John Brown took the road to California" may be similarly useful but false, and it is less context-dependent.)

Van Fraassen thinks that explanation is also heavily context-dependent. He begins from Bromberger's puzzle about explanatory asymmetries: the height of the flagpole explains the length of its shadow but not vice versa, though the two deductions may be structurally identical. (Actually, van Fraassen also tells a blue story which is meant to show—I do not think it does show it—that there are "contexts" in which the length of the shadow does explain the height of the flagpole.) The obvious solution to this puzzle is an appeal to causality: explanations exhibit causes, while nonexplanatory deductions do not. But this is to jump out of the contextual frying pan into the contextual fire, for which factor is picked out as "the cause" also varies enormously with the context:

> . . . [T]he salient feature picked out as "the cause" . . . is salient to a given person because of his orientation, his interests, and various other peculiarities in the way he . . . comes to know the problem—contextual factors. (P. 125)

Now, John Stuart Mill, who first drew attention to this kind of thing, insisted that only the entire constellation of factors, amounting to a sufficient condition for the event to be explained, is really entitled to be called the cause of it. And John Mackie, with Mill in spirit though more amenable to ordinary ways of talking, said that a cause is an insufficient but necessary part of an unnecessary but sufficient condition. Both Mill and Mackie can admit that contextual factors may influence *which event a person wants explained*. And this is enough to dispose of many of the usual examples adduced to demonstrate context dependence of the explanations given (car crashes, fires, and such like).[14]

Van Fraassen takes a different course: he accepts the context dependence of explanations and tries to make it more precise. An explanation "is an *answer* . . . to a why-question" (p. 134). Every why-question has a *topic* (if we ask, "Why P?" the topic is P), an implied *contrast-class* (what we actually ask is, "Why P rather than Q, R, etc.?"), and an implied relation of *explanatory-relevance* which determines what shall count as a possible answer to the question (pp. 142–43). The topic, contrast class, and relation of explanatory relevance all depend upon the context, in particular upon "a certain body K of accepted background theory and factual information," which in turn depends "on who the questioner and his audience are" (p. 145). This gets complicated. The upshot of it is that

> the discussion of explanation went wrong at the very beginning when explanation was conceived of as a relationship like description: a relation between the-

14. This is shown in John Worrall's forthcoming review article, referred to in note 2 above. Worrall also shows that we can dispose in the same way of van Fraassen's story about the length of a flagpole's shadow explaining its height.

ory and fact. Really it is a three-term relation, between theory, fact, and context. . . . So to say that a given theory can be used to explain a certain fact, is always elliptic for: there is a proposition which is a telling answer, relative to this theory, to the request for information about certain facts (those counted as relevant for *this* question) that bear on a comparison between this fact which is the case, and certain (contextually specified) alternatives which are not the case. (P. 156)

Is it *really* elliptic for all this? At the outset of his discussion, in order to combat "the increasing sense of unreality" the usual examples bring, van Fraassen sets forth three "workaday examples of scientific explanation" (pp. 101–3). And he manages to set these examples forth *without mentioning contextual factors at all*. There are why-questions all right, but there are no implied contrast classes, relevance relations, or anything else which depends "on who the questioner and his audience are." Contextual complications have little to do with explanation in science, if van Fraassen's own "workaday examples" are anything to go by.

Is it true to say that explanation was ever "conceived of as a relationship like description: a relation between theory and fact"? The orthodox account says that explanations are arguments in which three things figure: theories or general laws, initial conditions specifying the cause of the event being explained, and in the conclusion a statement of the event being explained. Explanations *contain* descriptions, but they are not *like* them.

Van Fraassen (his earlier examples notwithstanding) wants to drop the theories out of explanations and relegate them to the "context," to the "background information" relative to which why-questions are asked and answered. (Here I was reminded of the Wittgensteinians, who assimilate theories to rules of inference and insist that these rules do not figure as premises in the inferences constructed in accordance with them.)[15] Still, one might think, the theories must be true if the explanations proffered in the light of them (or constructed in accordance with them) are to be correct. Van Fraassen thinks not. Empirically adequate and empirically strong theories will do just as well as true ones. His argument seems to be as follows. The fact to be explained is always an observable fact. The facts cited in explanation of it are also always observable facts. So what the theory has to get right, to be a good explainer, are just the observable facts. And empirically adequate theories, by definition, do just that:

So scientific explanation is not (pure) science but an application of science. It is a use of science to satisfy certain of our desires; and these desires are quite specific in a specific context, but they are always desires for *descriptive information*. . . . in each case, a success of explanation is a success of *adequate and infor-*

15. For a discussion of this view, see my "Wittgensteinian Instrumentalism," *Theoria* 46 (1980), pts. 2–3, 65–105.

mative description [*of the phenomena*]. And while it is true that we seek for explanation, the value of this search for science is that the search for explanation is *ipso facto* a search for empirically adequate, empirically strong theories. (Pp. 156–57)[16]

I am not sure that I have got the argument right here—but, if I have, there is lots wrong with it. Sometimes we explain observable facts by citing other observable facts (and laws). But this is not always the case, though it tends to be the case in the usual philosopher's examples, which bring an "increasing sense of unreality" to the subject. The flagpole is a good example again: both its height and the length of its shadow are presumably observable facts. (I can scarcely bring myself to mention the "explanation" of why some bird is black, which consists in pointing out that it's a raven and they are all black!) But, in van Fraassen's own examples of scientific explanations, there are initial conditions such as: the specific heats of water and copper are 1 and 0.1, respectively; the earth's magnetic field at a certain point has a vertical component of approximately $5/10^5$ Tesla; the energy levels associated with stable electron orbits in hydrogen atoms take the form $E_n = -E_o/n^2$ where E_o is called the ground state energy. These initial conditions are *generalized* ones because the facts to be explained are *general* ones (with the possible exception of the second). But, setting aside the problem of their generality, they do not look much like observable facts, and their provision does not look much like providing descriptions of the observable phenomena. Nor, in science, is it always observable phenomena that we try to explain: we sometimes try to explain theories.

Van Fraassen not only thinks that explanation is a pragmatic affair in Morris's sense (context dependence), he also thinks that explanatory power is one of the *pragmatic virtues*, concerning which he says in general:

> In so far as they go beyond consistency, empirical adequacy, and empirical strength, they do not concern the relation between the theory and the world, but rather the use and usefulness of the theory; they provide reasons to prefer the theory independently of questions of truth. (P. 88)

If what I have said about scientific explanation is right, then the explanatory power of a scientific theory does depend on whether it tells the truth about the unobservable and therefore does go beyond empirical adequacy and empirical strength. But van Fraassen is obviously right when he says:

---

16. The italics in this quotation are mine. And I felt justified in adding "of the phenomena" to "description" because in the preceding sentence (not quoted here) the phrase "adequate description of the phenomena" occurs.

> Nor can there be any question of explanatory success as providing evidence for
> the truth of a theory that goes beyond any evidence we have for its providing an
> adequate description of the phenomena. (Pp. 156–57)

This is obviously right, because empirical adequacy is *defined* as correctness so far as the observable evidence is concerned. Explaining things cannot provide a special sort of *evidence* that theories are true rather than just empirically adequate. Realists, made of sterner stuff than constructive empiricists, still demand that a theory be true for the explanations in which it figures to be adequate. And realism carries this much metaphysical baggage: realists can point to no evidence over and above evidence of empirical adequacy that their sterner requirement has been met.

But there is excess baggage of a different kind in the constructive empiricist position. There is, above all, the philosophical excess baggage of defending an observable/unobservable distinction and giving it crucial epistemological significance. There is the excess baggage of providing an alternative to the obvious realist explanation of science's novel predictive success. And there is the excess baggage of a complex account of the pragmatics of explanation.

I suggested earlier that, in comparing constructive empiricism with scientific realism, we should assess the risks, penalties, and gains associated with each. The risks have been discussed, as have the penalties in the form of philosophical "excess baggage" of various kinds. As to the respective gains (or losses), I can only repeat a hackneyed point. The realist values theoretical science as an attempt to *understand* the world and sees continuity between commonsense and scientific knowledge. The constructive empiricist, browbeaten as much by the positivist emphasis on prediction as by esoteric problems in interpreting quantum theory realistically, jettisons understanding and seeks to drive a wedge between theoretical science and commonsense (taking his excess baggage aboard to do so).

Let me conclude by agreeing with van Fraassen that it would be a pity if scientific realism were to become a philosophical dogma. Bas van Fraassen's book certainly roused me from any dogmatic slumber to which I might have been prone. It has, in fact, given me sleepless nights! His antirealism is more viable than earlier antirealist positions. But, in philosophy of science as well as in science, viability directly depends on weakness. Constructive empiricism is weaker than earlier antirealist views in all kinds of ways, and correspondingly closer to realism. This is why I conclude, undogmatically I hope, that realism emerges a little bloodied but unbowed from its encounter with constructive empiricism.

10   What Can Theory Tell Us about Observation?

Mark Wilson

One of the more intriguing ideas in Bas van Fraassen's *The Scientific Image* is that scientific theories should be able to elucidate their observational content *internally*:

> If there are limits to observation, these are a subject for empirical science, and not for philosophical analysis. Nor can the limits be described once and for all, just as measurement cannot be described once and for all. . . . To find the limits of what is observable in the world described by theory *T* we must inquire into *T* itself, and the theories used as auxiliaries in the testing and application of *T*. (P. 57)[1]

This attitude contrasts greatly with old-fashioned empiricist attempts to graft observational properties onto scientific theories externally. Logical empiricists had a list of wholesome observational predicates, e.g., "is red," "is five feet long," and so forth, but unfortunately orthodox textbook systemizations of, e.g., physics, tended either to ignore these (e.g., "is red") or to leave their linkage to its more problematic portions unspecified. The empiricist sought to correct these lapses by welding on a lot of extraneous definitions, reduction sentences, and so forth. It is generally agreed today that such muscular tactics will not work; the proposed bridge principles were always too indelicate to be true. Beyond the simple failure of the program, one has the suspicion that the attempt was misguided from the beginning; the original, unsullied theory should be expected to delineate its observational content on its own, without need for positivist assistance. As an analogy, consider the quantity *temperature* and its historical relationship to Newtonian mechanics. In the early years after its first introduction, there was simply no other accepted way to measure temperature except by thermometer, and the concept's relationship to mechanics was certainly unknown. By the end of the nineteenth century, however, the development of statistical mechanics found a home for *temperature* within the Newtonian scheme. As a side effect of the whole process, the linkage between temper-

---

1. All page references are to Bas C. van Fraassen, *The Scientific Image* (Oxford: Oxford University Press, 1980).

ature and thermometer reading was broken, revealing that thermometers are often wrong even in cases (air temperature on a sunny day) formerly regarded as paradigmatic cases of correct temperature ascription. We can imagine some P. W. Bridgman of 1630 insisting that "thermometers measure temperature" be adopted as an irrevocable axiom of science, on the grounds that otherwise we'd have no means for checking the correctness of temperature claims. But, of course, mechanics *from the beginning* carried its own capabilities for determining what temperature is and how thermometers work. The 'irrevocable axiom" is likely to be simply inconsistent with the account the underlying theory will provide if left to its own devices. Logical empiricists seemed guilty of analogous impatience; they tried to pry the observational content of a theory loose by bridge principle crowbars, rather than letting it reveal itself in the natural course of its development. *Being red*, after all, is simply a property whose natural detection instrument is a human being rather than a thermometer.

At the heart of this rejection of traditional methods are at least two assumptions:

1. The problem of what is observable for human beings is to be treated in the same manner for any other detection device.
2. A fundamental scientific theory, such as physics (to which we shall restrict our attention in this paper), will solve this problem by the same means as would be applied to any other scientific problem. Thus, if we find a measuring instrument manufactured by the long-extinct Krell, we expect the question of what properties the machine measures to be settled by straightforward scientific investigation rather than conventional stipulation by an alert operationalist.

A program for investigating notions of "observation" and "observationality" in this spirit I shall call *internalism*. As it stands, my characterization is vague, and a better understanding of what it leads to can be reached only by consideration of examples. But, as a general programmatic, I believe van Fraassen would agree to the above. In fact, I think many readers will have found this view of "observation" one of the most appealing ideas to be found in the first half of *The Scientific Image*.

The reason I hesitate to straightforwardly attribute internalism to van Fraassen is that in the course of his discussion he makes a number of assumptions about observation's place in science that seem alien to the internalist viewpoint. For a start, there is a puzzling fluctuation in the degree of stringency required for observationality: on page 65 he seems to allow that such properties as the spin projection of a single electron count as "observable," whereas on page 60 he claims that even macroscopic mass, force, momentum, and kinetic energy are not observable properties in classical mechanics (nor, one would suppose by parallel reasoning, in quantum mechanics as well). Some of this apparent inconsistency seems generated by van Fraassen's desire to locate *empirical substructures* in a scientific theory to which data are supposed to correspond by "isomorphism" (p. 64). To make this a little clearer, van Fraassen proposes that we view fundamental physical theory as an attempt to delineate a class of structures to serve as potential models for reality. In its essentials, I agree with van Fraassen that this is often a more useful way to view theory than the proof-theoretical orientation of olden days. However, in order to explain how such modeling is confirmed or disconfirmed in practice, van Fraassen assumes that these potential models are characterized by substructures which serve as "candidates for the direct representation of observable phenomena" (p. 64). From an internalist viewpoint, I think it is fairly obvious that the hypothesis that "observational facts" appear as a "substructure" is too tidy to be possibly correct, just as logical empiricism's primary error lay in its attempt to be too fastidious in this same swampy area. For example, van Fraassen claims that the empirical substructures appropriate to classical physics consist in what he calls "motions," trajectories in frames where at least one physical object is at rest (p. 45). But there is a plausible sense of "observable" in which this account is in some respects too generous. In celestial mechanics, we generally observe motion only in the polar angles and not radially (except if we can use radar or triangulation); otherwise, the differences between Ptolemy, Copernicus, and Kepler could have been resolved much more easily. Likewise there would seem to be a problem in "observing" the motion of a portion of an opaque, homogeneous fluid. If we start with our notion of "observable data" something like the data Tycho Brahe collected, vulnerable in addition to stellar aberration and psychological oddities like the moon illusion, it is questionable that we will be able to locate the result as a "substructure" of a Newtonian model.[2]

This will probably seem like quibbling. Since we expect any eventual treatment of "observationality" to be messy, it will often by useful to sup-

---

2. This could probably be proven if the notion of "substructure" were suitably clarified (together with the relevant sense for "observational"). The orthodox logician's notion is:

$\mathcal{E}$ is a substructure of $\mathcal{T}$ if and only if
(1) $E \subseteq T$, where E and T are the domains of $\mathcal{E}$ and $\mathcal{T}$
(2) The interpretation of R on E = the interpretation of R on T restricted to E for every predicate symbol R in the background language $\mathcal{L}$.
(Cf. Donald Monk, *Mathematical Logic* [New York: Springer-Verlag, 1976], 328).

Obviously van Fraassen does not want $\mathcal{E}$ to interpret all R found in the language appropriate to $\mathcal{T}$. Moreover, none of the *primitive* terms appropriate to $\mathcal{T}$ may be appropriate to $\mathcal{E}$. $\mathcal{E}$ should somehow be formed from E and a new language $\mathcal{L}'$ of terms *definable* in $\mathcal{L}$. But it is unclear what sense of "definability" is required here. First-order definability is probably too weak, but overgenerosity can allow any theory to have "empirical substructures" adequate to any set of data. Van Fraassen seems to assume that this same language $\mathcal{L}'$ will form the empirical substructures in all the models of the theory, rather than having $\mathcal{L}'$ vary from model to model. My "Sirius" example casts doubt on the correctness of this assumption.

ply a schematic and somewhat provisional account of these matters. But then the delineation of such substructures will not be part of the *formal* task of presenting a theory: a simple presentation of its basic potential models should be enough. It is important to stress this fact, because there is a popular reconstructive program in the philosophy of science, represented most clearly perhaps by the work of Joseph Sneed, where the assumption of empirical substructures is motivated by what we might call "anti-internalist" considerations. Unfortunately, as we shall see in the course of this paper, van Fraassen himself tends to agree with far too many assumptions of this school.

Indeed, one must not assume that some single notion of "observationality" stands like the Holy Grail at the end of our internalist quest. The search will show, I think, that the modal considerations involved in the "-ality" side to "observationality" are far too complex to be unpacked in any univalent manner. This does not mean, as suggested by Grover Maxwell's famous debunking of "observationality,"[3] that study of this family of notions won't be of utility to philosophy of science. On the contrary, I believe that some quite sharp answers to important traditional queries about "observationality" can be obtained in this spirit, as long as one does not assume that the same notion of "observationality" will suffice in all instances. Some of the tensions and inconsistencies in van Fraassen's examples seem attributable to a determination to make a single notion perform too many philosophical chores.

Accordingly, my chief task in this essay will be to characterize *roughly* those features of *The Scientific Image* we can expect an internalist treatment of observation to ratify. Before commencing, a proviso should be entered concerning van Fraassen's account of theory as the delineation of a set of models. As philosophers of science, we often think of "classical mechanics" as a single theory, but in truth it is better viewed as a phalanx of loosely related forms of model building: the simple mechanics of point particles without constraints, the generalized versions of this permitting holonomic constraints due to Lagrange and Hamilton, mechanics of continuous media, mechanics of functionally determined relationships (systems of gears and other simple mechanisms), and much more. Today we tend to suppose that the simple point-particle mechanics should serve as a foundation for the rest, but this is an opinion that would not have been widely shared in the history of science. It represented the point of view of Boscovitch but not that of Maxwell, Hertz, Kelvin, or even Newton him-

3. "The Ontological Status of Theoretical Entities," *Minnesota Studies in Philosophy of Science*, vol. 3 (Minneapolis: Minnesota University Press, 1962). Likewise, although I am sympathetic to many of the complaints voiced in Michael Gardner's "The Unintelligibility of 'Observational Equivalence,'" in *PSA 1976*, ed. Suppe and Asquith (East Lansing, Mich.: Philosophy of Science Association, 1976), I am not so defeatist about the value of studies in this direction.

self. I shall argue later that one can easily form a distorted impression of the classical difficulties with *force* if one doesn't keep this historical situation in mind.

Indeed, this profusion of approaches without evident conceptual unity converted many scientists of the nineteenth century to vigorous antifoundationalism. I believe the popularity of instrumentalism among many Victorian physicists stems in no small part from this origin. Possibly these motivations for instrumentalism ought to be taken seriously even today,[4] because there is little doubt that a similar antifoundationalist methodology guides most research in quantum theory. It is important to note that van Fraassen's "anti-realism" is not of this type; the fact that he is willing to identify a theory with a set of models rather than with the nonunified bundle of the antifoundationalists makes him their opponent as much as the most devoted realist.

While on the topic of van Fraassen's treatment of theories, another point may be added. Sometimes the content of a scientific theory can be nicely boiled down to a set of equations, sometimes to a clear set of mathematical models; but often one will have fairly definite ideas about a given domain (e.g., an ether conception in electrodynamics) yet find that their translation into manageable equations or manageable models can be easily effected only at the cost of a host of special assumptions and approximations. The net results may completely resemble those of your action at a distance opponent across the channel although your original inspirations were quite different. It is usually a delicate question to decide whether such theories are really the same, although inessentially connected with two different "pictures," or that the "translation" of ideas into mathematics is yet incomplete. So, while I am as convinced as anyone of the great utility of studying scientific theories in a formalized setting, one must always recognize that an application to a historical situation is apt to be rather idealized because of simple mathematical difficulties, if no other. Otherwise, a profusion of "observationally equivalent, although distinct theories" in the historical record will appear as a simple artifact of one's idealized conception of theory.[5]

Let us now see how van Fraassen proposes to explicate "observation-

---

4. A serious "anticonjunctionalism" might be grist for the mills of Nancy Cartwright ("The Truth Doesn't Explain Much," *American Philosophical Quarterly* 17 [1980]) and Geoffrey Joseph ("The Many Sciences and the One World," *Journal of Philosophy*, Dec. 1980).

Incidentally, *contrary* to van Fraassen, Boyd, and others, I think it is a mistake to characterize "scientific realism" in terms of the "aims of science." Some of the greatest scientific discoveries don't contribute in any direct way to the description either of truth or "empirical adequacy" (e.g., Onsager's proof of a phase transition in a two-dimensional Ising model). A "realist" asks only that science be judged by the same epistemological yardstick as any other branch of knowledge. An inventory of wares usually constitutes a guide to truth, but the accomplishments of Billy Sol Estes remind us that this is not always the purpose of such accounts.

5. If the term "theory of classical mechanics" casts the net of "theory" too widely, van

ality" in internalist terms. We are left somewhat to our own devices, because we need to construct a notion of observationality for propositions, states of affairs, or the like, whereas van Fraassen typically discusses only the observationality of concrete objects like electrons. But explication of mere "object observationality" will not serve the purposes of *The Scientific Image*, which, after all, is intended as a brief for believing in only a restricted subset of currently accepted scientific doctrine (=the stuff we tend to believe if we are brainwashed by realist propaganda). But we are not encouraged to believe that observable objects possess *all* of their known properties, but only the observable ones. So the first notion we need to explicate is that of a *property* being observable.

When do we credit an instrument m with the ability to detect a property F? As a first approximation, this happens just in case m has two potential states S_1 and S_2 such that m would assume S_1 after interaction with a test object o possessing F but would assume S_2 if o does not possess F. The state S_1 can then be regarded as a "yes" answer as to whether o possesses F, and S_2 as a "no" answer. It should be apparent that this is but a rough sketch of what transpires in measurement: for example, a given device will accurately detect F only in a restricted range R of test objects. Qualifications such as this to our general model will assume some importance later in this essay. Van Fraassen clearly has this same general account of property measurement in mind:

> [W]e call system Y a measurement apparatus for quantity A exactly if Y has a certain possible state (the ground-state) such that if Y is in that state and coupled with another system X in *any* of its possible states, the evolution of the continued system (X *plus* Y) is subject to a law of interaction which has the effect of correlating the values of A in X with distinct values of a certain quantity B (often called the "pointer reading observable") in system Y. (Pp. 58–59)

Once we have accepted this account of property detection, van Fraassen suggests that we adopt the same general model for detecting the presence

---

Fraassen's own "state space" approach runs the risk of being too narrow. Thus, point-particle classical mechanics (or quantum theory) first delineates a collection of phase spaces (Hilbert spaces) and then locates its possible systems within these. Van Fraassen often writes as if a physical theory is to be represented by a single phase space rather than a collection of them. (Cf. his "A Formal Approach to the Philosophy of Science," in *Paradigms and Paradoxes*, ed. Robert Colodny, [Pittsburgh: University of Pittsburgh, 1972], 311). This would be merely eccentric if he didn't identify the observational properties with certain portions of the phase space. But there are plenty of obviously observable properties, e.g., particle count, which can be properly recovered only as subsets of the *class* of phase spaces (since such properties are constant within a given space). (This is one of the ways in which the term *observable* in quantum mechanics is misleading.) But, once the scope of allowable properties is enlarged in the way suggested, then traits like "is a system consisting of three electrons" become represented in the theory. Thus, arguments for the existence of a proper empirical substructure to quantum theory such as reported by Edward MacKinnon ("Scientific Realism: The New Debates," *Philosophy of Science*, Dec. 1979) lose their persuasiveness.

of a test object of a suitable type (p. 58).[6] Thus, a steady reading on a thermometer usually indicates that it is in contact with an object whose temperature it can measure. Assembling these two components we obtain

> *Device m can test the truth of the possible state of affairs Fa* if and only if *m* can detect *F*, *a* belongs to the range *R* in which *m* can detect *F* and *m* can suitably indicate the presence of *a*.

This account is easily extended to relational states of affairs. Finally we get

> *Fa is an observational state of affairs* if and only if a human being can test the truth of *Fa*.

Again I stress that these explications are meant to supply only a rough picture, since many essential details have been suppressed. Also, I have couched matters in terms of "possible states of affairs" rather than "sentences" or "propositions" in keeping with van Fraassen's desires to keep linguistic considerations to a minimum.[7] It is clear that a given observational property or state of affairs might be denoted by a quite "theoretical" predicate or sentence within a given language. Indeed the language might not have the capabilities for describing the state of affairs at all. Furthermore, a state of affairs may be "observational" in our newly defined sense for an agent who is totally unaware that he is testing such a fact. In van Fraassen's terminology (p. 15), he needn't to be able to "observe that *Fa*," a locution which requires the observer to possess some conceptual understanding of F-ness. Hence "*a* is largely H_2O" will count as an observational state of affairs for primitive people who totally lack any conception of H_2O. The internalist account is interested in a human being's ability to discriminate, not to understand.

Because of this fact, it should be clear that many properties will be counted as "observable" that traditional empiricism would not allow. It seems to me that many of the traditional classifications are based upon a rather primitive yet compelling metaphysics. Consider the property of *being red*, understood as a fairly permanent property of physical objects ("Roses are red even in the dark") and not as a property of sense-data,

6. Assume one has a region of space in which theory *T* says a group of electrons dwells. I assume that if an antirealist is forced to attribute a range of "observational" properties to *S* comparable in scope to the full panoply of properties attributed by *T*, then the battle between realist and antirealist is essentially over, even if one still wants to call the electrons "unobservable."

7. If one tries to mark which *sentences* of *T* are to count as "observational," one faces the difficulty that one's recommended set of chastened beliefs is not closed under logical implication. (Cf. Michael Friedman's review of *The Scientific Image*, *Journal of Philosophy*, May 1982.) Van Fraassen's response to this sort of problem seems evenly divided between insisting that the component predicates of the sentences be regarded as "fused" or "bracketed" (his term) or retreating to nonlinguistic objects of belief, as I have done here. It seems to me that there are great complications involved in either approach, but I will not pursue these here.

objects in peculiar lighting conditions or whatever. There is a very natural temptation to feel that one has a *complete understanding* of what this universal comes to as long as one has once upon a time had the right perception of the right kind of object. Hume somewhere opines that a person who has never tasted a pineapple cannot really understand the property *tasting like pineapple*. Or, to adopt Russell's more felicitous terminology, the pineappleless individual may know of the property in question by description but not by direct acquaintance. But, once we become properly acquainted with a property, we may meaningfully consider its applications even in locales where there is no obvious or convenient means for detecting its presence. Suppose cosmonauts were sent to some planet of Sirius. Since the frequency distribution of solar radiation is quite different there from on earth, objects will appear different colors, just as colors often look different here under fluorescent lighting. Unless our travelers had brought along a book of sample color strips from earth or other apparatus, it might be practically impossible to determine which Sirian objects were true red and which were not. That is, our interstellar friends could sort out a pile of what appeared to them to be red objects, but that collection could turn out to contain a mixture of colors if transported without change to earth.

This example serves to bring out a major tension between traditional views about observationality and our internalist account. According to the traditional view, the property of being true red will be an observable property of a Sirian ruby simply because *being red* is a property we have become acquainted with through observation. On the internalist treatment, however, we find that we cannot detect through observational means the presence of this property in its present circumstances. If we wish to count *being red* as an observational property of the ruby, it can only be through consideration of some counterfactual such as "If the ruby were brought to earth without change, it could be observed to be red" or the like. Moreover, on the internal account, symmetry considerations seem to dictate that our space patrol detect *some* legitimate colorlike property on Sirius, although that property will be different from any of our familiar color properties (call the trait our Sirian scouts can detect *being Sirian red*). In fact, we earthlings can observe the property *being Sirian red* here no more easily than our Sirian colleagues can observe *being true red* there. It is safe to claim that *being Sirian red* would never have been counted as an "observable property" on the traditional view. Even Hume would have probably supposed that a sheltered cosmonaut, who had never experienced red sense-data until he basked in the light of Sirius, would have thereby gained acquaintance with the property *being true red*, not *being Sirian red*. (What sort of rationale could be supplied for this prejudice I cannot say.)

I hope these brief remarks supply the reader with a sufficient sense of the acquaintance view of observationality. Perhaps the key difference between it and the internalist perspective is that the traditional account is based entirely on the agent's *understanding* of the property, whereas such considerations do not enter into the internalist account at all. If I train you to waggle your head in a certain way and report whether you've seen a yellowish image, you will thereby be serving as a fairly reliable detection device for the direction of polarization of the sky's light,[8] although you will probably be completely unaware that you are detecting any property of external objects at all. Of course, it may seem odd to say that we can "observe" such a property, but the internalist treatment of observationality is not intended as an analysis of ordinary usage (which is probably corrupted by the acquaintance view, anyway). And we shall see that, for many purposes for which we need a notion of "observationality," it is essential that we *not* insist that the agent always understand the properties he can observe.

It is fairly clear from *The Scientific Image* that van Fraassen has no truck with the acquaintance treatment of observationality. Nonetheless, it is important to remember that many traditional antirealist sentiments ultimately derive from these doctrines. Eddington's "two tables" problem in large measure stems from such a perspective: if one "fully understands" the property of *being red*, how can one come *to learn* that that property should be identified with an unexpected and complicated property involving the chemistry of colorless matter? I think Stebbing's famous cry, "If the table isn't solid, then what does 'solid' mean?"[9] expresses the acquaintance philosophy rather than some dubious paradigm case argument. If one is dissatisfied with British attempts toward appeasement between the kingdoms of ordinary life and science yet holds on to acquaintance, one is likely to adopt some form of instrumentalism. The temptation is even more overwhelming if one feels it is impossible to really understand the bizarre property that quantum theory assigns as correlate to *being red*, anyway.

After this digression on what sort of assumptions do *not* fit into the internalist program, let us return to our positive characterization of observationality within a theory. Just before we were sidetracked, we had given a sketch of when a state of affairs should count as humanly observable. We now want to bend this account to a physical theory, T, to determine which features of its proposed models count as observable. To do this, we need to know which subsystems of T are to count as (=be used to model the behavior of) human beings. Clearly, the formal presentation of T (say, as a

8. Example drawn from Feynmann, Leighton, and Sands, *The Feynmann Lectures on Physics*, vol. 1 (Reading, Mass.: Addison-Wesley, 1964), 36–37.
9. *Philosophy and the Physicists* (New York: Dover, 1958), 52. A good, although extreme, example of the classical line of thought I mean to characterize is the "pink ice cube" argument of Wilfred Sellars's "Philosophy and the Scientific Image of Man," in *Science, Perception and Reality* (London: Routledge and Kegan Paul, 1963).

set theoretic predicate) will not settle this question, which must instead be left to an empirical process of attempting to correlate human behavior with various subsystems of T. A specification of the observational will emerge only as a side product of the process of applying T to the world. It is clear that under the most ideal of circumstances we will never achieve this goal completely; at best we can hope to delineate an oversimplified class of structures which we can use to model the grosser aspects of human behavior. Accordingly, we should never expect a complete account of "observationality" within T, although a rougher sketch may be adequate for our purposes.

In the case of an empirically inadequate theory, such as Newtonian physics, one is faced with an additional difficulty. No one, except a truly rabid Duhemist, would suppose that classical physics possesses any subsystem which could remotely act like a human being—most of the facts we know about organic chemistry seem so deeply quantum mechanical in origin that it is totally dubious that similar effects be produced through classical means. But, in lieu of clear models for mankind, we must not expect classical physics to internally carve out its observational facts. At best, we can delineate what is observational by special Newtonian subsystems of a different nature. Sometimes studies of this type can be useful. Thus, in my own look at classical particle mechanics,[10] I assumed that any property detectable by a set of simple Newtonian "meters" was "observable," on the grounds that if human beings, *per impossibile*, were to exist in a Newtonian world, they would be able to read the pointer readings of my simple meters. But such a choice is doomed to be somewhat arbitrary and, except under special conditions, there will be no best way to supply a notion of "observationality" in an empirically inadequate theory. Accordingly, van Fraassen's formal explication of "observational equivalence" (p. 67) must be taken with a grain of salt, especially in trying to adjudicate a historical situation. One can extract quite different "observational content" from a theory, depending upon which pieces of the theory one wants to regard in a favorable light.[11]

Even if we possess a reasonably clear account of what the human sub-

10. "The Observational Uniqueness of Some Theories," *Journal of Philosophy*, May 1980. The particular choice of instrumentation in that context turned out to be unimportant because, under a generous method for specifying the other modal aspects to "observationality," the facts detectable by such meters turned out to be *maximal*. Hence one can certainly set some limits to unbridled claims of nonobservability.

11. This will generally happen if you decide to study the "observationality" of a discarded theory T by identifying some of its postulated traits with quantities held by *current* theory to be "observational" and then counting them as likewise observable in T. But such patterns of identification are by no means unique—a famous case is Carnot's use of "caloric." (Cf. Laszlo Tisza, "Evolution of the Concepts of Thermodynamics," in *Generalized Thermodynamics*, [Cambridge, Mass.: MIT Press, 1977.)

systems of T should be, we have so far been careless about some modal considerations buried in our notion of an observational fact. We have noted that almost all measuring instruments are *nonuniversal*—they can't detect a given property in all circumstances. In fact, most devices are *dependent* upon certain prevailing conditions to work correctly—a spring balance requires moderate temperatures and a homogeneous gravitational field to be able to measure mass, for example. Human beings are particularly touchy in this regard—they have a regrettable tendency to expire quickly if a lot of conditions such as the presence of oxygen are not met. Moreover, most instruments are *inconstant*, as well—they measure one or more properties in a given place, but yet others in different locations. Our scale fails to measure *mass* on the moon, although it will detect *impressed gravitational force* there. The "Sirian red" example displays our own fickle detection nature.[12]

In a T model which already contains a human observer, these questions are not crucial, because the specification of the model itself tends to resolve these worries. But this only settles what facts are *de facto* observed by the observers inside the model;[13] it does not settle the broader question of what facts are *observable*, i.e., could be observed by a human. I can think of at least three general programs for achieving this needed extension. The first is to simply say that a fact Fa in a model m is *observable* if and only if the property F is observed by a human in at least one other model m'. As I remarked before, this was the approach of the acquaintance school, with the proviso that this range of observational F be limited to those observed on earth. But, on an internalist account, this restriction is unwarranted; *being Sirian red* and *being true red* should count as equally *observable*. For the purposes of theory comparison, this generosity is wise; if competing theory T^1 could not produce an observational class with the same extension (under isomorphism) as theory T uses to model the behavior of the Sirian-red objects near Sirius, then we could obviously run a test to distinguish T^1 from T.

A more reasonable approach is to modify the original models of T to allow for observers.[14] Often in foundational studies, one assumes that model m and test instrument t are initially isolated and separated at infinity (so the interactive part of their joint Hamiltonian is negligible). Test instrument t is brought into interaction with m through spatial approach and

---

12. Some additional discussion of the submerged aspects to detection can be found in my "Predicate Meets Property," *The Philosophical Review*, Oct. 1982.

13. I am being purposely generous to an anticounterfactual stance. Actually, some modality, i.e., consideration of possible systems resembling the given system, is needed to guarantee that the observers aren't getting the right answer by accident. Van Fraassen alludes to this need in his call for *correlation* between systems Y and X in his description of measurement above. Moreover, this ensemble of similar systems is also needed to allow for occasional error.

14. I believe van Fraassen would favor this line of attack.

then withdrawn. Such idealization is fine for the measurement problem in quantum mechanics but constitutes pretty rough handling for human subjects (who do not thrive when completely isolated). Hence, if we want to evaluate the truth of "if object a in model m were observed by observer o, o could observe Fa," we need to specify how o is to be introduced into the neighborhood of a and how conditions around o are to be arranged. Example: a is a ruby buried in a mountain. Are we supposed to bury the unfortunate observer in the mountain or set o only as close as will sustain life? Is o allowed to chip away the mountain, to cart the ruby back to earth, etc.? It is easy to see that different ways of filling in these needed background assumptions will provide quite different accounts of what facts count as "observable" within a model and that no account possesses any inherent superiority. Accordingly, internalist considerations tend to splinter a univalent concept of "observationality" into a large family of notions, each appropriate to different situations. In addition, observationality explained in terms of outside systems with unlimited resources seems appropriate to classical and quantum mechanics but not to general relativity where the models in question are cosmological and only admit of inside observers with limited energetic resources. It would be better for philosophy of science, I think, if more of these essential details were fleshed out in discussions of "observational equivalence."

A third method which might be mentioned is to claim that Fa is observable in m only if there is an expanded model m^1 "sufficiently like" m in which an observer detects Fa. I think the problems and prospects for this approach are equivalent to those of the second strategy.

It is clear, then, that any reasonable explication of "observationality" in a model requires modal considerations of this nature. Van Fraassen is oddly ambivalent about this point. In the course of trying to establish that mass is sometimes (always?) a nonobservable property, he writes:

> For, consider, as [a] simplest example, a (model of mechanics in which a) given particle has constant velocity throughout its existence. We deduce, within the theory, that the total force on it equals zero throughout. But every value for its mass is compatible with this information. . . . [T]he example shows there are models of mechanics—that is, worlds allowed as possible by this theory—in which a complete specification of the basic observable quantities does not suffice to determine the values of all the other quantities. Thus the same observable phenomena equally fit more than one distinct model of the theory. (Remember that empirical adequacy concerns actual phenomena: what does happen, and not, what would happen under different circumstances.) (P. 60)

What puzzles me about this passage is that van Fraassen wishes to regard *any* property of his little one-particle system as observable. Why should he consider the particle's position and velocity to be observable, for example?

After all, the impoverished universe of the model does not contain any observers; and, if we allow an outside observer to interact with our particle sufficiently to check its velocity, we are certainly asking, "What would happen under different circumstances?" But what constraints keep our modal intruder from measuring the mass, e.g., on a beam balance he carries with him?

Van Fraassen's example looks more convincing than it really is, because it is embedded in a discussion which claims that, insofar as certain systems of classical mechanics go, it is a matter of complete convention what the relative masses of the particles are—hence *mass* could hardly be observable in such cases. Thus he claims, "In [George] Mackey's [version of mechanics] any two bodies which are never accelerated, are arbitrarily assigned the same mass" (p. 60). Some of this rests upon a misunderstanding; the only systems Mackey assigns arbitrary masses are those in which the bodies *couldn't* be mutually accelerated no matter how their initial positions and velocities were altered.[15] In short, they *can't* interact with each other (and, one would assume, with any outside system). Such particles, on collision, would pass through each other, more uncatchable than the most elusive neutrino! It is a mere curiosity of Mackey's account that such models be considered Newtonian systems at all; it is physically reasonable (although mathematically inconvenient) to insist that a system's potential energy function be non-null under close approach. Of course, if we refuse to allow any interaction between observer and particle, then *all* properties of the particle will be unobservable. But, in Mackey's other, nondegenerate systems, *mass* should be as detectable as anything else. Hence, Mackian pronouncements that "the universe is given to us exactly once" are bound to be misleading in a discussion of "observationality within a model"; all acceptable accounts known to me involve some considerations of a complicated, although benign, modal character.[16]

15. Perhaps the problem lies in the fact that Mackey's "Aj" is not a simple function of time but of phase, as well. (Cf. George Mackey, *The Mathematical Foundations of Quantum Mechanics*, [Reading: W. A. Benjamin, 1963], 4.) In general, one should always expect such degenerate systems to be found in any formalization of mechanics, for simple convenience if nothing else. For example, who wants to include the local repulsive interaction term corresponding to contact action in the Hamiltonian every time one does celestial mechanics? Moreover, an author in mechanics will usually frame his basic axioms so that most of its principles will carry over to generalized coordinates. Otherwise he will be chastised for inelegance.

16. There is an odd argument against counterfactuals on p. 62, which depends upon a *modus tollens* from the principle

If ψ and $R_{2\pi} \psi$ really represent exactly the same physical state, then the superposition $(K\phi + m\psi)$ represents the same state as $(K\phi + m R_{2\pi} \psi)$

and the evident falsity of the conclusion. But it is the *principle* which is false, not its antecedent. Multiplying a component of a superposition by a phase factor will generally alter its *relative* contribution to the whole but will be harmless if $K = O$. It is an accidental feature of 2π rotations on spinors that they shift phase; hence, "care must be taken in the formal group-theoretical manipulations" (J. M. Ziman, *Elements of Advanced Quantum Theory* [Cambridge:

One of the most restrictive, yet least argued for, assumptions that van Fraassen makes about "observation" is that in making an "observation" one apparently should not be allowed to employ instrumentation. Thus, "while [a] particle [can be] detected by means of [a] cloud chamber, and the detection is based on observation, it is clearly not a case of the particle's being observed" (p. 17). As a point about our ordinary usage of "observe," this is again probably correct, but that fact should be of little relevance to us. Van Fraassen wants to distinguish between "unmediated true observation" and "mediated detection through observation of an instrument." This mediated/unmediated distinction forms a natural part of the acquaintance view of observationality, but it seems completely foreign to our internalist treatment. *All* detections, at least by human beings, require a fairly extensive set of supporting conditions to facilitate the observation, e.g., presence of an electromagnetic field in low excitation (for sight), a homogeneous gravitational field (for judgments of "weight"), and so forth. Drawing a distinction between conditions which mediate a given detection and those which merely "facilitate" it seems a rather hopeless task. As a little test, consider the stone wolframite which has the interesting ability to reemit incident ultraviolet radiation in the visible region of the spectrum. This mineral, whose true color is brownish black, appears bright yellow under an ultraviolet lamp. This is caused by a permanent fluorescent property. Clearly, human beings can *detect* this property, but should they properly be said to "observe" it as well? Or should we hold instead that detection of the fluorescent property is "mediated" through prior observation of its property of *glowing yellow now*? Likewise, should we claim that *being Sirian red* is not really "observational" near Sirius, contrary to our earlier conclusions, but rather mediated by some more familiar property of the ruby?

However one wants to answer these questions, it seems that the desire to draw a mediated/unmediated distinction flows from some other philosophical spring than pure internalism. (I submit the contamination comes from the doctrine of acquaintance.) Moreover, whatever the merits of the distinction, it is *not* one we want to draw if we are trying to unpack notions such as "observational equivalence." If theory T tells us that in one of its

---

Cambridge University Press, 1969], 249). Van Fraassen continues: "What should we conclude from this? . . . If in one possible world, an isolated system is in state ψ and in another it is in state $R_{2\pi} \psi$, no amount of empirical information actually available can tell the observer which of these two worlds he is in. But there is a *counterfactual* statement we are inclined to make about the case: if the system had interacted with another one in such and such a way, the results would have been different in the two cases." But nothing stated above should incline us to say this. Two *isolated* systems in states $K\phi + m\psi$ and $K\phi + m\,r_{2\pi}\,\psi$ will generally behave in an observationally different manner from each other and from a system in yet another state, the eigenstate ψ (also describable as $R_{2\pi}\,\psi$). None of these states represents an interaction with another system.

models humans, aided by instruments, can sort out a class of objects α (as all having the property *mass of six grams*), then rival theory T^1 must likewise credit the observers of one of its models with an ability to detect some property forming an isomorphic set β; otherwise T and T^1 will be observationally distinguishable. It matters not a hoot whether the properties detected of the objects in classes α and β are "observed" in an unmediated way or not.

All of these factors indicate that the scope of "observational fact" within a theory will be fairly wide, as long as one fills in the missing modal parameters in a reasonable way. Van Fraassen typically supposes that the limits are much more narrowly constrained:

> In the context of [classical mechanics], and arguably in all of classical physics, all measurements are reducible to series of measurements of time and position. Hence let us designate as basic observables all quantities which are functions of time and position alone. These include velocity and acceleration, relative distances and angles of separation—all the quantities used, for example, in reporting the data astronomy provides for celestial mechanics. They do not include mass, force, momentum, kinetic energy. (Pp. 59–60)

It is important to realize that claims such as this, although they have a long and glorious philosophical tradition behind them, are *not* ratified in any particular way by the internal workings of classical mechanics. In the first place, it is not clear what the claim of "reducibility" means. One would initially suppose "derived through calculation from other quantities," for in celestial mechanics that is how we figure out the masses of the planets. However, this limitation in celestial observationality is a mere artifact of the difficulties in experimenting with the planets: it certainly seems possible to measure mass directly in local situations. Consider a typical measurement of mass on a bathroom scale. Are we really "measuring only change of position" here? Certainly the spring inside has contracted and the needle has shifted position, but have we *measured* these? The answer is clearly no; most of us are such bad estimators of length that we need to employ a ruler to determine how far the needle moved. In the case of the spring inside, we will usually be completely ignorant of its movements. The only properties we can readily measure in these circumstances are mass and impressed gravitational force. The project of explicating van Fraassen's sense of "reducible" seems both difficult and rather extraneous to an internalist viewpoint.

In this connection, I find it surprising that van Fraassen is willing to consider (linear) momentum as an "observable" in quantum theory but not in classical mechanics. This is odd because, generally speaking, quantum mechanics doesn't supply us with *new* means of measuring momentum unavailable in the classical picture (including electromagnetism); the uncertainty principle rather seems to exclude some possibilities. Thus, we cannot completely determine the momentum of an electron by measuring its velocity and multiplying by its mass; instead, we need to run the particle into a wall, say, and observe the recoil. In the classical framework, both measurement techniques are allowed. Hence I fail to see the motivations for van Fraassen's change of standards.

But what about the fact that *mass* and *force* are often called "theoretical" notions in classical physics, whereas *position* is not? Joseph Sneed writes:

> Examples of theoretical functions with respect to classical particle mechanics are mass and force. An example of a non-theoretical function with respect to classical particle mechanics is the position function. I claim that there is no application of classical particle mechanics in which the means of measuring masses and forces employed in the application do not presuppose that some application of classical particle mechanics is successful. In contrast, there are *at least some* applications in which the means of measuring position do not presuppose this.[17]

If T is particle mechanics, Sneed calls terms like *mass* "T-theoretical." Such terminology makes Sneed's dichotomy look like a nice theory—internal distinction. Certainly it is true that, if we supply a classical explanation of how a spring balance happens to measure mass or force, Hooke's law, the laws of gravitation, and other doctrines of Newtonian physics must enter the picture—this is Sneed's notion of "presupposition." But, if we consider how a Newtonian account of ruler measurement might go, again we find Newtonian elements trooping on stage—the particles of the ruler are bound by strong balanced forces which dominate gravitational or other distorting influences. In fact, since a theory of measurement describes the time evolution of test object plus measuring system, it is hard to see how much less than the full panoply of Newtonian ideas won't be needed in the treatment of every measurement.

The resolution of this riddle is that Sneed's notion of "T-theoretical" is not really derived internally from T but from T's place in a reconstructed series of theories of ever-increasing complexity. It is only because Sneed lists classical mechanics after what he dubs "classical kinematics"[18] that enables measurements of position to count as "non-theoretical" in particle mechanics. The motivations for this placement seem to me to be largely

17. Joseph Sneed, *The Logical Structure of Mathematical Physics* (Dordrecht, Holland: Reidel, 1977), 34.

18. The meaning of the term *kinematics* is hardly static in physics. Thus, one often uses "the kinematics of a system" to be limited to the properties used to form its state space; but, if those systems are under constraint, those quantities needn't (and often can't) be normal velocity and position but often "non-kinematical" traits such as the energy in a given vibrational mode. Moreover, a system of 37 weakly interacting particles has 111 degrees of freedom; if the 37 particles from into a rigid body, the answer is 6. Thus, the property of having a given "kinematics" is a classification of the forces acting on the system.

philosophical. Although such attempts to reconstruct scientific theory according to this pattern (which derives from Patrick Suppes's less doctrinaire recommendations) can be undeniably quite useful, I think it is a great mistake to suppose there is any unique answer as to how it should be done. If I were to provide a list of theories of mechanics in order of "logical priority," I might instead choose *statics*—the science of forces held in equilibrium, as in a balanced seesaw. It has been known since antiquity how to compare the magnitude of the forces on the arms (whether one chooses to call the property in question "force" or not). So I am puzzled when Sneed claims:

> All means of measuring forces, known to me, appear to rest, in a quite straight forward way, on the assumption that Newton's second law is true in some physical system, and indeed also on the assumption that some particular force law holds.[19]

But "$F = ma$" could hardly be very important here, because the seesaw is in equilibrium and total force F and acceleration vanish. And, of course, we have no idea at all of what "force law" in the Newtonian sense governs the contact forces exerted by the weights on the seesaw. (A good example of how statics can serve to ground classical dynamics can be seen in a standard derivation of Hamilton's principle from the principle of virtual work.) A search for a unique best reconstruction of mechanics in the Sneed/-Suppes framework seems as quixotic as the reconstructions of the positivist bridge principle builder and motivated by similar anti-internalist motivations. (I would like to stress that a good many important insights about theory confirmation are suggested by the work of this school, and claim only that their suggested infrastructure should not be regarded as a formal part of the theory studied.)

But if *length* is as "theoretical" as *force*, why all the fuss about the latter concept in the history of physics? The real problem, it seems to me, is not that classical forces can only be "theoretically calculated" in Sneed's sense, but in many cases they can't be determined by any rational means whatsoever. We noted earlier that "classical mechanics" really picks out a loose assemblage of diverse theories, and in some of these a place for Newtonian style *force* is not clearly indicated. As a simple example, consider impact between two atoms, which Newton theorizes in the *Optics* to be rigid and impenetrable spheres. When they collide, what forces change their motion? Immediately one sees that the formula "$F = ma$" cannot be applicable here, since each ball's acceleration will be undefined (the velocity may change from $+ v$ to $- v$ or o *instantaneously*, because the balls are

19. Sneed, p. 117.

rigid). Newton doesn't seem to have seen this problem clearly, partially because his treatment of the calculus obscured the difference between $m\Delta v/\Delta t$ and ma, but Leibniz used considerations of this nature to argue that a non-Newtonian conception of "force" was needed in mechanics. These arguments are fairly cogent, and a long history of attempts to clarify the situation followed.[20] One of the greatest advances lay in Lagrange's generalized mechanics where a large class of contact interactions (e.g., balls sliding on tables, etc.) could be treated through work or energetic considerations without needing to produce a Newtonian force law governing the forces between ball and table. This trick of *finessing* the problem of forces rather than directly answering it had an enormous impact on subsequent physical methodology, and many instrumentalist strains in nineteenth-century philosophy of science were inspired by the success of this clever antifoundationalist approach. "Forces" make a dual appearance in Lagrange's formalism—directly, as part of the potential, and indirectly, buried somehow in the constraints on the system. Likewise, in continuum mechanics, "forces" appear as stresses rather than as anything satisfying "$F = ma$." Given this distressingly protean nature to *force*, it was natural to hope that some other concept, e.g., energy, might provide a unity to this grab bag of tricks. Working out this energetic program can lead quite naturally to the various forms of antifoundationalism that we associate with nineteenth-century instrumentalists such as Ostwald. Of course, these quite concrete scientific considerations were frequently muddled together with traditional empiricist views that *forces* were nonobservable, noncomprehensible, or in need of operationalist definition. It seems to me that Sneed's and van Fraassen's judgments about the "theoretical" nature of *force* merely represent the somewhat ephemeral residue of an originally quite serious scientific problem.

Implicit in van Fraassen's discussions of examples like this is the assumption that, if someone obtains a value for *mass* through calculations from other quantities, he could not have observed the trait in question. Again as a point of ordinary language, this may be correct, but again it would seem to be a distinction we should not like to add to an internalist treatment of observation. After all, most sophisticated measuring devices used in physics laboratories today employ a computer to perform computations and control the experiment. Nonetheless, they are still good measuring devices, and it would obviously be hard to supply reasonable standards for when a given machine was "calculating" or not. Likewise, there are many types of mineral which can't be positively identified even by experts without performing a battery of auxiliary tests—cleavage,

---

20. A good account is Wilson Scott, *The Conflict between Atomism and Conservation Theory, 1644 to 1860* (London: MacDonald, 1970).

streak, hardness, solubility, etc. Nonetheless, *being hornblende* is a humanly detectable property, and all empirically adequate theories should be expected to model it. Like remarks hold for van Fraassen's own example of *being a VHF receiver* (p. 81); if we do not simply trust the authority of the manufacturer, it will take some tests to determine the wavelengths the device can receive.

I think we can now begin to see that van Fraassen has tried to perform too many philosophical jobs with a single notion of "observationality." One such task is to supply a reasonable account of the checkable content of a scientific theory. What limits does such a theory impose upon human knowledge? and so forth. The internalist treatment of observation, as we have sketched it here, seems a proper framework for attempting answers to such questions, although it also suggests that we need to decompose "observational content" into a family of related notions.

On the other hand, van Fraassen wants to use "observationality" as a guide to belief—he recommends that we believe only the "observational" parts of current scientific dogma. Here we shouldn't expect the internalist approach to be of much use. After all, the internalist account tells us what facts about the world we would be able to detect if we lived in a world modeled by theory T, given that T is true. But, of course, we needn't believe that we are detecting such facts if we are suspicious about T's truth. Typically, the more we trust a given theory, the more properties we will be willing to assign to the world around us on the basis of observation. But the internalist account, based as it is solely on the mathematical content of T, will not reflect these shifting standards of "observationality" simply because our current faith in T is not part of the content of T. At best, the internal treatment sets only a *maximal limit* to what we believe we can determine through observation. For purposes of theory comparison and so forth, it seems appropriate to deal with this limit, but what the internalist account marks as a *checkable fact* needn't be an especially *believable fact*.

On the other hand, the acquaintance view of "observationality" has often been linked to empiricist views about believability: one should only believe that objects have the properties that we are acquainted with. I hope I have persuaded the reader that many of the traditional assumptions about "observationality" that van Fraassen relies upon stem from this school and should not be accepted uncritically. Both van Fraassen and I reject the metaphysics of acquaintance, but van Fraassen's rejection leads to a puzzle which I'm sure that many other readers of *The Scientific Image* have felt. In the book, van Fraassen usually does not attempt to argue directly for limited belief in the current scientific corpus, but merely tries to establish his constructive empiricism as the rightful heir to the ancient House of Empiricism. There are, of course, many other reasons why one might succumb to instrumentalist temptations in the philosophy of science—I mentioned earlier some scientific difficulties that led many nineteenth-century physicists astray. Yet in *The Scientific Image* van Fraassen does not align himself particularly with analogous concerns (e.g., in quantum theory), but only with a purely philosophical tradition whose feet are firmly planted on "acquaintance"-based soil. Accordingly, I understand why many acquaintance theorists became antirealists (because scientific statements are not fully understandable), but I do not know exactly why van Fraassen wants to be one.

The main reason supplied explicitly in *The Scientific Image* constitutes what might be called an argument for agnosticism through the proliferation of empirically equivalent theories. The idea is that if two quite different theories T and T^1 are adequate to the known data (and one isn't obviously sillier than the other), one should be agnostic about any claims on which T and T^1 disagree. This seems like a fairly reasonable epistemological principle. Thus, Clerk Maxwell in his delightful encyclopedia article "Atom"[21] argued that, although we should believe in atoms, we should be agnostic about their other traits, even though in every concrete theory they were granted further properties (e.g., elasticity, inherent mass, etc.). The reason was simply that the numerous extant theories all disagreed about these qualities and none of the theories was clearly superior. But simple agnosticism does not in itself an antirealist make: a reasonable realist would have followed Maxwell's assessment. (Indeed, Maxwell's collection of theories were not empirically equally good—they were, rather, equally *bad*.) If we apply an internalist program to contemporary science, we find cases where distinct theories could be uncheckable by all conceivable evidence. The best-known example, cited by van Fraassen, are Glymour and Malament's results on our potential knowledge of the topological structure of a general relativistic universe.[22] But no scientific realist need postulate a global topology for the universe just to have a line on the matter. Our realist might like at least to know completely the laws which govern our universe, but the singularities of relativity suggest that even this hope is overly optimistic. So again one may have to settle for agnosticism. But in none of this do I see any betrayal of the realist's philosophic colors. What the antirealist needs to demonstrate is a more spectacular and unexpected arena for agnosticism than even these—e.g., a demonstration that all facts about the existence of forces are really uncheckable.

I won't be dogmatic (here!) about the outcome of such an investigation. However, it does seem to me that most writers who have written in this

---

21. In his *Collected Scientific Papers* (New York: Dover, 1958).

22. Clark Glymour, "Indistinguishable Space-Times and the Fundamental Group," and David Malament, "Observationally Indistinguishable Space-Times," both in *Minnesota Studies in the Philosophy of Science*, vol. 8, ed. Earman, Glymour, and Stachel (Minneapolis: Minnesota University Press, 1977).

vein have been far too cavalier in their presentations—both in terms of argumentation and delineation of examples. It should be clear from even as sketchy a study as this that the family of notions centered around "observational equivalence" constitute quite a complex lot, and it will usually take a fair measure of analysis to figure out what a given author might mean by the notion. And then it will take a further argument to show that the example should distress the honest scientific realist.

Part II

Bas C. van Fraassen
Replies to His Critics

11 Empiricism in the Philosophy of Science

Bas C. van Fraassen

It is a great honor and pleasure—though a scary one—to reply to this fair and careful critique of my version of empiricism in the philosophy of science.

The first draft of *The Scientific Image* took shape during my 1974/75 sabbatical; the book was published in 1980. Looking back to the beginning, the major change I see is in the sophisticated complexity of philosophical positions now arrayed with and against my own. Every contributor to this volume has a somewhat different position from all the others. I am glad that the situation was, or at least seemed, simpler to begin.

My reply will have two parts. Part 1 will outline my present position in epistemology. This is the area on which the critical papers mainly concentrated. I will be able to address some objections to *The Scientific Image* along the way, without any pretense, of course, that the answers were already implicit in the book. Part 2 will consist of short responses to further objections by the individual contributors, with references back to the first part. Inevitably, some subsections will summarize or duplicate parts of papers I wrote since the book, and I will indicate these.

Before I begin properly, I would like to mention some developments that I have found sympathetic. First of all, to my shame, I became only belatedly better acquainted with the writings of Larry Laudan. His collection of essays *Science and Hypothesis*, which appeared in 1981, gave me real new insight into certain philosophical issues, and I shall draw on it below. His antirealism and his arguments are different from mine, but closely allied. In Brian Ellis's book *Rational Belief Systems* (which I read and reviewed while preparing the final draft of *The Scientific Image*), he introduced the position of *scientific entity realism*. I think this is close to the modified realism adopted by Nancy Cartwright (whose *How the Laws of Physics Lie* appeared in 1983). What scientific entity realism rejects in realism, I think ought to be rejected. In his contribution to the present volume, Ellis makes it clear that he does not want to stop there; he advocates instead what he calls, with Putnam, *internal realism*. The thrust of his argument is that a scientific realist ought either to give up his realism or switch to the pragmatic theory of truth. That argument looks very convincing to

245

me. Finally, Arthur Fine (1984b) has attempted to find an intermediate position between realism and empiricism, which he calls the natural ontological attitude (NOA). The idea of NOA was that it should be acceptable to both camps and that each may see the error of going beyond it. As presently formulated, NOA is not compatible with constructive empiricism, but with minor modifications it would be (minor to me). Fine's advocacy of his position—like Laudan's, Ellis's, and Cartwright's of theirs—is accompanied by searching and persuasive refutations of previous arguments for realism, which I welcome as reinforcing my own.

I will not have the space to reply to every published comment or objection (though I will try to include reference to all I know in the bibliography). Omitted will be responses to, for instance, charges concerning sexism—in the "Tower and the Shadow" story—irresponsibility about explanation, and an unbelievably high price for the clothbound edition (though I will reply to a technical point in Friedman's review).[1] I mention these so as not to create the impression that I have unintentionally failed to give them all the attention they deserve. But I would like to share with the reader a letter I received from the Sierra National Forest in California. I quote parts only:

> *The Scientific Image* appeared on the "new books" shelf at the library here, so I thought I would read it and find out whether any advances had been made in the philosophy of science since I left school in 1961.
>
> The only thing I have to say to you is that the remark about Darwin on page 39 is puerile. . . . The idea that a mouse's enemy is a cat is as rank a distortion of evolutionary theory as would be the idea in physics that massless particles fail to be obedient to the law of gravity because they have no ears to hear the commandments of God.
>
> It is unDarwinian to think of mice and cats as enemies. Did you learn your biology from the Lake poets? Mice and cats are friends. Friendly mice feed cats. Friendly cats help mice control their population densities.

Point taken; cats and mice do not compete for the same resources. I have been trying to use this insight to construct a sociobiological Sixth Way for the "Gentle Polemics" chapter, but so far with no great success.

I. Sketch for an Epistemology

As I argued in *The Scientific Image*, and shall argue again below, acceptance is not belief. The distinctions I draw here entail a sharp division between epistemology and the study of scientific methodology, and a break

---

1. See part 2, "Ad Alan Musgrave."

with the characterization of science as merely the process of updating, re-fining, and improving our opinion of what the world is like.

But acceptance of a theory does involve some belief, and when it comes to opinion about empirical matters of fact—what the actual, observable world is, was, and will be like—it is reasonable only to turn to science, insofar as it goes. Hence, the central topic of epistemology, *rationality of opinion*, is never far away in our philosophical discussions of science.

So in this part I shall sketch my (tentative and developing) views on ra-tional belief. Paul Churchland has argued that belief and opinion, concepts of "folk psychology," are already anachronisms. In my review of his book (van Fraassen 1981), I argued to the contrary that, as epistemology is now developing, the refinement of those concepts is a proper part of the rise of scientific philosophy. He will find me in any case unashamedly tradition-alist in my epistemological concerns. The death of epistemology, preached in such different ways by Rorty and Churchland, is only the fire in which the phoenix is being reborn.

A quick preview first. The theory of belief—or, better, of opinion—which I consider right makes probability theory the logic of judgment (epistemic judgment, the judgments that constitute the state of opinion). This entails that the empirical adequacy of an empirical theory must al-ways be more credible than its truth. It also entails that a theory as a whole (no matter how explanatory, unified, or consilient, and no matter how many novel predictions it has led to) cannot be more credible than any of its subtheories. (From the semantic point of view, the statement that a the-ory is empirically adequate formulates one of its subtheories, the one we still sometimes conveniently but inaccurately try to identify in positivist phrasing as the sum of its "observable consequences.") And finally it en-tails that no virtue which consists in greater informativeness (among which I reckon explanation) can give an additional reason for belief. All these consequences I shall argue below, but this announcement should make clear that the theory of belief will play an important role in my arguments.

I wish to add immediately that it cannot provide the whole basis of my arguments. The reason is exactly that, in my view, acceptance is not belief, and the methodology of science is accordingly not covered by general epistemology.

1. Rationality and the Logic of Judgment

In his contribution, Ronald Giere compliments me on not taking refuge in Bayesianism (though he adds, "But virtue is not always rewarded"), and I would not happily forgo this compliment. So I will take care in this sec-tion to make clear (at least for readers familiar with Bayesian writings) where my agreements and disagreements with that movement lie. As Richard Jeffrey has been elaborating recently, the probabilism of De Finetti,

Ramsey, and Savage counts among its ancestry such great antecedents as Christian Huygens, the Port-Royal logicians, and Carneades; so I shall not be ashamed to admit to a large measure of agreement. My disagreement comes exactly on attempted Bayesian reconstructions of concepts of scientific methodology (see, for example, Paul Horwich's recent *Probability and Evidence*). In what follows here I shall only summarize my conclusions, and I beg the reader to look for supporting arguments in my recent articles.

Let us begin with *rationality*. Is it rational to believe in angels or electrons? I construe the term *rational*, as applied to opinion here, as a term of *permission* rather than of *obligation*. To say that you are rational in your opinions does *not* mean that your opinions are rationally compelled—that any rational person with the same experiences as yourself would have to agree. It is not irrational to "go beyond the evidence," and belief in angels or electrons or the truth of theories in molecular biology does not *ipso facto* make one irrational. The constraints or bounds of rationality leave much underdetermined—*rationality is bridled irrationality*.

What are the bounds of rationality? I am not too sure of even a general outline for an answer. They *include* the bounds of logic. But those bounds are, as it were, nonexistent; logic is empty. What I mean is, what the bounds of logic are is clear to you if (and to the extent that) you understand the language you are using. Secondly, for being rational, *ought* implies *can* (unlike, perhaps, in ethics). This I mean in the strong sense that self-policing must be possible, that you are not guilty of irrationality if you violate a standard you could not yourself have applied in time. Let me clarify this by means of the distinction between *reasonableness* and *vindication* in the evaluation of right action, right decision, and right opinion. Whether or not you were vindicated in a decision or action depends on the outcome it led to in the actual circumstances that obtained—much of which you could not have known or reasonably expected. Whether or not your present opinion about tomorrow's weather will be vindicated depends on tomorrow's actual weather. Lack of vindication can be a reproach, as Machiavelli pointed out, but it cannot impugn the rationality of the action or opinion. Whether or not that was reasonable depends on factors settled at the time and, in some sense, accessible. There is still a connection because the paradigm of irrationality is to form or organize your actions, decisions, or opinions so as to hinder needlessly your chance of vindication. If your opinion is logically contradictory, you have sabotaged yourself in the worst possible way—you have *guaranteed* that your opinion will not turn out correct—but milder forms of self-sabotage are easily envisaged.[2]

This short reflection on rationality gives the essential clue to the semantic analysis—and hence logic—of judgment. Consider the phenomena of

2. Reichenbach introduced this theme, which can be carried considerably further; see Van Fraassen 1983a.

judgment first. I express my state of opinion by saying that it seems likely to snow, more likely to snow than to rain, yet also almost certain to rain on the supposition that the temperature rises by fifteen degrees. I may further express my opinion by saying that it seems to me not unlikely, quite possible, that the theory of evolution is true, and certainly very likely that it will keep being corroborated by empirical findings and theoretical developments for the next century or so.

Sharp personal probabilities are notably absent, though I could have given natural-sounding examples ("It seems exactly as likely to me as not that it will snow"). But as everyone should know, the postulation of sharp (fully determinate) personal probabilities is neither needed nor presently common in the theory of belief. Thinking about the above examples, you will understand what it is for a probability function to *satisfy* or *represent* my opinions insofar as they go. To say that my opinions are vague and full of gaps means that, if one probability function jointly satisfies all my present judgments, so do many others. Let us call the class of all those functions my *representor*.

Bayesian Dutch book arguments, but also certain other arguments, try to show that rationality requires my representor to be nonempty (*coherence*). If my representor is empty, then I must be violating the probability calculus, and those arguments purport to show that, in that case, I am precluding vindication in some respect. The arguments do not go as far as one might wish, but that is not because we have counterexamples outside their clear domain of application—and they do go impressively far. That is why I say that probability theory is the logic of judgment.

The representor represents the class of judgments that are overtly mine. Each of my judgments limits the class further; so, for example, my judgment that A seems more likely than B limits membership in my representor to functions P such that $P(A)>P(B)$. Note that my lower and upper personal probabilities could still be entirely unconstrained: the infimum and maximum of the numbers $P(A)$, for members P of my representor, may still be *zero* and *one*. We may summarize this by saying that my vague personal probability for A is then the whole half-open interval $(0, 1]$.

In this last statement I really insinuated something new: that I am committed to judgments satisfied by all members of my representor. Implicitly I introduced a notion of *entailment*: judgment Q is entailed by judgments $P_1 \ldots , P_n$ exactly if all probability functions that satisfy $P_1 \ldots P_n$ also satisfy Q. In deference to this construal of the notion, I can say that, if I deny a judgment entailed by other judgments I make ("deny" in the strong sense of making a contrary judgment), then my opinion is incoherent (has an empty representor). So the entailment as construed is entailment on pain of incoherence. Once we accept this, there is a "supervaluation" relationship between "my" judgments and "my" representor.

This way of thinking about vague personal probability, via a class of probability functions, is part of the theories of belief developed by such otherwise divergent writers as Isaac Levi and Richard Jeffrey.

What about rational change of opinion? We sometimes see the conditional probability of A on B (*given B, on the supposition that B*) explained as the probability which A would have for me if I were given B as total new evidence. This explanation is totally flawed as an interpretation of conditional probability (in part because some propositions could never be anyone's total new evidence and in part because some which admittedly could be true could never be ones one could believe to be true). Worse, it has reinforced the idea that accepting probability as the logic of belief brings with it an implicit theory of rational change of belief. This is exactly the point where traditional philosophers of science feel threatened by the Bayesian line. What is at fault, however, is not the elaboration of consistency as coherence but the simplistic modeling of evidence and of rational belief change which accompanied it.

The basic picture of the deliverances of experience—the *revelation model of evidence*—is of a man in a glass booth with a ticker tape that prints out statements which he treats as divine revelation. Each time a new such statement appears, he becomes fully certain that it is true, and the only other thing he does is to adjust his prior opinions to accommodate it (i.e., he conditionalizes on this evidence). It is clear that, if the ticker tape has been delivering only nontheoretical statements, then his theoretical opinions derive mainly from the prior opinions he brought with him to this situation. No proofs about how the evidence can, in the long run, swamp any given prior opinion can take the edge of this damning point: that he can today have no significant theoretical opinions unless his prior opinions also had significant theoretical content.

This picture we must challenge in two ways. The first is that evidence does not come in the form of a ticker tape with the status of divine revelation. The second is that rational belief change is not restricted to mere accommodation of what one takes as evidence. In this I follow a traditional line: coming to believe something which is not established by the evidence (even relative to our previous opinions) may, though clearly fallible, be rational; *and* in what we take as evidence we are also fallible and not certain to accept only truths. To this I wish to add two typically empiricist concerns: (1) nothing except experience may be treated as a source of information (hence, prior opinion cannot have such a status, so radical breaks with prior opinion—"throwing away priors," as born-again Bayesians call it— can be rational procedure); (2) the deliverances of experience do fall within definite and strict limits, due to our factual finitude in several quantitative and qualitative respects. The problem is now to replace the challenged "revelation model" with something better. I have been working on this program in recent papers, and I do not want to go into its technical aspects here. But I shall describe the *voluntarism* which provides the philosophical framework and which I see as indispensable to the program.

Inevitably looking rather bare when presented without defense, here are the main points. First, when I express a judgment (express my opinion), I do not make an (autobiographical) statement but express an attitude. (Expressing an attitude is distinct from stating that I have it.) Important features are shared with such acts as promising or wishing and expressing intentions or commitments. (Thus, I am expected to "stand behind" my avowal.) In purporting to describe what is the case, I can at best express my opinion, and by doing this I have already ceased to be a passive onlooker and have become involved.[3] Second, it is impossible to find propositions that experience "gives" one as evidence. (Working statisticians, of course, are given propositions to be treated as evidential input by their employers, which may have affected Bayesian modeling.) As soon as I try to describe my experience ("I just saw a flying saucer"), even if only to myself, I am formulating a new belief *in response to* my experience. At this point one may be tempted to introduce a theoretical description of a process that determines this response. Let us leave that to scientists and not be our own armchair psychologists. The correct characterization of taking-as-evidence and its near relatives is this: I *accept a constraint* on my posterior opinions. This constraint I attempt to satisfy, and a critique is possible of how and whether I satisfied it in realizing my opinions. The acceptance of the constraint is itself also subject to critique, first of all by others (*"I* don't believe that you saw a flying saucer") and later by myself ("I now believe that I made a mistake, though at the time it seemed as clear as day"). And the constraint can take many forms—it may merely be to raise or lower a personal probability (or to make it sharper or vaguer), to change an expectation value, and so forth—and not just the form of a new, full belief (see Van Fraassen 1980b; 1981a).

Third, there is the rationality of coming to believe new hypotheses and other changes of opinion which cannot be justified as mere responses to the constraints we accept as having been imposed directly by our experience. While I shall discuss this distinction in more detail below (it is the demarcation-of-the-observable problem, obviously), everyone must surely make some such distinction. Even if "observation with instruments" had the same justification status as observation pure and simple, many hypotheses go far beyond the deliverances of experience so far. We are here in the

3. This is relevant also to the point that probability is the logic of judgments. The usual deductive logic and usual notion of satisfaction apply directly to autobiographical *statements*, and the statement that my personal probabilities for A and *not* A add up to more than one is satisfiable because I might not be rational. The semantic analysis of *judgments* is like that of commands (the analogous satisfiability concept for commands is that they can be jointly carried out, not that they can be jointly issued).

debate William James, in his "The Will to Believe," carried on with W. K. Clifford. The latter, who thought he was speaking for scientists, said that it is always wrong for anyone anywhere to come to believe something on insufficient evidence. This was an early version of the view that the only rational belief change is conditionalization on the evidence. The contrary view, that we may rationally come to believe hypotheses that go well beyond what any evidence could have shown—and justify this by appeal to consilience, novel predictions, and explanation—became part of the "received view" about theory acceptance that underlies much scientific realist writing. I consider William James to be basically right.[4] While acceptance of a theory is usually qualified, and properly so, I take acceptance as such to involve belief that the theory is empirically adequate—and not *ipso facto* irrational merely because that belief goes well beyond what the evidence has established.

The question will at once arise why I do not consider belief in the truth of the whole theory to be "equally rational." *Distinguo*: I do not think that such belief is involved in scientific theory acceptance, *but* neither do I think such belief to be irrational. Yet "equally rational" is the wrong phrase (even if it is intelligible, which I doubt). A person may believe that a certain theory is true and explain that he does so, for instance, because it is the best explanation he has of the facts or because it gives him the most satisfying world picture. That does not make him irrational, but I take it to be part of empiricism to disdain such reasons. I realize that this disdain is defensible only given certain views about explanation, for example, and shall further defend those views below.

2. Empiricism and the Limits of Experience

Addressing the philosophical clubs of Yale and Brown universities in 1896, William James distinguished his empiricism from what he called absolutism on the one hand and skepticism on the other. (James 1897, preface and 12–14). Rejecting skepticism, he asserted that we are certainly capable of arriving at the truth about what we ourselves and the world are like. But against the other extreme he argued for the fallibility of all human claims to knowledge—we cannot pretend to arrive at objective certainty or absolute security. The ground for both assertions lay in the thesis he identified as the core of empiricism: experience is the legitimate and only legitimate source for our factual opinions. Hence, all conclusions about matters of fact are liable to modification in the course of future experience.

4. See my "Belief and the Will," in which I try to demonstrate that Dutch book considerations themselves lead to a principle, going beyond the axioms of probability, which saves James's view from Bayesian refutation but which requires voluntarism for its own defense.

Half a century later, in 1947, Hans Reichenbach gave a presidential address to the American Philosophical Association and characterized his own logical empiricism in similar terms.[5] Since metaphysicians had claimed to reach objective certainty about matters of fact by demonstration, earlier empiricists had mistakenly thought that, to be successful, they must show that this objective certainty can be reached on the basis of experience alone. But Hume had proved this impossible once and for all. The correct response is neither the despair of skepticism nor the impossible ideal of an empirically based metaphysics. Instead, we should present an empiricist theory of knowledge and rational belief which entails both the possibility and fallibility of rational opinion about matters of fact, based on experience.

So, for James and Reichenbach, the core doctrine of empiricism is that experience is the sole source of information about the world and that its limits are very strict. For an example of such a limit we may take our temporal finitude: it is not possible to have a guarantee about the future on the basis of our experience so far—otherwise we could indeed have opinions today which are not liable to modification in the course of future experience. I explicate the general limits as follows: *experience can give us information only about what is both observable and actual.* We may be rational in our opinions about other matters—Augustine's "faith in things unseen," which, he rightly said, pervades even our everyday opinions about everyday matters—but any defense of such opinions must be by appeal to information about matters falling within the limits of the deliverances of experience.

As to what is observable, that also has both general and special limits. The most general limit is that experience discloses to us no more than what has *actually* happened to us so far. Hence, any observable structure is one which, in today's scientific world picture, fits inside the absolute past cone of some space-time point. In addition, the structure must be finite; indeed, on a cosmic scale, rather small. These are general limits that I take to apply regardless of who *we* (the epistemic community) are and which will therefore always remain. They are already sufficient to establish that scientific theories tell stories which go way beyond the limits of experience, even in the long run (partly because of modalities and partly because of questions of global space-time structure), and hence to establish the distinction between truth and empirical adequacy.

There are also much more special limits which derive from the *de facto* constitution of the epistemic community. What these limits are is accordingly so very much an empirical matter that we cannot be entirely sure that we know what they are and, even less, what they will be. When I give ex-

5. "Rationalism and Empiricism: An Inquiry into the Roots of Philosophical Error," in Reichenbach (1959), 135–50.

amples, they always presume some suppositions about what we are like. For example, I always assume that we (the epistemic community) are all humans, and no one of us is really a person from Krypton, like the comics' Superman, who could see Lois Lane's pink underwear when she was fully dressed. What our special limitations are I take to be an empirical question, and I think all my critics agree with this. I think they all agree also on the vagueness of observability and the irrelevance of exactly where the line is drawn. An electron is so unimaginably different from a little piece of stone—or the little grid Hacking can hold with tweezers—that minor adjustments would make no difference to the issues. Yet these special limitations provide the focus for much criticism. I shall first discuss those critiques which do no dispute the *de facto* limits of observability, but only the role I gave them in epistemology; then the remainder.

Gary Gutting's illuminating imagined dialogue between (proxies for) Wilfrid Sellars and myself leads right to the crux of the debate over observability. Why should its limits be given any epistemological role at all? Why should the range of accessible possible evidence play a role, as well as what the actual evidence is?

In my book I made only two directly relevant remarks, and they were not sufficient to answer this question. The first was that I did not see how we could deny the epistemological relevance of the range of accessible evidence, except on the basis of extreme skepticism or untrammeled, wholesale leaps of faith. The way I would put it now is this: If we choose an epistemic policy to govern under what conditions, and how far, we will go beyond the evidence in our beliefs, we will be setting down certain boundaries. I could not envisage a nonextreme rational policy that would make these boundaries independent of our opinions about the range of possible additional evidence. I can see additional boundaries—for instance, I do not see how the desire for a comfortable world view or philosophy of life could ever be an acceptable defense for believing something. (William James did not rule this out, but in another essay, "The Sentiment of Rationality," he is quite scornful of philosophers' desires for explanation as reason for belief.) And I agree, of course, that the truth of a theory taken as a whole is put in doubt as soon as experience tells against any of its consequences. But the theory's vulnerability to future experience consists *only* in that the claim of its empirical adequacy is thus vulnerable.

The second relevant remark was that it is not an epistemological principle that one may as well hang for a sheep as for a lamb. I am glad it was Musgrave who commented on this, for he seems to be in agreement with me in not taking features other than agreement with the evidence (features such as explanatory power and the like) as providing additional, independent reasons for belief. Thus he can say:

Epistemological or not, the principle that one might as well hang for a sheep as for a lamb is a pretty sensible one. Given two criminal acts A and B whose risks of detection and subsequent penalties are the same but where A yields a greater gain than B, the sensible criminal will do A. . . .

Suppose the realist tentatively accepts a theory as true, while the constructive empiricist tentatively accepts it as empirically adequate. The realist does take a greater risk. But he takes no greater risk of being detected in error *on empiricist grounds*. So, given strict empiricism (the principle that only evidence should determine theory choice), it seems that we ought as well be hung for the realist sheep as for the constructive empiricist lamb. (This volume, chap. 9)

Although the phrasing insinuates several views which I do not hold,[6] he describes the main situation just as I see it. If I believe the theory to be true and not just empirically adequate, my risk of being shown wrong is exactly the risk that the weaker, entailed belief will conflict with actual experience. Meanwhile, by avowing the stronger belief, I place myself in the position of being able to answer more questions, of having a richer, fuller picture of the world, a wealth of opinion so to say, that I can dole out to those who wonder. But, since the extra opinion is not additionally vulnerable, the risk is—in human terms—illusory, and *therefore so is the wealth*. It is but empty strutting and posturing, this display of courage not under fire and avowal of additional resources that cannot feel the pinch of misfortune any earlier. What can I do except express disdain for this appearance of greater courage in embracing additional beliefs which will *ex hypothesi* never brave a more severe test?

Not everyone will share my disdain, but perhaps more will agree that, as far as the enterprise of science is concerned, belief in the truth of its theories is supererogatory. Suppose that nothing except evidence can give justification for belief. However flexibly this is construed, it means that we can have evidence for the truth of a theory only via evidential support for its empirical adequacy. The evidence then still provides some reason for believing in the truth, *a infirmiori* so to say [as one might say that a reason for (A *and* B) is *a fortiori* a reason for B, so a reason to believe B might be called a reason *a infirmiori* for (A *and* B)!], but the additional belief is supererogatory. However, not everyone shares my view that what I called the pragmatic virtues provide no independent reason for belief, so I shall come back to that below.

There are, finally, the many independent challenges to where I "draw

6. It is philosophers, not scientists (as such), who are realists or empiricists, for the difference in views is not about what exists but about what science is. And, if theory choice is a choice of what theory to *accept* (as opposed to believe), then many features other than agreement with the evidence are relevant. The commitment involved in theory acceptance may be vindicated or come a cropper independently of its empirical adequacy or truth.

the line" on observability. Musgrave and Churchland give arguments that throw doubt on the very coherence of my treatment of this distinction. Musgrave says that "B is not observable by humans" is not a statement about what is observable by humans. Hence, if a theory entails it, and I believe the theory to be empirically adequate, it does not follow that I believe that B is not observable. The problem may only lie in the way I sometimes give rough and intuitive rephrasings of the concept of empirical adequacy. Suppose T entails that statement. Then T has no model in which B occurs among the empirical substructures. Hence, if B is real and observable, not all observable phenomena fit into a model of T in the right way, and then T is not empirically adequate. So, if I believe T to be empirically adequate, then I also believe that B is unobservable if it is real. I think that is enough.[7]

Significant encounters with dolphins, extraterrestrials, or the products of our own genetic engineering may lead us to widen the epistemic community. And this community could in any case have been different from what it was. I feel no great difficulty with thought experiments about such eventualities. But Paul Churchland has given them several new twists, so that they are not covered by my remarks about Grover Maxwell's.

First, there is the case of the man all of whose sensory modalities have been destroyed and who now receives surrogate sensory input electronically. This individual existence is simply irrelevant to *our* epistemic range, just like my scarred and thereby impaired left eye. Next, he asks us to envisage an epistemic community consisting entirely of beings in this predicament. Thirdly, Churchland imagines that we encounter a race of humanoids whose left eyes have the same structure as a human eye plus an electron microscope. Science tells us that virus particles and individual DNA strands are observable to them.

Upon reflection, it does not seem to me that the second example provides difficulties that really go beyond the third, so I will focus on that one. The example as given tempts us to confuse two cases. The first case has us suppose that we have accepted these humanoids as persons, as members of our epistemic community. In that case we have already broadened the extension of *us*, and what is observable to them is observable. The second case I think of as the "faithful dog and supersonic whistle" case, because that is how it is usually posed. In that second case, we call the newly found organisms "humanoids" without implying that they bear more than a

7. Foss (1984), Rosenberg (1983), and also Giere (this volume, chap. 4) have each suggested that there may be a circle, because I define empirical adequacy in terms of observability and point to science as the source of information about what is observable. I think the preceding paragraph answers this point if taken in conjunction with my remarks *Ad Mark Wilson* in part 2.

physical resemblance to us. Then we examine them physically and physio-
logically and find that our science (which we accept as empirically adequate) entails that they are structurally like human beings with electron
microscopes attached. Hence they are, according to our science, reliable
indicators of whatever the usual combination of human with electron microscope reliably indicates. What we believe, given this consequence drawn
from science and evidence, is determined by the opinion we have about
our science's empirical adequacy—and the extension of "observable" is,
ex hypothesi, unchanged.

Churchland tempts us to confuse the two cases by presenting himself as
an all-knowing and authoritative spectator who can tell us that the humanoids really are persons and that virus particles, etc., really are observable
to them. This is the authoritative stance he takes, part narrator and part
commentator, when he writes:

> The difficulty for van Fraassen's position . . . is that his position requires
> that a humanoid and a scope-equipped human must embrace *different* epistemic
> attitudes toward the microworld, even though their causal connections to the
> world and their continuing experience of it be identical.

But, on the one hand, it is not warranted to speak of their *experience* unless
they are already assumed to be part of our epistemic community. On the
other hand, while I do not know what, or indeed whether, factual conditions concerning causal connections suffice to make something a member
of that community, I do know that what an empirically adequate theory
implies about causal connections may not be true. We do not, in addition
to the science we accept as empirically adequate, have a divine spectator
who can tell us what is really going on. And, if we supposed that we had,
there would still be two cases: does the supposition include that He is a
member of the epistemic community or . . . ?

Behind the temptation Churchland so skillfully poses before us there
lies, I think, a *modal argument*:

> We could be, or could become, X. If we were X, we could observe Y. In fact,
> we are, under certain realizable conditions, like X in all relevant respects. But
> what we could under realizable conditions observe is observable. Therefore, Y
> is observable.

The crucial third premise, however, is justified by appeal to science (at
best). If we assume only that science is empirically adequate, we can justify
only the premise that we are, under realizable conditions, empirically indistinguishable from beings like X in all relevant respects. The conclusion
then derived is only,

Under certain realizable conditions, all the observable phenomena are as if we are observing Y,

which is often true although Y is unobservable. Only unconscious positivistic leanings would tend to lead us further. Of course, if the third premise is altogether omitted, we deduce only that Y could be or could have been observable, from which it would take a modal fallacy to draw the conclusion that it *is* observable.

It is clear from these essays and from most of the reviews that scientific realists tend to feel baffled by the idea that our opinion about the limits of perception should play a role in arriving at our epistemic attitudes toward science. This may be because the empiricist premise, that experience is the sole legitimate source of information about the world, leaves them cold. Or it may be because disdain for opinions inflated beyond what can run the gauntlet of experience strikes no chord in their hearts. But I suspect it must also be in part a different appreciation of just how unimaginably different is the world we may faintly discern in the models science gives us from the world that we experientially live in (the scientific image from the manifest image, the intentional correlate of the scientific orientation from the phenomenological life-world). This difference has been stretched by empiricists from the beginning—once atoms had no color; now they also have no shape, place, or volume.[8] (Except, that is, on certain hidden variable interpretations, which are in my view, at best, metaphysical baggage but which in any case engender paradoxes of their own.) There is a reason why metaphysics sounds so passé, so *vieux jeu* today; for intellectually challenging perplexities and paradoxes it has been far surpassed by theoretical science. Do the concepts of the Trinity, the soul, haecceity, universals, prime matter, and potentiality baffle you? They pale beside the unimaginable otherness of closed space-times, event-horizons, EPR correlations, and bootstrap models. Let realists and antirealists alike bracket their epistemic and ontic commitments and contribute to the understanding of these conceptual enigmas. But, thereafter, how could anyone who does not say *credo ut intelligam* be baffled by a desire to limit belief to what can at least in principle be disclosed in experience? Or, more to the point, by the idea that acceptance in science does not require belief in truth beyond those limits?

3. The Scientific Method Justified?

In this section and the next I wish to discuss scientific methodology, beginning with what a philosopher cannot and should not do. This will

8. See, for instance, Karplus and Porter (1970), sec. 3.10 (e.g., replacement of the size and shape by the "90% boundary surface").

prepare the ground here for a response (in part 2) to Richard Boyd's program. But the reader will quickly perceive that in these preliminaries I am much less ambitious, postponing consideration of the careful and insightful analyses of scientific procedure given by Boyd.

It is natural and appropriate for a philosopher to ask how the sciences proceed, and then why they should proceed in just that way. But, depending on how we take the "why" (i.e., on what we'll regard as a relevant and satisfactory answer), we may shed light or create new darkness. Here is what I see as a recipe for disaster:

Step 1. Assume that there is such a thing as *the scientific method* and that it has been best described, if only in part, by account X.

Step 2. Assume that this method is a method for arriving at true, or at least reliable, information about aspect Y of the world.

Step 3. Raise the problem of justification: to show or explain how the method described by X is indeed especially well designed to lead to true or reliable information about Y, or at least that it is better than alternative methods that might have been followed. Produce such a justification, Z.

Step 4. Analyze given justification Z; note that it rests on certain assumptions. Then claim that the success of Z in explaining the success of the sciences which proceed by method X is good support for the correctness of those assumptions underlying Z.

In support of my assertion that this is a recipe for disaster, you will have noticed certain pitfalls. To begin, science has certainly been successful; whether it actually proceeds in accordance with a methodology correctly described by X may be less incontrovertible. But, even if account X is correct, step 2 is not redundant upon step 1, since it may be other features of scientific procedure, besides those described in X, that have led to its success; or, equally worrying, it may have been special circumstances in which this procedure was followed that also contributed crucially to that success. If any of these doubts should be correct, the enterprise of step 3 will be based on a mistake. Unfortunately, these doubts can, it would seem, only be disarmed by that enterprise, which means in effect that they cannot be disarmed at all. Finally, should the enterprise succeed (in the only sense that it can, namely, without guarantee that the justified methodology really has the virtues that the justification explains), it can still be no more than the best available explanation of the postulated facts. So the claim made in step 4 is at best an inference to the best explanation. Hence we have a dilemma: either this pattern of inference is licensed by the methodology described by X or it is not. In the first case, part of the justified methodology was assumed in the justification. In the second, the person taking this step must reflect that his inference is neither part of what he takes to be the best description of scientific method nor incontrovertible among philosophers.

This general argument should suffice to banish such attempts from philosophy of science forever, but it would be optimistic to think that it will. Past examples are, of course, in general disrepute. The above recipe is followed in certain justifications of induction, which end with certain contingent premises. The author may proudly point out that even if those premises can only be supported by induction, that still shows a laudable internal coherence for the inductive method. Or he may proudly point to his premises as supported by the success of the inductive method they justify. From Gravesande's axiom of the uniformity of nature in 1717 to Russell's postulates of human knowledge in 1948, this has been a mug's game. A similar, perhaps related, sort of example consists in speculations on the convergence produced by systematic self-correction, from the time of Newton till today, examined by Larry Laudan.[9] Two contingent principles appear: that there is convergence in scientific hypotheses and that, if there is convergence, it is to the truth. At least, they are contingent if made in the context of an intelligible and nontrivial measure of convergence for hypotheses, if that exists. Peirce, Putnam, and now Ellis insist that the second principle can be backed only by a theory of truth which entails internal logical connections between truth and warranted assertion. They insist on this presumably because they do not want to be caught in the destructive dilemmas described above.

I cannot end this catalog of horrible examples without including the fastest and most facile—so-called *evolutionary epistemology*. I have learned about this subject not at firsthand, but from the critical analyses by William Lycan, Robert Van Gulick, Neil Tennant, and Massimo Piatelli-Palmerini. Nor are the sensible references to evolutionary mechanisms by Boyd, Churchland, Ellis, and Hooker in these pages to be faulted or found guilty by association. But the "pure case" of evolutionary epistemology apparently runs like this: We need methods of forecasting if we are to plan and hence to survive at all. The methods we have been following are the methods of the most successful organisms on earth, namely, us. So they clearly have survival value. It remains only to show how, specifically, each feature of that methodology (as described by some account X, say, a list that includes inference to the best explanation) contributes to our fitness for survival. As Popper apparently first said, that is epistemology for dinosaurs.

This is a parody of evolutionary thinking, though it looks surprisingly like some patterns of argument in sociobiology. If any pervasive characteristic in the population can only be explained by showing how it contributes to gene survival, does it not follow that every pervasive characteristic must be one that increases fitness?

In response (briefly), we can make two points. First, adaption is often

9. See, for example, his "Peirce and the Trivialization of the Self-Corrective Thesis," in Laudan (1981).

by *satisficing* rather than by *optimizing*. Secondly, evolutionary theory allows for many mechanisms leading to the persistence of anti- or non-adaptive traits (random fixation by limitations to population size or colonization of new areas by small subpopulations, genes "riding along" on the same chromosome as a gene being selected, developmental side effects of a gene being selected for a different reason).

Though not falling into the traps of vulgar Darwinism, Hooker and Churchland both like to refer to the evolution of our cognitive methods and strategies. Both make the point that selection mechanisms would not differentiate, through pressures on cognitive strategies, between adequacy to the observable and to the nonobservable. This seems very implausible to me even if we grant the applicability of these biological concepts here. Suppose the same green color on copper may indicate the presence of one of two chemicals, one poisonous and one harmless, the difference being manifest only when food cooked in such a copper pot is ingested. Any avoidance strategy that spreads through the population through natural selection will be of the green color, not of the green-color-when-indicating-poison. In any practical sphere, our hypotheses are tested only to within empirical equivalence, and selection pressures are entirely within the practical sphere.

Karl Popper (1981) correctly outlined, as far as I can see, the sole evolutionary parallel that can be drawn for the development of scientific theory. If there is such a parallel, scientific progress results from *instruction* (analogous to inheritance) and *selection*, and *there is no instruction from without nor selection from within*. First, the external world does not select for internal virtues—not even ones that might increase the chance of adaptation or even survival beyond the short run. Second, there is no directed variation: the distribution of variations is not such as to aid survival in the actual environment. For theories, these two points mean the following: Only through agreement or disagreement with the observed phenomena does rude reality take a hand, cutting down without any relation to explanatory power, simplicity, internal unification, and the like. Secondly, the creative intuition that provides novel hypotheses is not especially in tune with the world. There is no subjective difference between *inspiration* and *guess*, nor between *faith* and *stubbornness*—these distinctions are made with hindsight, sorting out the lucky who were vindicated from the unlucky who were not.

I see no way in which an evolutionary epistemology could possibly beat the pitfalls and dilemmas of the recipe for disaster I outlined. But evolutionary *analogies* take a quick grip on the imagination, and it is easy to see how realists would be tempted by them. To show that two can play the game, I'll now present two antirealist evolutionary analogies, with rival pictures to bewitch your mind.

In a period of normal science, such as that from Newton till the late nineteenth century, we see a great uniformity. There is a period of sustained successful problem solving in which a single paradigm theory is extended into new areas and directions. Is the best explanation not the thesis that, in some major if approximate way, that paradigm theory represents the discovery of a truth and that its victorious extensions are due to the theory's exploitation of (partial) truths about real underlying mechanisms?

That "best" explanation is an intensional, *adequatio ad rem* explanation; now here is a Darwinian one. There was a famous puzzle in evolutionary theory, seen as a threat by Darwin himself: the apparent Cambrian explosion of life. Most of the major phyla of invertebrate animals appeared then in the span of a few million years—four billion years into earth history. Although some Precambrian fossils have now been found, the sudden development of multicelled animals was still apparently explosive. The solution (i.e., evolutionary hypothesis) is that the sudden diversification was triggered by a very small development, the appearance of a unicellular predator or cropper, (see Gould 1973, essays 14 and 15). This may seem counterintuitive, but, in the absence of such violence, diversification is inhibited. This is similar to my lawn: in the absence of such croppers as sheep or lawn mowers, the dandelion is taking over. The appearance of a cropper results in fast diversification.

Analogously, the period of normal science is the period during which the paradigm science encounters few serious reverses or seriously recalcitrant phenomena. In the absence of such violence, the imperialism of this theory is unchecked. Then come the violent reversals, everybody theoretically gallops off in a different direction at once, there is a new intellectual renaissance—a thousand flowers bloom. Nature, producing observable effects that burst the theoretical confines, suddenly turned predator.

For a second gambit, consider the glorious history of the atomic hypothesis. How is it that certain highly theoretical features of scientific models, initially far from the experimental arena, can play such an essential role in novel success? Must they not have been early discoveries of reality? The atoms of the early nineteenth century looked like metaphysical fictions and played no significant role in empirical prediction. Slowly it appeared that these hypotheses, initially all but totally untestable, held the clue to success—the essential clue to the saving of the phenomena found or created in the laboratories at the end of that century. If contribution to empirical success were the only touchstone, could the atomic hypothesis ever have grown up? Must it not have been an insight into reality, recognized as such by scientists on the basis of those clear signs of atunement to reality such as explanatory power, unification, simplicity, and the like?

A clear case of preadaptation! So consider the analogous puzzle for evolutionary theory. Today's human eye needs no praise; it is so clearly adapted to its function. But how did it ever evolve? Did the earliest form of this organ, say, the first five percent of the eye to evolve, really contribute to the organism's fitness for survival? What good is five percent of an eye in red-toothed jungle? Again, Darwin already reacted to this problem. The Darwinian solution is that *preadaptation exists only in hindsight*. No, the early forms of the eye were not selected because they would eventually allow visual perception; they were selected for *some* advantage they gave the possessor at the time or because they "rode along" on some adaptive trait. Slowly this rudimentary feature, perhaps non- or even antiadaptive or serving some other function altogether, took on the function for which we now say, with hindsight, it was so eminently suited.

For the epistemology of science, the philosophical justification of scientific method is a morass, a dead end, a false ideal, and a scandal. If anti-realists can neither justify nor explain the success of scientific methodology, they should know when they are well off.

4. The Scientific Method Classically Described

Does the scientific method exist, and, if so, what is it like? It is easy enough to give a trivializing answer; but, if the term is taken in some trivial sense, then the existence of this scientific method may be obvious but its relation to the practice and success of science so minimal as to make the subject banal. What worries me is that I am still not sure, for example, whether the assertion that the method of the sciences is hypothetico-deductive entails any more than the rule that accepted theory should be consistent with the evidence. That rule is irrelevant to those who don't believe the putative evidence, and merely the advice to be logically consistent to those who do.

In this section I mean only to state my own perspective on how philosophical views of scientific methodology developed. Laudan's *Science and Hypothesis* helped me greatly in reaching a coherent overview. Since I am no historian of science myself, I shall rely on his chronology as an outline for my sketch, without claiming more than a didactic advantage. The first stage of my rough history is the debate over the rival accounts of *the methods of hypotheses* and *the inductive method*, between the (roughly classified) heirs of Descartes and Newton, respectively. The second stage is the *synthesis* of the two accounts, generally credited to Herschel and Whewell. (If we wish the title of "hypothetico-deductive method" to be honorific, it is probably best to identify it with their account, despite the barren wastes of later textbook presentations under that heading. I don't know the venue of the term itself.) And the third stage is the *breakdown* of that grand synthesis in the last hundred years or so—the breakdown to which we today still continue to respond and contribute.

In proposition 204 of part 4 of the *Principles of Philosophy*, Descartes states that, concerning things which our senses cannot perceive, it suffices for science to explain how they *can* be. This is done by introducing a hypothesis about how they are, in the service of an account of how the observable phenomena are produced. Being given such a hypothesis about, say, atoms and a deduction of some chemical laws from the atomic hypothesis, the primary avowal of success is, yes, this is indeed how they *can* be. It includes no assertion of how they are. But proposition 205 adds that, nevertheless, we have a moral certainty that all the things in the world are indeed how (Cartesian) science has demonstrated that they can be.

His defense of these two propositions is a beautiful cameo of future centuries of debate. First, he accepts (perhaps only for the sake of argument here) that the one source of information about the world that science can appeal to is sense perception and that there will always be many ways in which unobservable things could be that are compatible with how observable ones are. His response to this point has two parts. The first I take to be truly empiricist. The whole point of the sciences, he says, including mechanics and medicine, is to arrive at the truth about how manipulation and interaction on the observable level lead to observable effects. Hence, we can have all we want in science without caring about the truth or falsity of our hypotheses with respect to the unobservable.

The second part comes in proposition 205: we can have moral certainty that the hypotheses are true, sufficient for practical purposes and the conduct of life. As example he imagines a message apparently in cipher; if we work out a code by which it has a meaningful content, we are by all practical standards guaranteed that it was written in that code.

Are they really two different replies? If the connection between "moral" and "practical" is too close, moral certainty that A is true may be no more than certainty that A's consequences relevant to practical concerns are true—and, by the earlier statement, those consequences concern only the observable. Alternatively, moral certainty may be certainty pure and simple when morally defensible. If the defense needs to refer only to practical matters, then an opinion substantiated by no more than reasons to believe in what it entails about the observable may indeed by defensible—regardless of its extravagances on other matters. I don't know how Descartes meant this nor what could be made of it.

In their historic rivalry with Cartesianism, Newton and his followers gave severe criticisms of the method of hypotheses; they are familiar enough now. They can be summarized as a charge of *arbitrariness*. The grounds are several. It is admitted that for all observable phenomena there are in principle many different hypotheses about the unobservable that will account for them. Whether a hypothesis becomes part of accepted science (by Cartesian procedure) depends on several historical accidents—namely, that it was thought up (formulated by someone) and that the other ones thought up happened to be among those which were inferior to it by some standard. Secondly, in this evaluation, of these standards one is indeed objective and incontrovertible, namely, agreement with the observed phenomena. But, when several meet *that* standard, recourse must be had to controversial criteria such as simplicity. To justify the idea that the victorious hypothesis is likely to be true or even likely to be right about future phenomena requires, therefore, the unwarranted assumption that hypotheses actually thought up are more likely to be true than the noncontestants and that the ones we prefer for reasons like simplicity are more likely to be true than their rivals.

It is easy to see that these objections are not so much against Descartes's defense as to the idea that hypotheses about the unobservable should be introduced into science at all. However, they apply equally to hypotheses about the future and distant past which introduce or relate to no unobservables at all. Underdetermination by the evidence is a general problem for theory evaluation. Newton believed that he had the solution to this problem in the *method of induction* (credited to Bacon). Since the Newtonian held his spectacularly successful theories to have been arrived at and supported within the confines of induction, they saw no need for Cartesian hypothesis-mongering.

That method of induction, as formulated in Newton's *Rules*, was a paradigm of epistemic conservativism. Rule 3, the centerpiece, formulates the policy of straight exploration, to ascribe to all bodies those qualities which have been found to belong within the reach of experiment. Even prior to Hume's critique, it must have been clear that a sophistical choice of quality description could lead to unwarranted generalizations under that rule. So rules 1, 2, and 4 set limits. The first and second insist on great economy in causal explanation: the doctrine of the *vera causa*. However Newton may have meant it, his followers interpreted this as allowing recourse only to those causes which are demonstrably real and as eliminating any introduction of explanatory hypothetical entities. Rule 4 explicitly rejects positing of alternative hypotheses unless forced by new recalcitrant phenomena and in accordance with rule 3. Specifically, the propositions inferred by induction are to be regarded as accurate or very nearly true, despite any contrary hypotheses that can be imagined, for as long as they remain unrefuted by new phenomena. "This rule we must follow," Newton adds, "that the argument of induction may not be evaded by hypotheses."

This was all very well as long as one could remain convinced that the great successes of mechanics had been gained within the confines of these rules. The pleas of various ether theorists, on behalf of the explanatory power of their hypotheses, could be rejected with sanguinity. There must exist a method for finding the truth with objective certainty, the sciences ought to follow that method, and Newton's success is adequate reason to believe that he had discovered the principles of that method. (Does *this* argument accord with the rules?)

The turning point for the official description of scientific methodology came early in the nineteenth century, with the triumph of the wave theory of light. Recalcitrant phenomena defeated its Newtonian rival; moreover, these phenomena were found just because the hypothesis of light being constituted by waves in the hypothetical ether led to their discovery. Actual scientific developments had burst the bounds of the official inductive methodology. The inductive method, through the extreme conservativism of its principles, had escaped the searing critiques that applied to the method of hypotheses—but now this conservativism had to be seen as a straitjacket science could not afford. Yet in its old formulation the method of hypotheses could not escape those critiques.

The synthesis due to Herschel and Whewell can best be seen, in my opinion, as the advent as a central principle of the thesis that *evidential support must be independent support*. The inductive method can be regarded as a very conservative version of the method of hypotheses (as Le Sage had indeed argued and as Herschel had in effect made plausible through his distinction between *context of discovery* and *context of justification*). What was needed was less stringent but still substantial strictures to limit the method of hypotheses. In the Fresnel-Poisson experiment that was crucial to the triumph of the wave theory of light, we see the paradigm example: Poisson showed that Fresnel's model entailed the possibility of a new phenomenon, never yet observed and with small prior plausibility—and that phenomenon was indeed disclosed by further investigation. Since it had not been known before nor was entailed by anything that Fresnel's model had been tailored to fit, this constitutes *independent support*. The Newtonian disdain for the explained facts as support for an explanatory hypothesis could therefore be accommodated: that disdain is apt, unless the cited agreement is novel and independent of what was known in the context of discovery (or, as we should really say today, context of theory construction).

This idea was generalized into Whewell's notion of *consilience*. If the diagnosis of the preceding paragraph is correct, those who equate consilience with varieties of explanatory power, unification of theories, internal coherence, and the like are looking at the wrong side altogether. The advance was not to recognize the legitimacy of explanatory relations but to raise to primacy the notion of *independence* in the analysis of support. I do not deny that Whewell was to some extent guilty of the same confusion. The two forms of independent support stressed by both Herschel and Whewell were:

1. prediction of novel, unexpected phenomena,
2. explanation of facts, possibly previously known, but in novel areas for which the hypothesis had not been constructed,

and Whewell added his characteristic emphasis on *variety of evidence*:

3. explanation of two or more independent classes of facts.[10]

10. Though today we draw an automatic distinction between fitting the facts and properly explaining them, I see no evidence of such a distinction in these authors.

What exactly is independence? This is the point where the content of the context of discovery enters the context of justification. The independence must be relative to the knowledge and accepted theories of the context of discovery. (The old charge of arbitrariness lurks nearby, of course, for this is to some extent a matter of historical accident. But it is perhaps a new version of the epistemic conservativism advocated in Newton's rule 4, where previous inductive inference has squatter's rights.)

The triumph of the wave theory did not seriously undermine the enormous optimism of writers on scientific methodology. As the archaic "context of discovery" terminology reveals, the conviction that there exists a method of discovery of the truth leading us to objectively warranted certainty (now even with respect to the unobservable) is, if anything, intensified. It looks rather disingenuous now, but Whewell (1847, 2:283–90) claims the new method of hypotheses with frills to be the true explication of Newton's method of induction.[11] By this methodology he argues that we do isolate the *vera causa*, "but . . . its verity ceases to be indistinguishable from its other condition, that it 'suffices for the explanation of the phenomena.'" He adds, pointing to the required consilience, that, "when the validity of the opinion adopted by us has been repeatedly confirmed by its sufficiency in unforeseen cases, so that all doubt is removed and forgotten, the theoretical cause takes its place among the realities of the world, and becomes *a true cause*." The historical claim he made was that *no truly consilient theory has ever later turned out to be false*, (p. 67 and p. 286). Newtonian mechanics was, of course, his great and paradigmatic example of such a truly consilient theory (!).

So the new and undoubtedly much more adequate description of the scientific method came to us yoked unto what was undoubtedly an epistemology for realists. Today I don't think even any realist would hold that there is an inductive method that leads us to knowledge of the truth with objective certainty, in which "all doubt is" rightfully "removed and forgotten." So the question for epistemology is, how *shall* we regard the described method—as tending to lead to truth in the long run or to probable approximate truth or to more probable approximate truth than any alternative that has been formulated (an easy one, since we have formulated no serious alternatives!) or probable empirical adequacy or what? All these alternatives sound at first hearing as if the method is a theory-producing procedure. It isn't. But, if we carefully rephrase it as a procedure leading to recommendations of acceptance, the alternatives remain *mutatis mutandis*.

Before launching ourselves into this problem thicket, let us gain some perspective on what the scientific enterprise is like, if Herschel and Whewell's descriptive account is more or less correct.

11. My quotes are from pp. 284 and 286; see also p. 67.

Bracketing every judgment concerning the value of the method, the procedure that best instantiates it in concrete form is this: Imagine a huge cooperative of creative writers—a sort of literary Bourbaki—engaged in the task of writing a large-scale historical *roman fleuve*. This historical novel is to cover the fortunes of a dynasty in, say, central Europe founded in some small earldom early in the Dark Ages. They agree that historical accuracy shall always be the rock-bottom criterion of adequacy. Thus, every new paragraph is the subject of various rival proposals: any reason to think that the proposed paragraph leads or will lead to later conflict with the historical evidence is a major point against the proposal. Very often, as a literary cul-de-sac is entered or historians produce new relevant evidence, earlier chapters need to be amended or revised or jettisoned altogether. Desirous of fame and competitive, the writers not only try to think up new plot lines but work in libraries and archaeological digs to find evidence that may refute rival proposals. Sometimes no proposals are needed to advance the story: historical research itself writes the next line in the story (in the sense that it rules out all but one possible rival continuation of the story produced so far). Conversely, new creative story ideas stimulate research, and a writer is happiest when he can claim vindication in the new documentation or artifacts uncovered thanks to the line of (archival or archaeological) excavation suggested by his own literary proposals.

As analogy, this fits science in some aspects very nicely. The ideal of the unity of science—the science that the sciences are building is one—has its analogue here as an immediate corollary to the decision to write one historical novel together. The dual nature of the theory-experiment relation that I described in chapter 4 of *The Scientific Image* has its neat parallel in the last few sentences preceding this paragraph. The crucial role in practice of many standards relating to internal coherence and satisfaction of the audience's desire for intelligibility—as described for tragedy in Aristotle's *Poetics*, for example—and also their relative independence from the rock-bottom criterion of checkable accuracy are clearly paralleled. The reader is invited to meditate on possible epistemic attitudes toward this great historical novel.

5. Scientific Method: Another Perspective [12]

Although Herschel's distinction between contexts of discovery and of justification stopped some of the more blatant armchair psychologizing of the inductivists, it could not well be made *too* sharp. As I pointed out, reference to the context of discovery is made in justification when evidence

---

12. This section incorporates two sections, essentially unchanged, of my paper "Theory Construction and Experiment," in *PSA 1980* (1980), vol. 2, pp. 663–67. Reproduced by permission.

is called *novel* or facts are classified as mutually *independent* (one virtue of the advocated theory being that, relative to it, they are not independent). Moreover, although there may not be a logic of discovery, to think of scientific advances as the offspring of romantic inspiration, dreams, or images seen in the flames is to ignore the systematic aspect of theoretical and experimental exploration. In chapter 4 of my book I addressed this topic, introducing the "Clausewitz doctrine" of experimentation (the continuation of theory construction by other means). In this section I shall further address the methodology of theory construction. The next section will turn to the other (justification or evaluation) aspect, incuding testing and evidential support.

New theories are constructed under the pressure of new phenomena, whether actually encountered or imagined. By "new" I mean here that there is no room for these phenomena in the models provided by the accepted theory. There is no room for a mutable quantity with a discrete set of possible values in the models of a theory which says that all change is continuous. In such a case the old theory does not allow for the phenomenon's description, let alone its prediction.

I take it also that the response has two stages, logically if not chronologically distinguishable. First, the existing theoretical framework is widened so as to allow the possibility of those newly envisaged phenomena. And then it is narrowed again, to exclude a large class of the thereby admitted possibilities. The first move is meant to ensure empirical adequacy, to provide room for all actual phenomena, the rock-bottom necessary condition of success. The second move is meant to regain empirical import, informativeness, predictive power.

It need hardly be added that the moves are not made under logical compulsion. When a new phenomenon, say X, is described, it is no doubt possible to react with the assertion that, if it looked as if X occurred, one should conclude only that some familiar fact or event Y had occurred. A discrete quantity can be approximated by a continuous one, and an underlying continuous change can be postulated. From a purely logical point of view, it will always be up to the scientists to take a newly described phenomenon seriously or to dismiss it. Logic knows no bounds to *ad hoc* postulation. This also brings out the fact of creativity in the process that brings us the phenomena to be saved. Ian Hacking put this to me in graphic terms when he described the quark hunters as seeking to create new phenomena. It also makes the point long emphasized by Patrick Suppes that theory is not confronted with raw data but with models of the data and that the construction of these data models is a sophisticated and creative process. To these models of data, the dress in which the debutante phenomena make their debut, I shall return shortly.

In any case, the process of new theory construction starts when described (actual or imagined) phenomena are taken seriously as described. At that point there certainly is a logical compulsion, dimly felt and, usually much later, demonstrated. Today Bell's inequality argument makes the point that certain quantum mechanical phenomena cannot be accommodated by theories which begin with certain traditional assumptions. This vindicates, a half century after the fact, the physicists' intuition that a radical departure was needed in physical theory.

Of the two aspects of theorizing, the widening of the theoretical framework and its narrowing to restore predictive power, I shall here discuss the former only. There we see first of all a procedure so general and common that we recognize it readily as a primary problem-solving method in the mathematical and social as well as the natural sciences, anyplace where theories are constructed, including such diverse areas familiar to philosophers as logic and semantics. This method may be described in two ways: as *introducing hidden structure* or "dually" as *embedding*. Here is one example.

Cartesian mechanics hoped to restrict its basic quantities to ones definable from the notions of space and time alone, the so-called kinematic quantities. Success of the mechanics required that later values of the basic quantities depend functionally on the earlier values. There exists no such function. Functionality in the picture of nature was regained by Newton, who introduced the additional quantities of mass and force. Behold the introduction of hidden parameters.

The word *hidden* in "hidden parameters" does not necessarily refer to lack of experimental access. It signifies that we see parameters in the solution which do not appear in the statement of the problem.

We can "dually" describe the solution as follows: The kinematic relational structures are embedded in structures which are much larger— larger in the sense that there are additional parameters (whether relations or quantities or entities). *The phenomena are small but chaotic; they are treated as fragments of a "whole" that is much larger but orderly and simple.* This point could, I believe, be illustrated by examples from every stage of the history of science. When a point has such generality, one assumes that it must be banal and can carry little insight. In such a general inquiry as ours, however, perspective is all, and we need general clues to find a general perspective. (That this particular point may be productive of more specific insights is in any case not an unreasonable hope. The most spectacular recent theoretical development may well be the deduction of Maxwell electrodynamics, Einstein geometrodynamics, and the Yang-Mills quark-binding field dynamics from the requirement of embeddability in space-time by Hojman, Kuckar, and Teitelboim.)

In order to illustrate the general view of what theorizing is like, which I have just presented, and to try and persuade you that it is a reasonable one,

I shall now describe some recent activities in the foundations of quantum mechanics (see also Van Fraassen 1982).

The whole point of having theoretical models is that they should fit the phenomena, that is, fit the models of data. So we need to look at what the latter must be like in general. Hence the development of Randall and Foulis's "empirical logic"; Mackey, Jauch, and Piron's preliminary discussions of experimental questions; Ludwig and Mielnik's filters; and so forth. These authors write sometimes as if their program is one of transcendental deduction: study what the data models must be like, deduce what structure theoretical models must have if the data models are to be embeddable, demonstrate the basic axioms of quantum mechanics as corollaries to this deduction. Since the theory has clear empirical content, success can be at most partial, but it is astonishing how much can be achieved in this way. Moreover, the very fact that success must necessarily be partial is what gives the approach its value for the future: it brings within our ken alternatives to the extant theory that can rival it.

In Suppes's description, the experimentalist brings to the theoretician a small relational structure, constructed carefully from selected data. The examples Suppes mentions are specific, and the little structures are algebras; hence he calls them "empirical algebras." The authors in quantum mechanics point to such small structures that represent data, and they are not always algebras; they are more generally partial algebras or just partially ordered sets with some operations. Let us see how this happens.

In the typical sort of experiment discussed in connection with the Einstein-Podolski-Rosen paradox and Bell's inequalities, we have two apparatuses, L (for left) and R. Each has (say) three settings or orientations; let, for instance, L1 be the proposition that L has been given the first setting. The experiments have each (say) two distinct possible outcomes, which we may represent by the numbers zero and one. Let, for instance, L30 be the proposition that L has the third setting and outcome zero.

When we carry out a particular run on this dual apparatus, we can give a score of T or F to some of these propositions. For example, the first time we do it, each apparatus was placed in the first setting; L had outcome 1 and R had outcome 0. An experimental report looks, in part, as follows:

Proposition	Score
L1	T
L2	F
R1	T
L10	F
L20	No score
R10	T

We note that L20 received no score. It could have been given F simply on the basis that L had not been given the second setting. But this is useless information and does not appear in the experimental report.

This single report is not likely to come to the theoretician's desk. What reaches him, rather, are reports of the form:

(a) With initial preparation X, the probability of outcome Lia, given setting Li, equals r.

(b) For all initial preparations, the probability of (Lia and Ria), given settings Li and Ri, equals zero.

There was an extrapolation before these conclusions were reached: the extrapolation from found relative frequencies to probabilities. This is no different from the extrapolation of data points on a graph to a smoothed curve.

But the report that comes in forms (a) and (b) leads us to a mathematical structure that may properly be called the *data model*. The important relation stated between Lia and Ria in (b) is that, when they can receive an informative score at all (i.e., when the preconditions Li and Ri obtain), they cannot both receive the score T. We then call Lia and Ria *orthogonal*. It also means that $Ri1$ must receive score T when $Li0$ does (*modulo* probability zero) and again when the informative scoring conditions obtain; and we call that *implication*. The latter is a partial ordering, and so we have here a partially ordered set (*poset*) with an orthogonality relation.

Reflection on this form of representation leads to assertions of the form: all data models can take the form . . . A popular way to fill in the dots is to say "ortho-poset" (i.e., poset with orthocomplement). A. R. Marlow, whose work I am about to take as a special example, used the more general characterization *dual poset*; that is, a partially ordered set with zero element and equipped with a single operation, *duality* ($x \neq x'$; $x = x''$; and x *implies* y only if y' *implies* x').

In the world of mathematical entities there are many dual posets. Widening our theoretical framework will consist in the provision of models that can have very strange dual posets of experimental propositions embedded. But the embedding must be good; that is, we must be able to see in the theoretical model all the significant features. I mentioned parts of the experimental report labeled (a). They start, "With initial preparation X . . . ," and then they mention probabilities. These probabilities characterize what is called the *state* prepared by procedure X. And these states (they look like fragments of ordinary probability functions, in that they assign probabilities only to propositions for which the informative scoring conditions obtain) must be "visible" in a certain sense in the computational structure of the theoretical model.

Here is Marlow's theorem.[13] It requires two preliminary definitions. A *probability function* on a dual poset is any function f with the properties $f(0) = 0$, $f(x') = 1 - f(x)$, and $f(x) \leqslant f(y)$ if x implies y. A *base* for the dual poset is any set B of elements which does not contain the zero element, nor does it contain two orthogonal elements [here defined by the relation $(x \text{ implies } y')$], but does contain either x or x' for each x. (Remark: every set with the first two properties can be extended to one that has all three. Note, also, that any set of elements that have all received score T on a particular occasion is intuitively required to have the first two properties.) The theorem says now that, if we have a dual poset and a base, we can embed the poset in the algebra of projection operators of a Hilbert space in such a way that duality becomes orthocomplementation, the partial ordering is preserved for elements within the base, and each probability function on the poset can be associated with a vector and becomes calculable by means of the familiar trace computation used in quantum mechanics.

This is an extraordinarily general result. Marlow takes the result as justification for his project to write space-time theories in Hilbert space formalism. The result provides good reason, after all, to write all physical theory in that mathematical framework. Of course, he realizes that in some ways the theorem is less than totally general (the implication order is preserved only within the base!) and in some ways less than informative (there are enormous Hilbert spaces with room to embed almost anything), but the postulates he adds, and intends to add, will narrow down this *embarras de richesses* to recover empirical content.

Let us look, however, as second main illustration, at a line of thought born from dissatisfaction with the way phenomenal structures are embeddable in the Hilbert space framework. I refer to *operational quantum mechanics*, associated with Ludwig, Mielnik, Davies, and Edwards.

To explain how data models can take the form of ortho-posets, or more generally dual posets, I already gave a brief sketch of a quite typical experimental setup, of an intermediate degree of complexity. Let me now start an alternative sketch; the two will not be incompatible. In a typical simple test, a system is *prepared* in a certain way, an *operation* is performed on it, and a *question* is asked. In the simplest case, the question has yes or no as possible answers. We may keep count, as we repeat the test, of how often we receive yes as answer. We visualize the situation by imagining a *source* which sends out a beam of particles that encounter a *barrier*, and a *counter* on the other side of the barrier that clicks every time a particle reaches it. This is a good picture, for it has all the general features indicated; we appear to lose no generality if we focus attention on it.

---

13. "An Extended Quantum Mechanical Embedding Theorem," in Marlow (1980).

The barrier affects the intensity of the beam; for example, if the barrier were not there, the counter might be clicking twice as rapidly. The sort of barrier determines the sort of question being asked. When the situation, so conceived, is embedded in the mathematical apparatus of Hilbert space, each source has an associated statistical operator W (representing the *prepared state*), and each barrier an associated projection operator P (representing the *question asked*); the probability of a single particle passing the barrier (the factor by which the beam intensity is diminished) is calculated by the Born rule as $p = Tr(PW)$.

Many situations of the general sort described can indeed be modeled in this way. But others have to be treated in more roundabout fashion. Intuitively, we say that there is an *observable* which is measured in such a setup. Whether that is so in the sense now instilled in us by quantum theory has a simple criterion: there must be a state such that the yes answer becomes certain, that is, receives probability *one*. That state is then called an eigenstate of the observable being measured.

It is certainly possible to find examples where it looks intuitively as if we are measuring something but where the criterion is not met. Suppose we place an atom of radioactive substance near a Geiger counter and ask, "Will the counter click within four to five minutes from now?" Is that not a simple form of measurement, of something we could call, say, the decay time of the atom? Yet we cannot prepare the atom in a state so as to make the yes answer certain.

A clearer example, first introduced into the literature, I think, by Shimony and discussed especially by the Dutch authors de Muynk, Cooke, and Hilgevoort, is pictured in figure 1. We have a battery of three Stern-Gerlach apparatuses testing for spin. They are so arranged that particles exiting from the first, along the top channel, encounter the second, while those exiting along the lower channel encounter the third apparatus. Now, the relative orientations of these three apparatuses determine the transition probabilities, that is, the probability that a particle exiting in the first top channel will exit along the second top channel, and so forth. We may choose the orientations so that those transition probabilities equal 1/2. As is easily seen in the diagram, there is no initial state which makes certain an exit along the second top channel, for the probability of that exit equals $p/2$, which has maximum value 1/2. Here we have a question for which the yes answer *cannot* be certain. We can at most represent it by means of a sequence of questions of the directly representable sort. Perhaps we could say that a 'derived observable' has been measured.

Operational quantum mechanics may be thought of as enlarging the theoretical models so as to allow the embedding of all such empirical structures in the same way. The questions asked by our battery of apparatuses

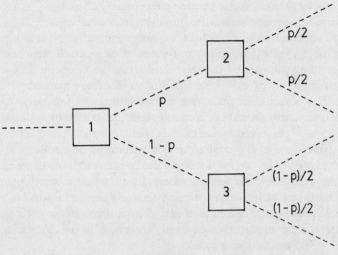

Figure 1

are treated on a par with those asked by a single apparatus. This enlargement proceeds as follows: We represent the (ensemble prepared by the)
source by means of two parameters: the statistical operator W and a number t which represents the production intensity. The barrier then affects
both state W and intensity t, changing W to some other state W' and reducing the intensity t to rt (with $0 \leq r \leq 1$). We may set the initial intensity
equal to 1, so that t itself must belong to the interval [0,1]. Referring to the
statistical operators W—positive definite Hermitean operators with unit
trace—as the "old states," we represent the new states by their multiples
rW. These are not in general themselves statistical operators, for, if W has
unit trace, then rW has trace r.

Mathematically the enlargement proceeds in several steps. We begin
with the "ball" S of old states; generate the real cone

$$B+ = \{rW : W \in S \text{ and } 0 \leq r \leq 1\},$$

and then enlarge that cone by closing it under the usual linear operations,
thus forming a real vector space B, a Banach space with the trace as norm.
The physical operations can now be represented uniformly by considering
their effects on "old state" and trace/intensity.

These two longish examples I take to reveal philosophically especially
significant aspects of scientific theorizing. Brian Ellis is quite right to see
my view of theory structure as originating in those studies of time and
space with which I began. My dissertation was on the foundations of the

causal theory of time, and it became part (mainly chapter 6), in a less technical form, of my first book, *An Introduction to the Philosophy of Time and Space*. The program of the causal or more generally relational theory of space-time, as I saw it especially in Russell, Reichenbach, and Grünbaum, was first of all to describe the relational structure of events in the world, constituted merely by collisions, exact coincidences, genidentity, and light signals. It was very clear that, without exceedingly strong empirical postulates, this structure would not coincide with the spatiotemporal order. The reason, clearly perceived by Leibniz, is that the actual is only a fragment of the possible—not every geodesic is the path of an actual particle or light ray. At this point we can call metaphysics to the rescue: either we can postulate the existence of space-time itself, as an arena in which events take place, or else we can give a possibility (namely connectability, as opposed to actual connection) the status of fact. The former course gives up on the relational theory; the latter saves it only by giving up the empiricism which was for me its main motivation.

The solution, as I saw it, is that space-time is an "ideal entity" (in Leibniz's phrase). It is a mathematical model which guides all our thinking in spatiotemporal terms, but its relation to reality is that the structure of actual connections, *whatever it may be*, must be embeddable in it. It has thus the same status as Wittgenstein's prime example of a logical space. For the color spectrum is a mathematical structure—part of the real line—in which we postulate all colored entities to be locatable in a way that reflects their color relations and which provides thus the picture that guides our thinking and discourse about colors. Reading an old book by Evert Beth (*Natuurphilosophie* [1948], which I obtained in 1965), I began to see how phase space and configuration space in classical mechanics and the Hilbert spaces of quantum mechanics play exactly the same role. They represent the complete general form that all phenomena allowed by the theory fit in a fragmentary way. Phenomenal reality need not be fragmentary in itself, but its chaotic nature vis-à-vis human understanding forces us to treat it, conceive of it, as fragmentary.

Once we see that this is what we are doing in science, we can do it in good conscience without requiring a metaphysical justification. We do not need to postulate that there are elements of reality corresponding to all elements of the model.

6. Acceptance Is Not Belief

The new perspective I wish to provide for the context of justification is coded in the above title, which is the conclusion of this section.

The ideal we inherit from the past is a notion of confirmation or evidential support which is:

(a) *objective*: it is a relation solely between theory and total body of evidence, independent of the context of evaluation;

(b) *comparative*: not only whether the evidence on the whole supports the theory but also whether it supports one theory more than another, or supports a theory more than alternative evidence would have, is objective;

(c) *unique*: the propositions to be believed on the basis of the evidence are a determinate and logically consistent set (for instance, the disjunction of all the theories maximally supported by the evidence), and rationality requires that, with given total evidence, one believes all and only those propositions.

The idea of an inductive logic or of an organon or canons of induction is exactly the idea of a systematic description of the relation (evidence—to—propositions to be believed when this is the total evidence) described under (c). It is the idea of *an inductive consequence relation*, analogous to the familiar deductive one. Whewell apparently believed that he had incorporated the method of hypotheses into the deductive method under such strict control that the synthesis still had features (b) and (c). Sometimes Whewell talked as if (a) remained also; but, as I pointed out, certain features of the historical context enter theory evaluation through the notions of novelty and independence. In our century, Carnap and Hempel sought to restore (a). Today, despite all the puzzles and criticisms, *Inference to the Best Explanation* (with or without frills, and under various names) remains a widespread and popular conception of something that can fill the role. Glymour especially, in his discussion of the role of explanation in theory evaluation, insists on (a). In his words, there are "objective unpragmatic relations between theories and statements of empirical phenomena, relations which when apprehended produce understanding and give grounds for belief." [14]

To focus this discussion, let us distinguish two sorts of judgments we can make about theories (see also Van Fraassen 1983d). The first is the *judgment of opinion or credence*: that we believe a theory more than we did formerly or more than we do another theory or perhaps more idiomatically, that it is more credible to us. The second is the *judgment of evidence*: that we have more or better evidence for a theory than we did formerly or than we have for some other theory. In the traditional picture, the former is mainly about us (our attitudes, which may be historical accidents), but the latter can be evaluated equally and objectively by anyone who knows what our evidence is and what the theories are.

14. I quote part of a negative sentence, but the paragraph of which it is part and the next two pages make clear that he advocates this conception and does not shy away even from such terms as "inductive principles" and "inductive logic."

The well-known Bayesian "explication" of confirmation as enhancement of credibility ("evidence E confirms T for us exactly if T is more credible to us on the supposition that E than it is alone") links the two sorts of judgment. It is predicated on the idea that judgments of evidence do not depend solely on the relation between theory and evidence but express a three-term relation between theory, evidence, and us. To call this an explication is, of course, a travesty, for it is just the advice to jettison the idea of objective evidential support altogether. At the same time, it is persuasive because it retains the idea that what the evidence supports is credence or credibility, belief that the theory is true. Evidential support is still identified with grounds for belief. So it is important to see that, if we do take the advice and jettison the traditional ideal of an objective confirmation relations, we shall still not necessarily become Bayesians, as long as we do not equate methodology with the study of grounds for belief.

Returning to our two sorts of judgments, let us consider two theses concerning their objectivity. The first is the thesis that, *sub specie* rationality, the evidence exerts total control over opinion or credence:

TCC: If two agents or communities are rational, and equally aware of logical connections, and have always taken exactly the same propositions as evidence, then they have exactly the same state of opinion (credence).

This thesis is most easily identified in Carnap's program, where the criteria of rationality were meant to single out a unique probability function as the correct prior before any proposition has yet been taken as evidence. But TCC is also implied by the idea of a correct inductive method satisfying (a) − (c), transposed into our terminology of judgments of credence and of evidence. There are heirs to Carnap's program and to inductivism today, but I don't think even they subscribe to TCC. The second thesis, weaker and more interesting to attack, is that the evidence exerts total control over the judgments of evidential support:

TCE: If two agents or communities are rational, and equally aware of logical interconnections, and have always taken the same propositions as evidence, then they have exactly the same judgments of evidence.

This is not tautological, because judgments of evidence are comparative. To arrive at the judgment that our total evidence supports one theory more than another, for example, requires that we first decide which evidence is relevant and then gauge the extent of the support it provides for the two theories. The criteria of relevance and standards of comparison are not written in the evidence. They could only come from the criteria of rationality alluded to in the first clause. So in the way that TCC makes precise part (c) of the old ideal, TCE does that for part (a)—and it implies that the standards of comparison are context-independent.

Without embracing the Bayesian final solution, however, I must still say that their critiques of the theory of confirmation (such a hopeful name!) make faith in TCE extremely difficult to defend. In general terms, we must distinguish *conforming* from *confirming*. Confirmation by evidence must be gotten, if at all, from observed facts conforming to the theory: facts that have a home in some model of the theory. But conforming is still a far cry from confirming. For the very same facts that conform will show that the theory cannot be held in conjunction with some other theory (which may have as much to recommend it) or may, in the presence of certain background information or beliefs, seriously lower the given theory's credibility.[15]

Reading the above, not only a Bayesian but also a Popperian may feel relatively sympathetic. For the Popperian can read all the above as throwing doubt on the ideal of the inductive method. However, it is in my opinion merely trivial, indeed banal, to say that the evidence exerts "negative control": a theory seen to conflict with the evidence is ruled out. This point is simply irrelevant to those who do not believe the evidence, and it follows trivially from the principle to maintain ordinary logical consistency for those who do. (It does so even on empiricist grounds, if acceptance of a theory entails belief that it is empirically adequate!) So, if that point were the core of methodology, emperor Popper would be wearing no clothes. That, of course, is why Popperians have written so much about theory comparison and verisimilitude; without going into specifics, I would say that they allow pragmatic elements into this which would tend to refute or undermine (a) and (c) and hence both theses of objectivity, despite their terminological obeisance to objective rationality.

Newton-Smith argues that "the failure of Popper's endeavors clearly establishes the indispensability of inductive argumentation in science" (1981, 52). This would be a cogent line of argument if there existed a viable inductive method, inductive logic, or set of canons of induction for Popper to reject. Newton-Smith clearly does believe that some such thing exists, and so does Glymour; and I shall come back to this question below. But I suspect that writers on the subject vacillate between the tenable and the audacious. When I maintain that there is no inductive method or inductive logic, I mean that there is nothing satisfying (a)–(c), nor anything approaching that. But when Newton-Smith reiterates, as a fact about us or about background philosophical discussion that "we are so totally committed to the thesis that the character of our experience provides inductive

---

15. For details, see Van Fraassen 1983d, and, for similar points with respect to Glymour's theory of confirmation, via testing, 1983c. See also my reply to Boyd in part 2.

evidence" (p. 64), it seems to me that he must be referring to something much more banal. And that is the fact *that we do form expectations of the future, which are reasonable and yet do go beyond what is now evident to us— and that this is all within the bounds of rationality.* It does not follow from this that we are engaged in anything meriting the name of inductive inference. For the terms *inference, logic, rule, canon* are simply misused in this context when desiderata (a)–(c) are not fulfilled.

How is reasonable expectation possible? One answer, which must surely be part of the truth, is that it is reasonable to base our expectations on the scientific theories most worthy of acceptance. The superlative must refer to the results of evaluation and comparison in the light of our total evidence. The conclusion to which we are driven, however, is that this evaluation does not and cannot satisfy the ideal that (a)–(c) encode. And once the evaluation depends on historical, contextual, pragmatic, or even personal and communal peculiarities or choices, it becomes (by my lights) impossible to equate these reasonable expectations with rationally *compelled* belief.

My thought on this point found its focus at a symposium on scientific explanation in which I participated with Wesley Salmon and Clark Glymour. I noted that Salmon and I agreed that to give a scientific explanation must consist at least in part in giving relevant information on the topic. (We disagreed in our characterization of relevance, which I argue to be context-dependent.) On the other hand, I agreed with Glymour that explanatory value or power is a major respect in which theories are comparatively evaluated. Conjoining the two points, I conclude that, *ceteris paribus*, theories are better if they are more explanatory, which in turn requires them to be more informative. But there are other reasons, besides the weight given to explanation, to conclude that, *ceteris paribus*, the better theories are the more informative. We go to science to have our questions answered about the empirical world, for many purposes—not just explanation, but practical control, mere factual curiosity ("thirst for knowledge," if you like), suggestions for new directions of research, and perhaps others. When a theory can putatively satisfy such needs, we consider that a *virtue*, a reason for acceptance.

At the same time, we worry that the information given by the theory may be false. So we also compare theories with an eye to the possibility of falsehood, their credibility. However this comparison is carried out, and whether the criteria are objective or context-dependent or subjective, *credibility varies inversely with informativeness.* This is most obviously so in the paradigm case in which one theory is an extension of another: clearly the extension has more ways of being false. So our two models of evaluation are in radical tension, conflict—as I put it then: *informational virtues are not confirmational virtues.* Indeed, the two desiderata cannot be jointly maximized. They conflict; they detract from each other (see Van Fraassen 1983b).

If some reasons for acceptance are not reasons for belief, then acceptance is not belief. And indeed some reasons for acceptance hinge crucially on the audacity and informativeness of the theory. So acceptance is not belief.

At the end of section 4 I suggested that, in the Herschel-Whewell synthesis, the process of science has as best analogy the cooperative creation of a historical novel. One feature there is the acceptance of new additions and revisions by a quasi-political process, in which only *local comparisons* of alternatives (not global comparisons of The Book with other possible books they could be writing) are to the point. In the analogy, too, many reasons for acceptance have little to do with likelihood of truth, although some definitely do, because freedom from conflict with evidence is the bottom line. This quasi-political process of decision, in which much depends on interests that are just then in the valuational limelight and also on just what alternatives have actually been devised by the participants, is within the bounds of reason. It is perfectly reasonable to go ahead on the basis of its outcome. Thus it is with science, too, even if the process of evaluation and comparison is much tighter and more logical and more objective than the analogy suggests. The Newtonians' criticisms of the arbitrariness inherent in the method of hypotheses must be regarded as correct descriptions of scientific procedure once that method (in a not too constrained form) is seen as part of that procedure. But, if acceptance is not identical with belief, that may not be so bad.

What is there in acceptance besides belief, and how much belief does it involve? In practice, acceptance will always be partial and more or less tentative. In the ideal limit, where the acceptance is unqualified, I suggest that the expectation of observable phenomena is entirely based on the theory, no more. In addition, the acceptance involves a commitment to maintain the theory as part of the body of science. That means that new phenomena are confronted within the conceptual frame of the theory, and new models of phenomena are expected to be constructed so as to be embeddable in some models of that theory. It should go without saying that, even when acceptance is unqualified, it need not be dogmatic; fervent and total commitment still need not be blind or fanatical. If we construe acceptance in this way, then the commitment to the unity of science—*the science that science is building, is one*—allows for the use of choice criteria that are logically independent of truth, empirical adequacy, or likelihood thereof. Yet the construal also explicitly disavows those criteria as yielding legitimate sources of belief and hence remains consistent with the empiricist tenet that experience alone has that legitimacy.

II. Replies to Individual Authors

In this part I shall reply to some specific points and objections that were not covered along the way in part 1. I shall intersperse references and cross-references, some to authors and papers not included in this volume. The incompleteness of my replies will be glaringly obvious, so I needn't dwell on it; every one of these essays deserved a reply equal to its own length.

Ad Richard Boyd

That there are enormous disagreements in approach between Boyd and myself will have been apparent from part 1, section 3, in which I criticized a class of possible programs in philosophy of science to which his surely belongs. It is true that my criticisms there apply directly to older and less sophisticated versions, and I shall not pretend that extrapolation to Boyd's work is straightforward. But I believe that they include general arguments for the futility and lack of cogency of that sort of program.

When Boyd reacts to my brief use of an evolutionary parallel, he first suggests that I have not been paying attention to this century's synthesis of Darwin's theory of natural selection with genetics. That sounds as if he wants to say that the parallel is there but, if pursued, favors the scientific realist. It appears quickly, however, that he wants to say just the opposite: features of biological evolution that Darwin needed to rule out to banish appeal to teleological or purposive forces in biology (Boyd's perhaps some-what one-sided description of what Darwin's success *was*) are really there in the development of science. In other words, the details show that the parallel does not exist.

In each of his separate points, Boyd's argument turns out to hinge on appeal to a point meant to be established in the first, constructive, half of his paper. I shall accordingly take up only his first point, then turn to a critique of the details of Boyd's own program.

His first point is that Darwin needed to insist on the empirical claim that there is no directed variation in nature. It is a fact that the color of moths changed in Britain during industrialization, but not (not even in part) because smog-plagued moth parents had a larger proportion of dark offspring when they encountered the signs of change in the environment. We must begin with a distinction on the biological side. Of course, there are asymmetries in variation; it is not random. For example, most dan-delion offspring look predominantly like dandelions. In theories, too, there is such "instruction from within": we strive to retain the success of earlier theories in our new ones. The newly proposed theories are designed to have the same success in those areas where their progenitors were per-ceived as successful. In addition, there is direction in the variation that comes to maturity. The standard dandelion is spreading in my backyard and taking over territory so far occupied by grass and other weeds. In the absence of a cropper (whether sheep or lawn mower), variation is tending toward zero. Boyd's claim must be that, *after* discounting normal inheritance and *before* the effect of confrontation with new phenomena, the range and variation of new theoretical proposals favor the chance of success in the new sciences.

Boyd claims (a) that it is so, and (b) that the only reasonable explanation is in terms of the correctness of the theoretical and conceptual riches of successful background theories—their correspondence to reality over and above correspondence on the level of observable fact. Put conversely, this is very appealing—*if* the theoretical structure enjoys *adequatio a rem*, *then* hypotheses exploiting and building on that structure can be expected to do well or better. But claimed fact (a) is no fact to me—a touching faith at best. Since Boyd claims support from his section 1, with reference to his study of projectability, we must look there.

Since I have already explained above my general objections to Boyd's sort of program, I shall restrict myself here to details. The first step in the recipe I outlined was to choose some account X of the scientific method. Justificationists in the past have been rather cavalier in this choice. For his, Boyd goes in the first place to Goodman's discussion of projection to provide a framework for his descriptive account of methodology. In my own view, he could hardly have done worse.

In the third and fourth chapters of *Fact, Fiction, and Forecast*, Goodman (1965) gave a new and devastating critique of numerical induction.[16] The idea of inductive logic, shown in principle mitaken, was to be replaced by that of the theory of confirmation, already well begun by Hempel. Unfortunately, his "grue" and "bleen" examples also effectively destroyed the idea of instance confirmation, which Hempel had not only preserved from older ideas about induction but had made the paradigmatic core of his account of confirmation. Goodman offered a diagnosis and a programmatic solution. Some predicates are *projectible*, and some are not. This is the diagnosis of the problem that apparently some hypotheses are supported by their positive instances and some (like Goodman's counterexamples) are not. The diagnosis entails or assumes that this apparent difference is due to a really important underlying difference between sorts of predicates. In a postscript written in 1964 to his "Studies in the Logic of Confirmation," Carl Hempel endorsed both the diagnosis and the program of delineating comparative projectibility of predicates.

But now we know, through the excellent Bayesian critiques of I. J. Good and Roger Rosencrantz, that the idea of instance confirmation was bank-

16. These chapters were given as lectures in London in 1953.

rupt all along, and for very different reasons. Bayes's theorem was originally used to "explain the success" of the old rule that a hypothesis is confirmed by verifying one of its consequences. If $P(E|H) = 1$, and evidence E and hypothesis H are neither certainly true nor certainly false, then $P(H|E)>P(H)$. Isn't that beautiful—a logical justification of the keystone of the hypothetico-deductive method! And so useless. For, usually, the evidence is considerably stronger than what the hypothesis entails, and, in addition, the evaluation cannot be without reference to background information. These two points, which are almost always relevant, crop up especially in the case of instance confirmation. The hypothesis that all crows are black entails that, if x is a crow, then it is black. But the actual evidence is the stronger proposition that x is a black crow (a positive instance) and comes against a background of statistical facts about the distribution of crows and black things among birds and among animals in general. In striking example after example, the fact that $P(H|E\&E'\&B)$ is smaller than $P(H|B)$ shows the silliness of the old saws and, even more, of their putative justification (see Good 1967; Rosencrantz 1982; Van Fraassen 1983d).

There is a good deal more to be said about projectability and the cluster of real and pseudo problems among which it was born. I shall restrict myself to a summary of my opinion: it was an illusory solution to an illusory, misdiagnosed problem. There is no need for a justification of scientific method which shows how this method enables scientists to distinguish projectable from unprojectable predicates, for the distinction is unreal.

Ad Paul Churchland

In part 1, section 2, I have tried to counter Churchland's specific arguments concerning the demarcation of the observable. Now I must place that dispute in the larger context of his paper. It was his intention to put me in a difficult dilemma. By asking me to consider various sorts of beings that we might have been or could become, he initiated a scrutiny that could have vitiated any significant distinction between observation and detection or else forced recourse to "superempirical" theoretical virtues as reason for belief. I hope that I escaped the dilemma by offering one in return: are we to assume that the envisaged beings are (already) part of the epistemic community or not? In the first case, the assertion that they have certain perceptual capabilities implies that we (as epistemic community) have them (despite the limitations of any one of our subcommunities or individual members). In the second case, the assertion that there are certain causal processes that lead to their correct description of (redundantly, to us) unobservable structures is at best implied by our evidence in conjunction with a theory that we accept. Hence, in the second alternative, we are simply brought back to the dispute over acceptance.

But I can't be let go so lightly. Much of Churchland's paper is devoted to the epistemological role of those superempirical virtues (simplicity, coherence, and explanatory power he lists explicitly). His theme is developed further by several other authors here, as well as by a number of writers and reviewers not represented here. So, in addition to my other remarks (in part 1, section 6, and in "Ad Clark Glymour" especially), I shall add here a line of criticism which I began a little while ago in correspondence with J. J. C. Smart.

The superempirical virtues are sometimes summarized as simplicity, sometimes as explanatoriness. Let me concentrate here on the former. To begin, we must distinguish two ways in which simplicity may be attributed. A continental Cartesian would have described Newton's achievement more or less as follows: He has provided us with a mathematically elegant, simple theory of a world full of inexplicable cosmic coincidences and lacking entirely that simplicity and coherence of design to be found in the natural causal order. (In Cartesian terms this is fair, for Newton's theory postulates instantaneous effects at any distance, however great, that cannot be derived from any possible mechanism of action by contact.)

If it is now suggested that simplicity provides a reason for belief, we must ask which attribute of simplicity, of the theory or of the depicted world, is at issue. In the first case the claim appears to be that, among the products of our theorizing, the simpler ones are more likely to correspond to the facts. Since there is no obvious connection between the two kinds of simplicity, this claim does not obviously have any status different from a claim that the more beautiful or inspirational or mathematically more tractable or more visualizable are more likely to be true. I see no ground for such ideas.

In the second case, the claim presumably rests on the conviction that the world is simple. While I have no prior conviction that this is so, many writers have expressed a contrary sentiment. *But suppose I grant it.* What reason does it give me to take, among depicted structures, the simpler as the ones more likely to correspond to reality? The structures described by proffered theories are proper parts of the world as a whole, and a simple whole can have parts which are, considered in and by themselves, exceedingly complex.

To put it briefly, there must be lots and lots of simple false theories, even among those that accommodate our evidence. Even if the world is simple, or even if there exists an exceedingly simple true theory among those that cover *all* aspects of the world, there need be no predominance of true theories among the simple ones, nor does the proportion (or some infinitary analogue of proportion) of true ones need to go up with increased simplicity. We can use simplicity as a basis of theory choice, but not on the grounds that this will increase our chances of metaphysical or even empirical success.

Ad Brian Ellis (See also part 1, section 5, and "Ad Clifford Hooker")

Both Ellis and Hooker, in each case at the beginning of their papers, describe what they take to be empiricism and discuss where my position falls in the empiricist tradition. Let me briefly take issue with both at this point.

Like Ellis in his first two pages, I identify empiricism with the epistemological thesis that experience is the sole legitimate source of information about the world. Therefore, if the differences between Galileo and Bellarmini were as Ellis says, he and I would both classify Galileo, and not Bellarmini, as an empiricist. Historical questions aside, I would say that the empiricist critique of knowledge undercuts all grounds for scientific realism. Thus, Galileo was perhaps an empiricist who delighted in leaps of faith (which I do not consider irrational, though groundless) or else one who did not see the implications of his own position. Bellarmini, on the other hand, was a very acute philosopher who spelled out such implications correctly (on the supposition that the phenomena were as he took them to be). This may have been the Anselmian technique ("The fool sayeth in his heart . . .") of convicting the proponent on his own ground. Or he may have held to a restricted empiricism, which recognizes other sorts of information only for matters outside the domain of natural science. I do not know. As I indicated at various points in my book (though I did not insert the qualification at every appropriate point), I do not consider leaps of faith or belief in things unseen, arrived at for whatever reason, necessarily irrational—only the pretense that we are rationally compelled (e.g., through arguments concerning explanatory value) to embrace more than strict empiricism prescribes.

In philosophical practice, the dividing line between empiricists and others does indeed appear in connection with explanation. Ellis sets out the crucial point very well in his introductory section. When it seems that two theories fit the phenomena equally well, what criteria shall we use to choose between them? The answer will have to cite properties and relations going beyond considerations of empirical adequacy. Here explanation takes center stage.

But suppose we accept some such proffered criteria and take their satisfaction to be reasons for belief. Then we have identified something new to treat as a legitimate source of information about the world. And then, by the account of empiricism discussed above, we shall not be empiricists anymore. For what is the best available explanation, and what it is like, depends on such factors as what theories we have been able to dream up and (I but perhaps not Ellis would add) also on our interests and other contextual factors giving concrete content to the notion of "better explanation." These features of the participants and other aspects of the context of discovery, all independent of what experience has disclosed about the relevant phenomena, are then given a role in shaping our expectations of the future. This is in direct conflict with the empiricist thesis that experience is the sole legitimate source.

Now, it is indeed true that such considerations, beyond evidence directly bearing on empirical adequacy, play a role in theory evaluation and choice, hence in rational acceptance. Therefore, we face a dilemma.

On the one hand, we can profess to have reasons for believing these other considerations to yield reliable indications of truth. For example, the theories we do come up with are only a few among the ones we could have produced after inspection of the same evidence, but it might be held that we are especially prone to "get it right." (This is how the reliability of intuitive abstraction that goes beyond mere numerical induction was regarded in the Aristotelian tradition, the mind being supposed to have a special ability to abstract the essential features of the observed objects.)

On the other hand, we can attempt to remain empiricists and conclude that these extra considerations which legitimately bear on theory acceptance are not legitimate reasons for belief. This includes the corollary that acceptance is not to be identified with belief. The remainder of this story I have told in part 1, section 6, and in "Ad Clark Glymour."

Quite independently, I cannot agree that Ellis's formulation (3) of the aim of science should be acceptable to empiricists and realists alike. The best possible explanatory account may not be true. And, on my view of explanation, what is the best possible explanatory account for one community of scientists may not be that for another. So his (3) does not seem to describe, either on my view or on that of various realists, a worthy aim for science. If "explanatory" is to be given some meaning so that it implies empirical adequacy plus some other ahistorical feature, then it would, I think, bear little relation to actual claims of explanation. But I have no new arguments for this contention, beyond those in *The Scientific Image*.

At the end of part 1, section 5, I agreed with Brian Ellis that my view of theory structure is a generalization of the conclusions I reached about space-time theories. In his paper he argues especially that I was wrong to generalize in this way to models which add physical microstructure in which to embed the phenomena. He might admit that the nineteenth-century molecular models of gases were like that. We may add that Hertz's reconstruction of mechanics, in which each body was represented as an aggregate of point masses, was eminently like that. But these were also cases where ideal entities are introduced to model the real ones *without causal claims* for the elements of postulated structure. When a theory instead offers a causal explanation, its criterion of success must, according to

Ellis, include the reality of the causes. This distinction has been urged similarly by Nancy Cartwright and Ian Hacking.

I assert to the contrary that to see causal models in the same way is not only possible but gives us the best hope of eliminating metaphysics from our interpretation of science. Causality in the philosophical interpretation of science is just a *deus ex machina*, whose very saving powers depend on his essential lack of involvement in the real problem situation. "If you cure the illness, the fever disappears; but, if you merely alleviate the fever, the patient may still be ill." We perceive that the simple story in which this illness and fever are strongly correlated, a symmetric relation, leaves out some crucially important facts that doctors need to know. We can point the doctor in the right direction by adding, in our common universal shorthand, that the illness causes the fever. But is that more than shorthand? Does it say more than the sentence with which I began this example? That there is more to be said is clear, but does any general theory of causation say it? I submit that introduction of the notion of cause never gives better than a spurious and merely verbal unification of an incompletely conceived and ill-defined class of explanatory accounts.

But, through the works of Reichenbach, Salmon, Suppes, and others, we do have some necessary criteria for this class; more precisely, for what constitutes a causal model. These criteria I have utilized in my discussion of Bell's inequality (see "The Charybdis of Realism") to prove that certain phenomena (at least conceivable and probably actual) cannot be embedded in any causal model. In that discussion I follow throughout the point of view that to give a causal account is to display an embedding in a certain kind of model and nothing else. I see absolutely no inadequacies of principle in this way of looking at the matter.

If this is correct, then causal terminology gets its meaning not from any reference to aspects or elements of reality but from the structure of that (only partially defined) type of model. This is, of course, of a piece with my general thesis that at least at an elementary level (or first approximation) the language of science is to be viewed as a semi-interpreted language. That means that the initial interpretation of the nonlogical terms is by correlation with aspects of a certain logical space (i.e., with aspects of the general structure of the relevant sort of models). Reference to reality derives only mediately from the fact that some aspects of that logical space (of the comprised models) are in turn correlated with real physical systems.[17] I shall not pursue this topic here; it concerns the relation between

17. Giere explained this approach in chapter 5 of his *Understanding Scientific Reasoning* (1979) and elaborates a slightly modified version in a forthcoming book. As he notes in his article here, the approach goes back in part to Suppes's innovations in the fifties, which were developed in different directions by Sneed on the one hand and by Frederick Suppe and myself on the other. (For an early survey, see Suppe 1974, Introduction.)

the semantic view of theory structure and the philosophy of language. Since I believe that this point of view can be carried through consistently, I do indeed think, as Ellis states (in his section 2), that "the postulated entities of causal process theories have no more claim to be considered real existents than the theoretical constructs of model theories."

Ad Ronald Giere (see also part 1, section 2)

Analysis of the content and structure of scientific theories, both in general and in the foundations of particular sciences, is a meeting ground for realists and antirealists. There our common interest comes to the fore. When such an analysis is given, it can, of course, be framed in the concepts of one or other side, and the suspicion may grow that another participant's philosophical methodology is not adequate to the same purpose. I place a great value on cooperation in such common areas of concern. Ron Giere and I see in the semantic view of theory structure a framework for the analysis of science that both empiricists and scientific realists can use in good conscience (see further Van Fraassen [forthcoming]). By concentrating on models rather than on linguistic formulations of theories, we can escape the internal problems of logical positivism. If, in this liberation from positivist shackles, empiricism and realism have both been set free, as Giere says, so much the better.

Both sides may nevertheless indulge some secret glee. I want to say, "See, the theory of science can be developed without the realists' metaphysical baggage!" Giere wants to say, "See, the theory of science can be developed without the constraint of the empiricists' epistemological straitjacket!"

There are small differences in formulation among those who choose the semantic approach. I learned from Giere the value of beginning with a clear distinction between the definition of the class of models and the hypotheses that relate real systems to these models (*theoretical definition* and *theoretical hypotheses*). I still wish to take exception to his new proposal for the explication of "approximate." So as to head off excursions into the morass of *verisimilitude* and related (pseudo?) concepts, I want to insist on a simple basic point. To say that a proposition is approximately true is to say that some other proposition, related in a certain way to the first, is true. To say that a model fits approximately is to say that some other model, related in a certain way to the first, fits exactly. These remarks do not explicate "approximate," but they introduce a general form for any admissible explication. Giere suggests this should be rejected because to say that some Newtonian model fits approximately does not entail that any *Newtonian* model fits exactly. This is correct, of course, but is too narrow a construal of the proposed form. I think that his own proposed form can easily be put into mine ("related in a certain way to the first" becomes "similar to the first in specified respects and to specified degrees"). In any case, his formulation and my own both have the virtue that they do not suggest (as various realists have suggested) that we need a theory of approximate truth or verisimilitude or approximate fit as a subject in semantics on a level with the theory of truth. His formulation, moreover, makes very clear that assertions of approximate correctness are context-dependent, requiring for completeness a separate specification of respects, degrees, and criteria of similarity. In practice, I think, the specifications in mind are largely a function of interest, which affects not only the degree to which, but also the respect in which, the models must be similar to constitute "good" approximations.

Giere also says that, if we replace *observability* in the characterization of constructive empiricism by *scientific detectability*, "one major difference between constructive empiricism and constructive realism is removed." This entails, I think, a systematic divergence between his constructive realism and other versions of realism on a large variety of issues. He indicates that he would have sided against Newton in the disputes over absolute space (independently of the ultimate empirical failure or success of Newton's physics). I think he must similarly side against sentiments about absolute space-time that have been expressed by Earman, Friedman, Field, and others. For differences in the global structure of space-time are not detectable; yet, if space-time is a real, independent entity, questions about that global structure must be entirely objective. (I assume here that, even on the broad characterization which Giere gives, any detectable structure must lie entirely within the past cone of some space-time point or be subject to similar spatiotemporal limitations.) In quantum mechanics we find also the "problem of identical particles": several bosons can be in the same state. The total mass is that of several bosons, and so forth, so there cannot be just one. But not one of them can be part of a causal chain that the others are not. Nor is there any observable distinction associated with their permutation. Suppose two bosons are respectively in distinct states ψ_1, ψ_2; then both in the same state φ, then later in distinct states φ_1, φ_2. The assertion that the particle now in φ_1 is identical with the particle that was in ψ_1 must be, for most realists, objectively true or false.[18] Yet the theory itself tells us that there is even in principle no measurement whose outcome could reflect that supposed objective fact. If I understand Giere's position, constructive realism sides with constructive empiricism in placing such questions entirely outside the domain of our epistemic attitudes to science. Quantum theory plus that assertion (or plus a principle that rules on such assertion) is empirically equivalent to quantum theory plus some contrary

18. Reichenbach's alternative solution that bosons are unlike fermions in not being genidentical (in his terminology) does not work without either introducing another hidden variable or violating the identity of indiscernibles.

assertion (or contrary principle) on both Giere's and my criteria of equivalence. For both of us, therefore, the difference is irrelevant to science; truth, as such, does not matter.

I also wonder how easily Giere's constructive realism and modal realism lie together. We can detect only what is actual. Detection, like the narrower mode of observation, can reveal only simpler, actual facts. It can reveal the presence of a particular six-toed sloth but not the necessity of all sloths being six-toed (nor even the truth of all these animals, throughout world history, being six-toed). If we wish to settle the truth of a modal statement, the best we can do is to investigate some related nonmodal ones. The relation, however, can only be given by a theory, for it is a highly theoretical matter. If a theory is correct about everything that is scientifically detectable, it does not follow that it is right in its implications of what could or would have happened (or, have been detectable) under nonactual conditions. We are here, however, at a point of epistemological dispute, and I can only offer my reasoning here as a challenge.

The main dispute between Giere and me is about the nature and justification of acceptance. First of all, Giere concentrates on the belief aspect of acceptance. The alternatives he considers are that acceptance is belief that the theory is true and belief that the theory is empirically adequate. His discussion of justification entirely concerns the justification of the belief in question. This puts at least a terminological distance between us in that, for me, the question whether acceptance of a theory is justified, warranted, rational, or rationally permissible involves the evaluation of a pragmatic aspect—a certain commitment—as well as of a belief.

Giere uses a classical rather than Bayesian decision theory to furnish an account of theoretical deliberation. The temptation is to respond at once that introduction of values into deliberation about belief threatens to bring in ulterior motives. He guards against this threat, however, by using only the value judgment that believing a true proposition is better than being agnostic, which is in turn better than believing a false one. The utility of believing truths and of not believing falsehoods cannot count as ulterior motives here.

But the decision model he uses seems to me dangerously oversimplified, and I'm not sure that its oversimplifications can be removed without detriment to the argument. To begin, it is true that we normally have only a few rival hypotheses before us. This is the focus of one of the Newtonian criticisms of the method of hypotheses. For a choice from the proffered batch may give us merely the best of a bad lot, and that may be worse than remaining agnostic. By considering a choice between H and its denial, Giere looks at a case where one of the lot must be true. This can always be done, but only on pain of introducing the other classic problem, namely, that of determining the probability of conforming evidence conditional on the de-

nial of the hypothesis. In another paper, Giere (1983) has made a new proposal to solve this problem, but I remain unconvinced that the solution is
successful. We could fill the gap with a personal probability, against Giere;
this is a Bayesian move, though adopting it would in itself certainly not
make us full-fledged Bayesians.

The decision rule Giere discusses also has its problems. In a choice between H and its denial, the rule to accept H if the evidence conforms to it
and reject it otherwise is not symmetric between H and its denial. The
reason is that the evidence might conform to both—if H is false, the evidence could still be as it is. Introducing probabilities, if we use only those
conditional on H, we still do not arrive at a symmetric rule. [The reason-
able candidate is: for total evidence E, accept H if $P(E|H)$ is very high,
reject H if it is very low, remain agnostic otherwise. This still tells us im-
plicity to accept H if $P(E|H)$ and $P(E|\text{not } H)$ are both high.] Thus, we
have to bring in the contentious probability conditional on the denial of H
if we are to arrive at a reasonable decision procedure.

It is exactly here that Giere sees the advantage going to the realist side.
The argument is reminiscent of Boyd's (1973) *Noûs* article. The realist may
already believe some background theories which sharply limit the range of
admissible hypotheses. Given that belief, the denial of H amounts to a disjunction of quite determinate possibilities, and, conditional on that disjunction, the probability of evidence conforming to H may indeed be (objectively determinable as) very low. No personal probability with little
warrant is needed to fill the gap. This is the import of the Watson-Crick
example:

> What the empiricist cannot do, however, is appeal to knowledge of the inter-
> nal causal structure of nucleic acids and their interactions with X rays in order
> to bolster the judgment that the predicted pattern is indeed unlikely if H is
> false. (Giere, this volume, chap. 4, sec. 5)

But I think there is a delicate vacillation here between what the background theory *says* and what its realist adherent *believes*. The probability
of the evidence conditional on H is objective if H explicitly gives the
chance of the experimental outcome which the evidence records. The
probability of Halley's comet returning, conditional on Newtonian celestial
mechanics plus Halley's assumptions about that comet (which were extensive), was *one*, because theory plus assumptions entailed that the comet
would return. There was no objective probability, but only a probabilistic
measure of ignorance, for the comet's return conditional on the theory
alone (induced by the probabilities of correctness for Halley's various assumptions, which were not objective chances but measures of opinion).
Similarly, the probability of finding vertical polarization for a photon,

given quantum theory and the premise that the photon is in given state ψ, is objective; but it is not objective conditional on the theory alone.

Suppose now that we want to reach an epistemic decision concerning H in the context of background theory T. Relative to T, the denial of H may be logically equivalent to a hypothesis H'. It may well be that if H bestows an objective chance on conforming evidence E, so does H'. In that case, the decision problem has no gaps, for we must use those two objective chances as probabilities. But it may also be that H' entails a disjunction of chances for E, and T gives no probability distribution over that disjunction. Then there is a gap which cannot be filled by objective calculation of probability. Suppose, for example, that quantum theory entails that, in the described macroscopic situation, the produced photon must have one of two states, ψ, ψ', but entails no more (due, of course, to incompleteness in our described knowledge of that macroscopic situation). Then, unless the chance of finding vertical polarization for the photon is the same in both states ψ, ψ', we have no objective probability for that finding, even if we do believe the quantum mechanical description of microstructure to be literally true. We may have personal odds for state ψ over ψ', deriving from our personal odds on certain incompletely determined macroscopic features of the experimental situation—but that is all.

This objection does not require personal probabilities to be subjective à la Bayesianism. They could be frequencies in selected reference classes, while that selection itself is not uniquely determined by the total evidence. (And that, in my opinion, as it was in Venn's, is the usual case; see Van Fraassen 1983a).

We can put this discussion in better perspective if we go back to how I characterize acceptance. If we have accepted background theory T already, we are committed to limiting our acceptance of new hypotheses to ones that "cut down" the family of models of T. Hence, if we try to decide between H and not-H, we are really considering a cutting down either to those models of T in which H holds or those in which it does not. The latter subclass may indeed exhibit a much more limited variety than does the class of all structures in which H does not hold—a class so poorly delimited that it is not even guaranteed to be a set. But the expectation value of E on the supposition that A, relative to T, is the weighted sum of the chances of E in the models of T in which A holds, the weights being the probabilities that those models fit the real situation. Those latter probabilities cannot be provided by T, since it cannot talk about its own models. The destructive effect of that nonobjectively (or at least extraneously) supplied weighing can only be offset, in the decision process, by an internal connection between A and E, relative to T.

The situation is thus exactly the same, as far as objectivity and justification go, for those who believe that some model of T mirrors reality and those who merely believe that some such model fits reality empirically. In the case of Watson and Crick, if the background theory entails objective chance odds for various possible causal mechanisms, these will enter the decision problem in both the realist and the empiricist account. If it does not entail such objective odds, they can be supplied from outside, with no obvious claim of a scientific source, for either account.

Ad Clark Glymour (see also part 1, section 6)

Since writing his book *Theory and Evidence*, Glymour has been developing a theory of explanation to complement his account there of confirmation by testing. In the paper he presented to the American Philosophical Association in the spring of 1980 (in a symposium with Wesley Salmon and myself), he affirmed that explanation of the phenomena is a reason to believe the explanatory theory and is different from and additional to the sort of reason provided by its passing a test (Glymour 1980a). Specifically, it is an additional virtue to be credited to the theory when it explains the results of the tests they survive.

In my commentary I argued as I have here in section 6 of part 1: the virtues of explanation, insofar as they go beyond description, may indeed provide reasons for acceptance of the theory, but not for belief (Van Fraassen 1983b). And the grounds of the argument were simple. To be more explanatory, the theory must be more informative. This is a point (as I noted) on which Salmon and I agree, although Salmon and I disagree specifically on whether the relevance of the information is objective or pragmatically determined. For both the forms offered by Glymour for explanatory achievement, I argued that this is so: both explanation by *identification* (elimination of contingency) and by *unification* (elimination of theoretical chaos) are achieved at the cost of greater information content. But to contain more information is, to put it crudely, to have more ways of being false and, hence, to be no more likely to be true and, hence, to be no more worthy of credence. I assume here that no one can coherently call one hypothesis less likely to be true than another while professing greater credence in it. This assumption holds not only on Bayesian or Carnapian accounts of credence, but on much more tolerant ones as well.

It is clear from his present article that Glymour stands by his earlier opinion. His reply to my main objection comes in his section 5. It will be clear from my summary above that this reply is independent of any questions concerning whether explanation is a straight theory-fact relationship and objective or whether it is context-relative, pragmatic, or otherwise affected by subjective considerations. Hence, Glymour carefully isolates my main argument (the second one he considers there). His response is that it is irrational not to believe the inductive closure of what one believes, in a sense "akin" to not believing its deductive closure. Sections 3–6 of part 1, in which I give my views on induction must therefore be the main part of my reply. To summarize their upshot: What was valuable in the literature on induction were its more or less careful accounts of reasons for acceptance of theories. Of these, reasons to believe are a proper subclass, and the great tragedy of this subject, which turned the idea of induction into an impossible ideal, was to confuse reasons for acceptance *überhaupt* with reasons for belief. In my view, as long as we try to maintain that conflation, we cannot make sense of either scientific practice or methodology.

Of course, we want theories that are more informative in various ways— and also simpler, more unified, with more easily calculable solutions, or at least approximations to solutions, lending themselves more easily to application, and so on. But not because they would *thereby* satisfy our other desire, to make us more secure that we believe only what is true. The one is bought at a cost to the other. Now, belief and practical decision are so closely tied that we cannot use a theory to reach practical decisions except on the basis of our belief (or, more accurately, gradated credence) in the truth of its practical consequences. But our practical commitments require no credence in the more theoretical parts of the theory, and our commitments to research directions or to a direction for further theoretical exploration also need not involve that.

What about Glymour's response, then, about inductive closure? Of course, he has set out exactly the ideal of inductive logic: the ideal of rules that would lead us to "consequences" which are at the same time (a) more informative than the premises and (b) irrational not to believe when the premises are believed. This was a philosophers' invention, meant to throw light on scientific methodology as well as on ordinary, everyday practice. Instead, it mired its subject in a morass of philosophical difficulties.

Inductive logic is a make-believe theory; no one has ever written its principles. Attempts to do so have always landed in incoherence or fallen afoul of hilarious counterexamples.[19] But, reluctant to admit that they were only talking about a gleam in their eyes, philosophers always pretended, at least in terminology, that there is such a discipline as inductive logic. (Glymour does not pretend this; he offers a candidate for an inductive principle and conjectures that it, together with certain unspecified other principles, will provide a powerful basis for this discipline-to-be. I doubt it.)

There was a conviction, also, that there *has to be* such a thing as inductive logic, because rational belief change must surely proceed according to

19. Some readers have apparently thought that, in chapter 2, section 3 of *The Scientific Image*, I assert inference to the empirical adequacy of the best explanation to be a correct inductive principle. Not so; I *exhibited* this putative rival principle as part of a demonstration that we can have no good evidence for the psychological hypothesis that people do in fact follow the rule of inference to the best explanation.

some rules and does involve arriving at more informative bodies of opinion. Now, in the twentieth century, we have a powerful rival view, the strict Bayesian model of rational belief change through conditionalization on evidence, in which no permissible move, except strict incorporation of evidence, ever increases information. I do not say that this "Carnap's robot" model is correct, but its mere existence has completely disarmed the conviction I described in the initial sentence of this paragraph. (I do hold that the model must at least be liberalized so as to allow for formation of new belief and more generally for moves to more informative states of opinion unwarranted by the mere logical effect of evidence—but that is exactly a process which, according to me, does not proceed in accordance with rules and is not rationally compelled. It is only that, within certain limits, leaps of faith are rationally permitted.) So there is no longer any good reason to say that *there must be* a logic of induction, even if we have not yet formulated it.

Having argued so strongly against the epistemic role Glymour wants to give to explanation, I must add that we are comrades as laborers in the vineyard of philosophy of science. Testing and explanation are both central topics there, and we recognize the same tasks. "A philosophical understanding of science should . . . give us an account of what explanations are and of why they are valued, but, most importantly, it should also provide us with clear and plausible criteria for comparing the goodness of explanations," Glymour says near the beginning of his paper, and I agree entirely. For this task, as for that of the account of theory testing, Glymour has provided concrete proposals, grounded in historical case studies. In both cases, the features he describes seem much more to get at the heart of the matter than any relations of credence, enhancement of credence, or credibility could ever do. The actual structure of tests, the actual structure of explanations—about these I can accept his insights and we can agree, although we differ on the connections with general epistemology.

Ad Gary Gutting (see also part 1, section 2)

When I received Gutting's dialogue in the fall of 1981, it immediately seemed to me that he had definitively isolated the crucial hinge or focal point of the epistemological controversy. Why should the limits of observability, rather than the mere factual limits to what has been observed, be given a role in determining our epistemic commitment? I feel very sure of two points: that this is how it is in fact in science and that this is how an empiricist ought to say that it ought to be. It is true, of course, that we have in principle no better certainty of what those limits are than of any other factual matter. But our present opinion about them can play a role in how we fashion our new expectations in response to both experimental and theoretical developments.

Since then a number of other writers have raised this point, though never as eloquently and persuasively or in such effective literary form. It is not a point that can be simply answered and dismissed; the answer has to be an integral part of one's epistemological position. I have tried to make it so and refer especially to the middle of part 1, section 2, where I refer to Gutting and Musgrave.

Ad Ian Hacking

A main objection to my stand with respect to the demarcation of the observable is found in Ian Hacking's paper. I have been greatly aided in my appreciation of it by the opportunity to study his book-length manuscript *Representing and Intervening* (1983) and our correspondence and discussion of these issues over the past few years. In addition, Wesley Salmon has allowed me to read the manuscript of his new book on explanation, which has a section devoted to my views on empirical adequacy and this paper of Hacking's.

Hacking is a master of style as well as of his subject matter, and most particularly of the polemical use of linguistic presuppositions: his prose attacks my position on many levels simultaneously. When a realist gives a consciously and deliberately naïve description of what happens in an experiment or observation, I do not, of course, want to dispute a single one of his assertions on its own ground—but I want to stand back and ask *what* his language is doing. Can we deconstruct this text? Well, only (just as in literature) if it makes a good beginning of deconstructing itself.

"Do We See through a Microscope?" begins by explaining with admirable clarity the radical difference in status (internal to science) between even the ordinary light microscope and our familiar magnifying glass, thus illuminating the opinions of certain scientists and standard texts on microscopy that we do not *see* through a microscope. But the real issue is of course whether the information "naturally" or "uncritically" gathered by its means has the same status as what we get by mere looking—whether we should believe that the structures apparently revealed are really there. The nearest to this in scientific discussion is the internal distinction between reality and artifact in the apparently revealed structures. The test for this distinction is whether other modes of access lead to the same structural feature. As example Hacking chooses the appearance of the dense bodies in red blood cells, and says, "It would be a preposterous coincidence if, time and again, two completely different physical processes [electron transmission and fluorescent reemission] produced identical visual configurations which were, however, artefacts of the physical processes rather than real structures in the cell" (this volume, chap. 7, sec. "Truth in Microscopy").

Imagine I have several processes which produce very different visual images when set in motion under similar circumstances. I study them, note certain similarities; as I repeat this, I discard similarities that do not persist and also build machines to process the visual output in a way that emphasizes and brings out the noticed persistent similarities. Eventually the refined products of these processes are strikingly similar when initiated in similar circumstances. Now I point to the similarities and say that they are too striking to be there by coincidence, though, of course, the discarded dissimilarities were mere idiosyncrasies of the individual processes. What is the status of my assertion? What principle of reasoning could support it? Since I have carefully selected against nonpersistent similarities in what I allow to survive the visual output processing, it is not all that surprising that I have persistent similarities to display to you. What is the status of my interpretation, i.e., my procedure for assigning them significance?

Hacking himself examines and rejects any global justification through such putative canons of inductive reasoning as inference to the best explanation or the demand for common causes. The section "Coincidence and Explanation" undercuts, with four incisive distinctions, any such justification. Unfortunately, the rephrased conclusion at its end, "it would be a preposterous coincidence if two totally different kinds of physical systems were to produce exactly the same arrangements of dots on micrographs," is much too weak. We refer to two different sorts of instruments, so the sameness in the outputs must be attributed principally to similarities among the inputs. But no one doubts that it is in each case *blood samples* and not different kinds of physical systems that were fed into the machines. This conclusion warrants no inference about the reality of the imputed unobservable structure. So Hacking attempts to find a more instructive example and gives "The Argument of the Grid." While more striking, however, the new example is not more probative. It is no argument to say, "I know that what I see through the microscope is veridical because we *made* the grid to be just that way," since the premise needs to imply what is under dispute (that we *successfully* made the object to be that way). To add that agnosticism on this point requires belief in a Cartesian demon of the microscope reveals only the unstated premise that the persistent similarities in the relevant phenomena *require, must have,* a true explanation. But reliance on that premise is exactly what the previous section denied.

Wesley Salmon has discussed this sort of example consistently in terms of the notion of common cause. As I pointed out in *The Scientific Image* (p. 123), however, there is a crucial ambiguity when common-cause arguments are given in support of scientific realism. In the technical sense, two correlated event (-type)s have a common cause when, in a significant proportion of the correlated pairs, the histories can be traced back to an intersection of a third event (-type), bearing certain relations to the first two.

The correlation between heavy smoking and lung cancer has a common-cause explanation if we can point to a past history of smoking which is present in a significant proportion of those individual lung cancer patients who are heavy smokers, and so forth. (Imagine what you would think if it turned out that nonsmokers usually begin to smoke after contracting lung cancer.) The similarities in experimental outcomes that Salmon pointed to in connection with Avogadro's number, and Hacking for the visual images produced by different sorts of microscopy, are not a case in point. If we pick a *specific* pair exhibiting the similarity in their *particular*, precisely noted, results, we shall rarely find a significant intersection (such as a specific blood sample taken from a particular patient) of their histories. The explanation or inference is not by, or to, a *common* cause but a *similar* cause. If I see a facial resemblance between two beauty queens in different countries, I give a *common-cause* explanation if I point to a common ancestor and a *similar-cause* explanation if I point to fads and fashions influencing their selection.

Salmon has now, I think correctly, diagnosed the pattern of argument at work as one of *analogy* (see Salmon [forthcoming]). We see observable causes $C_1, \ldots, C_n$ all of a certain sort producing similar observable effects $E_1, \ldots, E_n$; then we attribute a similar but unobservable cause C to a further observable effect E by analogy. And what is the status of analogy, that now sadly neglected but once favorite inductive device? In *The Foundations of Scientific Inference*, Salmon espoused an "objective Bayesianism" in which arguments by analogy are one of the sources of prior probabilities.

This view has much to commend itself. First, if we think that Bayesian models are applicable at all—and I have explained above how and to what extent I think they are—we do need to look for sources of prior opinion. For the judgments we make before inspecting the evidence play a crucial role in the interpretation of that evidence (though, *if they are precise and empirically concrete enough* they can be swamped by a manageable amount of evidence). This idea raises the hackles of subjective Bayesians, but they may agree that asking a person to draw or look for analogies is exactly an efficient way to ferret out his prior probabilities. The difference between the subjective and objective Bayesian at this point must be just this, that the latter thinks of the prior judgment as being somehow *supported* if we can "back it up" by something that looks like an inference by analogy.

This is where we shall have to part company. Inspiration is hard to find, and any mental device that can help us concoct more complex and sophisticated novel hypotheses is to be welcomed. Hence, analogical thinking is welcome. But it belongs to the context of discovery, and drawing ingenious analogies may help to find, but does not support, novel conjectures.

Hacking himself may well reject any attempt to support his conclusions with reference to analogical inference. But at this point I can see no other support for his conclusions that could in principle appeal to an antirealist.

Ad Clifford Hooker (see also "Ad Alan Musgrave")

In the section "Ad Brian Ellis," I began by describing exactly what I take to be empiricism in order to focus my reply to his discussion of empiricism. I shall rely on that description again here to state my perspective on the initial part of Hooker's paper.

Hooker's sketch of the empiricist tradition in his sections 2 and 3 is not history but polemics—the sort of redescription of his predecessors that almost every philosopher since Aristotle has put to good rhetorical use. But here the bias is very great, in my view, in part because it concentrates almost entirely on the past half century or so. The empiricists to whom I feel most akin are, in the recent past, those French and German writers on science in the half century before logical positivism and, less recently, the nominalists of the fourteenth century, who destroyed the Aristotelian metaphysical tradition from within. Yet some of Hooker's main points generalize well to the whole tradition. For example, I think he is entirely right when he describes the main tactics of empiricist response to criticism of empiricism: recourse to pragmatics (with the denial that pragmatic considerations are cognitively relevant) and to new developments in logic. This is true not only of Carnap but also of Poincaré and of Ockham. It is also true that I accepted major realist criticisms of logical positivism and reinforced them sympathetically by some of my own. However, it is by focusing on the very recent history of empiricism, dominated by logical positivism, that Hooker achieves the polemical clout in his section 3. Without this bias in his history, I do not think he could have concluded that I deflected realist criticisms of empiricism by simply incorporating them into my own position. I enjoin us all to return to an empiricism unconstrained by logical positivist shackles.

In part 1 I have tried to deal at length with the epistemological concerns raised by Hooker in his section 4; these are most clearly and most closely related to arguments found in the other papers and in reviews. My view on his evolutionary naturalistic realism is, of course, the same as that on other metaphysical positions, however closely aligned with science: they involve the impossible ideal typical of pre-Kantian metaphysics, of knowledge beyond the reach of experience. But, to my surprise, we find in section 7 a telling attack on the *scientism* usually characteristic of such metaphysical positions today. Not only that, scientism seems to be laid here at the door of empiricism. It seems to me to be exactly the realist who "leaves no distinctively ethical-spiritual, social, or rational structure to extrascientific life, after all the major substance of life." For the realist's world, taken to be in principle exhaustively described by science, is as chock-full as an egg. Empiricism does not pretend to supply such structure as Hooker seeks, to be sure, but leaves room for it. And it also leaves room for the freedom to be found in a life *without* a world picture.

Ad Alan Musgrave

I have already replied to Musgrave on two points, both in section 2 of part 1. Both of these concern the truth/empirical adequacy distincton and its epistemological role. Here I have the satisfaction of seeing, in Musgrave's relatively robust attitude to matters of evidence and belief, at least an inclination (*malgré lui?*) to go my way—as when he says (this volume, chap. 9, sec. 3), "Not that realists would be too happy with an explanation of quantum mechanics which was demonstrably empirically equivalent with it: such an explanation could have no *independent* evidence in its favor." Realists there are, less empirically minded than Musgrave, who would take the fact that the proffered (hidden-variable) theory explained quantum mechanics *to be* the independent evidence in its favor!

In a footnote, Musgrave repeats John Worrall's (1984) quote, which now rather embarrasses me, from an early paper of mine. It is from my article on Beth, which introduced my version of the semantics of physical theories and concerns the relations between syntactic and semantic accounts of theories:

> There are natural interrelations between the two approaches: an axiomatic theory may be characterized by the class of interpretations which satisfy it, and an interpretation may be characterized by the set of sentences which it satisfies. . . . These interrelations make implausible any claim of superiority for either approach." (*Philosophy of Science* 37 [1970], 326.)

The first statement is true; the second, despite its vagueness, does not follow. I was at that time overly impressed (but perhaps Patrick Suppes was the only philosopher of science who wasn't) by the completeness theorem for quantificational logic (cf. Van Fraassen [forthcoming]).

Suppose that I present a scientific theory in what the semantic view takes to be the typical fashion: by describing its set of models, i.e., the set of structures it makes available for modeling its domain. Call this set M. Suppose in addition that, in doing so, the first object I mention is the real number continuum—also not all that unlikely. Now, you or I may formalize what I have done in some carefully chosen or constructed artificial language. Part of the formalization will be to select a set of axioms, from among the sentences of that language, to express what I meant to convey. If we now look at the models of that language—in this sense, a model is a structure plus a function that interprets the sentences in that structure— we can correlate the chosen set of axioms with the class of models of the language in which these axioms are satisfied. Call that set N. If we have been lucky, every structure in M now occurs in N, attached to one or more interpretations of the syntax. But the real number continuum is something infinite. The Löwenheim-Skolem theorems then tell us at once that N contains many structures not isomorphic to any member of M. So we certainly have not come up, in our axiomatic theory, with a new and precise and more tractable, etc., etc., description of M.

Of course, I am not saying that a scientist's presentation of his theory cannot be formalized. But I do say that if, after formalization, we discard the original and concentrate solely on the isolated abstract pattern (the formal distillate, so to say), we shall have lost the theory.

Originally I intended to include some such passage as the above in chapter 3 of *The Scientific Image*. I decided not to do so because I reasoned that, to most readers, these technical points would be irrelevant or obscure, while those who would easily understand them would not need to have them spelled out. I forgot how I myself had begun under the spell of the sort of logical model theory we find in standard logic texts. And I underestimated how strongly the logical positivists had succeeded in indoctrinating subsequent generations to think in terms of their first-order language picture of theories.

Thus, Michael Friedman's review of *The Scientific Image* asked at a certain point, if a theory is to be identified with the set of its models, is that set an elementary class or not? This question makes sense only if we construe "models" as referring to the models (in the second usage found in the preceding paragraph) of some particular language. In that case, an elementary class of models is one for which there exists a set of sentences which they and they alone satisfy. The class M is not a set of models in this sense; the class N is, and it is a sort of image of M produced through the lens (which may be more or less limiting or distorting) of the specific chosen language. The set N contains a more exact image M^* of M, namely, the set of those members of N which consist of structures in M accompanied by interpretations therein of the given syntax. But, moreover, as the preceding paragraph makes clear, M^* is not an elementary class. So I wish Friedman had called me on the telephone with that question, for he proceeds to assume that a theory is to be identified with an elementary class and thus is able to make some very powerful irrelevant points.

I should add two remarks. The first is that I am now, as much as in 1970, inclined to see as the main virtue of the semantic approach its naturalness and closeness to scientific practice. The extra technical leeway is there, so it has richer resources to draw on than the axiomatic approach—but, even in cases in which the theory can in principle be axiomatized (in the logician's rather than the scientist's sense), the superiority of its closeness to science and lesser degree of abstraction are marked. Not that the logistical approach was barren—it engendered many entertaining red herrings, such as Craig's theorem, the raven paradox, Ramsey sentences, and so forth.[20] Suppes's slogan, that the tool for philosophy of science is mathematics and not metamathematics, saves us from this miscreant brood.

The second remark is that I am a nominalist. Of course, one can be a nominalist only in the way Saint Paul held one could only be a Christian, namely, in the sense of trying to be one. I do not really believe in abstract entities, which includes mathematical ones. Yet I do not for a moment think that science should eschew the use of mathematics, nor that logicians should, nor philosophers of science. I have not worked out a nominalist philosophy of mathematics—my trying has not yet carried me that far. Yet I am clear that it would have to be a fictionalist account, legitimizing the use of mathematics and all its intratheoretic distinctions in the course of that use, unaffected by disbelief in the entities mathematical statements purport to be about. Within mathematics, the distinction between structures of different cardinalities and the nonisomorphism of real number continuum and natural number series are objective. I cannot spell this out further. But I feel sufficiently clear on what the antirealist strategy must be to resist as irrelevant disputes within the philosophy of mathematics while I cooperate in the task of an account of science, of its content, its structure, or its methodology.

Ad Mark Wilson

Mark Wilson presents an admirably clear analysis of observability from a point of view which sounds, to begin, close to my own. As his paper progresses, the deep underlying differences become progressively more apparent.

For both of us, observation is perception. But even without either of us delving deeply into the theory of perception, it is clear that we start with different ideas on that subject. Thus, I understand Russell's distinction between acquaintance and description well enough in ordinary examples: when he wrote *The Problems of Philosophy*, Russell knew Ottoline Morell by acquaintance and the emperor of China merely by description. But I would accordingly have thought that the aborigine's perception of the tennis ball is a case of acquaintance and that his failure to perceive that it is a tennis ball is due to his lack of command of the terms or concepts needed for that description. Wilson, however, employs these notions with reference to universals (or properties). When the aborigine sees the tennis ball,

20. I wish I could have made this sentence as crisp as its model: Poincaré's exclamation, upon hearing of the Russell paradox, "So logicism is not barren after all—it engenders contradictions!"

he does so despite his lack of acquaintance with the property of being a tennis ball. While it may be possible to be acquainted with an object without knowing or understanding much about what it is, acquaintance with a universal appears to require such understanding. I do not see the need to mobilize this much ontology in a discussion of the limitations of perception, and I am inclined to think that it must lead one into a morass of metaphysics. Suppose that the aborigine is acquainted with the universal, the property of being water. Suppose in addition, just for now, that water is H_2O and that this is necessarily so, a not uncommon doctrine today. Does it follow that the aborigine is acquainted with the property of being H_2O? If we say yes, then for universals, as for tennis balls, acquaintance with them does not require knowledge and understanding of what they are, not even of what they are necessarily. If we say no, we are in the domain of "fine-grained" ontological distinctions, where even necessarily coextensional properties need not be identified. This is not absurd; the distinction between features "which even God could not separate" was explored by the medievals under such headings as "formal distinction" and "distinction of reasoned reason." It seems to me that to join the issue here would lead us quite far afield, and I shall try to address the differences between us while skirting around such questions of ontology. This should be possible since, despite the differences, we both are willing to think of processes of observation as a subspecies of processes of measurement.

The general description of measurement is an important subject in foundational studies. Questions of consistency and internal coherence arise when a theory is used to predict measurement outcomes but is also rich enough to contain models of the measurement process. In Poincaré's writings on classical mechanics, we see these questions beginning to come forward; they arrive at center stage in philosophical inquiries into relativity theory and play starring roles in fundamental work in quantum mechanics. The description of what measurement is needed progressive refinement even in its most general form as the discussion went on. Surprisingly important turned out to be attention to the notion of the ground state—crucial to the measurement preparation—and I found that, to disown an offspring of the Einstein-Podolski-Rosen paradox (in an objection posed by Jon Dorling), it seems necessary to define what a measurement process is by first defining what counts as a measurement apparatus with a given ground state (see Van Fraassen 1974). While I agree with some main points Wilson makes with his Sirius example, I think that adequate attention to the notion of ground state will dispel the impression that it raises philosophical difficulties.

But, coming now to the difficulties, I want to draw a sharp distinction between the use of science to help delineate what is observable, and Wilson's program, which can issue only (if I understand it) in a theory-relative notion of observability. As I see it, what is observable is just as much a matter of objective fact as what is real or, to use a previous comparison, what is breakable or portable. At one point Wilson suggests that human observations are such complex processes that no Newtonian model can contain them. Would it follow that no Newtonian model can contain humanly observable things? As I understand that, the answer is no, because billiard balls are observable and there are Newtonian models containing billiard balls. But, as Wilson understands the question, it becomes, does it follow that there is no Newtonian possible world in which there are humans observing billiard balls? The answer to that would automatically be yes. But, of course, once we do make the supposition that a certain phenomenon has no Newtonian models, we cease to look for information about that phenomenon in Newton's theory. If for a moment we are allowed the contrary supposition, however, we do not conclude that billiard balls in a world devoid of humans are not observable. Again the questions sound different to me than to Wilson: *Suppose an ordinary billiard ball existed but no humans; would that ball be observable?* versus *Does a world which contains a billiard ball but no humans contain humans observing a billiard ball?* I would say yes to the first and no to the second but consider the first to be the relevant question. Perhaps there is an ambiguity in "theory-dependent," but in its significant sense a feature does not become theory-dependent just because one theory describes it as being different from the way another theory describes it.

References

Note: Besides the items cited in my text, I have included here the other reviews and critical studies of *The Scientific Image* which have come to my notice.

Beth, E. 1948. *Natuurphilosophie*. Gorinchem: Noorduijn.
Boyd, R. 1973. "Realism, Underdetermination, and a Causal Theory of Evidence." *Noûs* 7 : 1–12.
———. 1985. "*Lex Orandi Est Lex Credendi*." This volume, chap. 1.
Brown, J. R. 1982. "The Miracle of Science." *Philosophical Quarterly* 32 : 232–44.
Cartwright, N. 1982. "When Explanation Leads to Inference." *Philosophical Topics* 13 : 111–21. Reprinted as chap. 5 of Cartwright (1983).
———. 1983. *How the Laws of Physics Lie*. Oxford: Oxford University Press.
Churchland, P. 1979. *Scientific Realism and the Plasticity of Mind*. Cambridge: Cambridge University Press.
———. 1982. "The Anti-Realist Epistemology of van Fraassen's *The Scientific Image*." *Pacific Philosophical Quarterly* 63 : 226–35. Revised version in this volume, chap. 2.

Demopoulos, W. 1982. Review of van Fraassen (1980). *Philosophical Review* 91:603–7.

Derksen, A. A. 1983. Review (in Dutch) of van Fraassen (1980). *Bijdragen, Tijdschrift voor Filosofie en Theologie* 44:457–58.

Devitt, M. 1982. Review of van Fraassen (1980). *Australasian Journal of Philosophy* 60:367–69.

Earman, J., ed. 1983. *Testing Scientific Theories*. Vol. 10 of Minnesota Studies in the Philosophy of Science. Minneapolis: University of Minnesota Press.

Ellis, B. 1979. *Rational Belief Systems*. Oxford: Blackwell.

———. 1985. "What Science Aims to Do." This volume, chap. 3.

Fine, A. 1984a. "And Not Anti-Realism Either." *Noûs* 18:51–65.

———. 1984b. "The Natural Ontological Attitude." In Leplin (1984).

Foss, J. 1984. "On Accepting van Fraassen's Image of Science." *Philosophy of Science* 51:79–92.

Friedman, M. 1982. Review of van Fraassen (1980). *Journal of Philosophy* 79:274–83.

Galavotti, M. C. 1981. Review (in Italian) of van Fraassen (1980). *Lingua e Stile* 16:633–36.

Gauthier, Y. 1981. Review (in French) of van Fraassen (1980). *Dialogue* 20:579–86.

Giere, R. 1979. *Understanding Scientific Reasoning*. New York: Holt, Rinehart, Winston.

———. 1983. "Testing Theoretical Hypotheses." In Earman (1983), 269–98.

———. 1985. "Constructive Realism." This volume, chap. 4.

Glymour, C. 1980a. "Explanations, Tests, Unity and Necessity." *Noûs* 14:31–50.

———. 1980b. *Theory and Evidence*. Princeton: Princeton University Press.

———. 1985. "Explanation and Realism." This volume, chap. 5.

Good, I. J. 1967. "The White Shoe Is a Red Herring." *British Journal for the Philosophy of Science* 17:322.

Goodman, N. 1965. *Fact, Fiction, and Forecast*. Indianapolis: Bobbs-Merrill.

Gould, S. J. 1973. *Ever Since Darwin*. New York: Norton.

Grimes, T. R. 1984. "An Appraisal of van Fraassen's Constructive Empiricism." *Philosophical Studies* 45:261–68.

Gutting, G. 1982. "Scientific Realism versus Constructive Empiricism: A Dialogue." *Monist* 65, no. 3. Reprinted in this volume, chap. 6.

Hacking, I. 1981a. "Do We See through a Microscope?" *Pacific Philosophical Quarterly* 62 (1981):305–22; reprinted in this volume, chap. 7.

———. 1981b. *Scientific Revolutions*. Oxford: Oxford University Press.

———. 1983. *Representing and Intervening*. Cambridge: Cambridge University Press.

Hanna, J. F. 1983. "Empirical Adequacy." *Philosophy of Science* 50:1–34.

Hanson, P., and E. Levy. 1982. Review of van Fraassen (1980). *Philosophy of Science* 49:290–93.

Hausman, D. M. 1982. "Constructive Empiricism Contested." *Pacific Philosophical Quarterly* 63:21–28.

Healey, R. 1984. Review of van Fraassen (1980). *Philosophical Books* 23:100–102.

Hesse, M. 1981. "Anti-Realist Philosophy of Science." *Nature* 289:207–8.

Hooker, C. A. 1985. "Surface Dazzle, Ghostly Depths: An Exposition and Critical Evaluation of van Fraassen's Vindication of Empiricism against Realism." This volume, chap. 8.

Horwich, P. 1982. *Probability and Evidence*. Cambridge: Cambridge University Press.

James, W. 1897. *The Will to Believe*. New York: Longmans, Green, and Co.

Karplus, M., and R. N. Porter. 1970. *Atoms and Molecules: An Introduction for Students of Physical Chemistry*. New York: W. A. Benjamin.

Krueger, L. 1981. "Empirismus oder Realismus—Eine Alternative in der Wissenschaftstheorie?" Paper presented at German Philosophical Association Congress, Insbruck.

Largeault, J. 1982. Review (in French) of van Fraassen (1980). *Revue Philosophique* 173:572–75.

Laudan, L. 1981. *Science and Hypothesis*. Dordrecht, Holland: Reidel.

Leplin, J., ed. 1984. *Scientific Realism*. Berkeley: University of California Press.

Marlow, A. R., ed. 1980. *Quantum Theory and Gravitation*. New York: Academic Press.

McKinnon, E. 1979. "Scientific Realism: The New Debates." *Philosophy of Science* 46:501–32.

Musgrave, A. 1982. "Realism and Constructive Empiricism." *Philosophical Quarterly* 32:262–71. Revised version in this volume, chap. 9.

Newton-Smith, W. 1981. *The Rationality of Science*. London: Routledge and Kegan Paul.

Peacocke, C. 1981. "Not Real but Observable." Review of van Fraassen (1980). *Times Literary Supplement*, 30 January, p. 121.

Popper, K. 1981. "The Rationality of Scientific Revolutions." Hacking (1981b), 80–116.

Putnam, H. 1981. *Reason, Truth and History*. Cambridge: Cambridge University Press.

Reichenbach, H. 1959. *Modern Philosophy of Science*. New York: Humanities Press.

Rosenberg, A. 1983. "Protagoras among the Physicists." *Dialogue* 22:311–17.

Rosencrantz, R. 1982. "Does the Philosophy of Induction Rest on a Mistake?" *Journal of Philosophy* 79:78–97.

Salmon, W. 1967. *The Foundations of Scientific Inference*. Pittsburgh: Pittsburgh University Press.

———. (forthcoming.) *The Structure of Scientific Explanation and the Causal Structure of the World*. Princeton: Princeton University Press.

Smart, J. J. C. 1979. "Difficulties for Realism in the Philosophy of Science." In Abstracts of Section Six, 2–3. *Proceedings of the Sixth International Congress of Logic, Methodology, and Philosophy of Science*. Hanover.

Suppe, F., ed. 1974. *The Structure of Scientific Theories*. Urbana: University of Illinois Press.

Thornton, M. 1981. "Sellars' Scientific Realism: A Reply to van Fraassen." *Dialogue* 20:79–83.

Trigg, R. 1983. Review of van Fraassen (1980). *Mind* 92:291–93.

Van Fraassen, B. C. 1970a. *An Introduction to the Philosophy of Time and Space*.

New York: Random House. 2d ed., New York: Columbia University Press, forthcoming.

———. 1970b. "On the Extension of Beth's Semantics of Physical Theories." *Philosophy of Science* 37: 325–34.

———. 1974. "The Einstein-Podolski-Rosen Paradox." *Synthese* 29: 291–309.

———. 1980a. *The Scientific Image*. Oxford: Oxford University Press.

———. 1980b. "Rational Belief and Probability Kinematics." *Philosophy of Science* 47: 165–87.

———. 1980c. Review of Ellis (1979). *Canadian Journal of Philosophy* 10: 497–511.

———. 1981a. "Discussion: A Problem for Relative Information Minimizers in Probability Kinematics." *British Journal for the Philosophy of Science* 32: 374–79.

———. 1981b. Review of Churchland (1979). *Canadian Journal of Philosophy* 11: 555–67.

———. 1981c. "Theory Construction and Experiment: An Empiricist View." In *PSA 1980*, edited by P. D. Asquith and R. N. Giere, vol. 2, pp. 663–78. East Lansing, Mich.: Philosophy of Science Association.

———. 1982. "The Charybdis of Realism: Epistemological Implications of Bell's Inequality." *Synthese* 5: 25–38.

———. 1983a. "Calibration: A Frequency Justification of Personal Probability." In *Physics, Philosophy, and Psychoanalysis*, edited by R. S. Cohen and L. Laudan, 295–319. Dordrecht, Holland: Reidel.

———. 1983b. "Glymour on Evidence and Explanation." In Earman (1983), 165–76.

———. 1983c. "Theory Comparison and Relevant Evidence." In Earman (1983), 27–42.

———. 1983d. "Theory Confirmation: Tension and Conflict." In *Epistemology and Philosophy of Science: Proceedings of the Seventh International Wittgenstein Symposium*, edited by P. Weingartner and M. Czermak, 342–52. Vienna: Hoelder-Pichler-Tempsky.

———. (Forthcoming.) "The Aim and Structure of Scientific Theories." In *Proceedings of the Seventh International Congress of Logic, Methodology, and Philosophy of Science*, Salzburg, 1983.

———. 1984. "Belief and the Will." *Journal of Philosophy* 81: 235–256.

Whewell, W. 1847. *The Philosophy of the Inductive Sciences*. 2d ed. London. Reprinted by Johnson Reprint Corp., 1967.

Wilson, M. 1985. "What Can Theory Tell Us About Observation?" This volume, chap. 10.

Worrall, J. 1984. "An Unreal Image." *British Journal for the Philosophy of Science* 35: 65–80.

Contributors

Professor Richard N. Boyd
The Sage School of Philosophy
Goldwin Smith Hall
Cornell University
Ithaca, NY 14853-0205

Professor Paul M. Churchland
Department of Philosophy, B-002
University of California, San Diego
La Jolla, CA 92093

Professor Brian Ellis
Department of Philosophy
La Trobe University
Bundoora, Victoria
Australia 3083

Professor Ronald N. Giere
Department of History and
    Philosophy of Science
Goodbody Hall 130
Indiana University
Bloomington, Indiana 47405

Professor Clark Glymour
Department of History and
    Philosophy of Science
University of Pittsburgh
Pittsburgh, PA 15260

Professor Gary Gutting
Department of Philosophy
University of Notre Dame
South Bend, IN 46556

Professor Ian Hacking
Institute for the History and
    Philosophy of Science and
    Technology
Room 316, Victoria College
University of Toronto
Toronto, Ontario
Canada M5S 1K7

Professor Clifford A. Hooker
Department of Philosophy
University of Newcastle
Newcastle, N.S. Wales
Australia 2308

Professor Alan Musgrave
Dean of Arts and Music
University of Otago
Box 56
Dunedin, New Zealand

Professor Mark Wilson
Department of Philosophy
University of Illinois at Chicago
Chicago, IL 60607

Professor Bas C. van Fraassen
Department of Philosophy
1879 Hall
Princeton University
Princeton, NJ 08544